Rethinking Environmentalism

Linking Justice, Sustainability, and Diversity

Strüngmann Forum Reports

Julia R. Lupp, series editor

The Ernst Strüngmann Forum is made possible through the generous support of the Ernst Strüngmann Foundation, inaugurated by Dr. Andreas and Dr. Thomas Strüngmann.

This Forum was supported by the
Deutsche Forschungsgemeinschaft
(German Science Foundation)

Rethinking Environmentalism
Linking Justice, Sustainability, and Diversity

Edited by

Sharachchandra Lele, Eduardo S. Brondizio,
John Byrne, Georgina M. Mace, and Joan Martinez-Alier

Program Advisory Committee:

Eduardo S. Brondizio, John Byrne, Sharachchandra Lele,
Julia R. Lupp, Georgina M. Mace, and Joan Martinez-Alier

The MIT Press

Cambridge, Massachusetts
London, England

Series Editor: J. R. Lupp
Editorial Assistance: M. Turner, A. Ducey-Gessner, C. Stephen
Photographs: A. Eskin
Lektorat: BerlinScienceWorks

The book was set in TimesNewRoman and Arial.

This book is freely available in digital form at https://esforum.de

Library of Congress Cataloging-in-Publication Data

Names: Lele, Sharachchandra M. (Sharachchandra Madhukar), editor.
Title: Rethinking environmentalism : linking justice, sustainability, and
    diversity / edited by Sharachchandra Lele, Eduardo S. Brondizio, John
    Byrne, Georgina M. Mace, and Joan Martinez-Alier.
Description: Cambridge, MA : The MIT Press, 2018. | Series: Strüngmann
    Forum reports ; #23 | Includes bibliographical references and index.
Identifiers: LCCN 2018008760 | ISBN 9780262038966 (paperback : alk.
    paper)
Subjects: LCSH: Environmentalism--Social aspects. | Environmental
    justice. | Sustainability. | Human ecology.
Classification: LCC GE195 .R48 2018 | DDC 363.7--dc23 LC record avail-
able at https://lccn.loc.gov/2018008760

Contents

The Ernst Strüngmann Forum

Science is a highly specialized enterprise—one that enables areas of enquiry to be minutely pursued, establishes working paradigms and normative standards, and supports rigor in experimental research. Some issues, however, do not fall neatly into the purview of any one discipline, and for these topics, specialization can serve to hinder conceptualization or limit problem-solving approaches. The Ernst Strüngmann Forum was created to address such topics.

Founded on the tenets of scientific independence and the inquisitive nature of the human mind, the Ernst Strüngmann Forum convenes "intellectual retreats" to address problems confronted in research. These gatherings are not meetings comprised of lectures or presentations, they are carefully crafted forms of interaction designed to promote synergy between diverse areas of expertise—an extended discourse that aims to identify knowledge gaps, to explore novel ways of conceptualizing pressing issues, and to delineate trajectories for future research.

Having participated in an earlier Ernst Strüngmann Forum on land use, Thomas Sikor sought our help to provide an open platform for discussion. As he stated:

> The urgent need for a Forum arises from the increasingly apparent impasse in environmental research across disciplines. Since the beginning of the 21st century, ecologists, environmental chemists, economists, social scientists, and researchers from other disciplines have all worked on the key environmental problems. Interdisciplinary spaces and frameworks have been created to address environmental issues from a broad viewpoint, and particular aspects of these problems have indeed been illuminated. However, the initial burst of enthusiasm for these new interdisciplinary spaces and frameworks has given way to fragmented thinking. Research and thought on the environment now takes place in environmental economics, environmental sociology, restoration ecology, and environmental chemistry. What is lacking is a "science of the environment" in its broadest sense.
>
> There is also a real danger that disciplines may slip back into more fundamentalist and naturalistic positions about what environmentalism means, thereby losing sight of the social dimensions of environmental change and reducing attention to the multiple meanings attributed to the environmental in the Global South and North in public and practice. This tendency is reflected in a striking disconnect between the centers of scholarly expertise and those who actually experience environmental harm on the ground and act against it, in particular in many communities in the Global South. It also gets exposed in ideas like global tipping points, which may make sense within the assumptions of natural science models but fail to capture how large-scale environmental processes are mediated by economic, political, and cultural processes, and thus experienced differently. This fundamentalism and naturalism poses a major challenge to environmental research across various disciplines.

To get beyond the current impasse, environmental research and thinking require a multidimensional framing of and re-engagement with the environment
as a simultaneously natural and social problem. Only then will new theoretical
and conceptual frameworks emerge that are encompassing enough to accommodate the wide range of environmentalisms, yet also sufficiently specific to
guide future research, facilitate cross-disciplinary dialogue, and link academic
scholarship to praxis.

In short, this Ernst Strüngmann Forum was envisioned as an opportunity to
reflect on the critical underpinnings of environmental research. It's overarching goal was to promote transformative environmental research that would
contribute to social learning in a meaningful way.

Sharachchandra Lele joined us in bringing this topic to the attention of the
Forum's Scientific Advisory Board, and it was subsequently approved for further development. Shortly before the Program Advisory Committee was due to
meet, however, a medical condition precluded Thomas's further involvement.
Thanks to the commitment of Sharad Lele and the other committee members—
Eduardo Brondizio, John Byrne, Georgina Mace, and Joan Martinez-Alier—
development continued.

This volume summarizes the resulting discourse from the 23rd Ernst
Strüngmann Forum and contains two types of contributions. First, background
information is provided on key areas. Drafted before the Forum, these chapters
have been revised to reflect input from the Forum as well as the peer reviews.
Second, the discussions of each group have been captured in "reports" (see
Chapters 4, 7, 10, and 12) to communicate the essence of this dynamic discourse, from the perspectives of all involved. These chapters are not consensus documents: they expose areas where opinions diverge and highlight topics
where future enquiry is needed.

An endeavor of this kind creates its own unique group dynamics and puts
demands on everyone who participates. Each invitee played an active role and
for their efforts, I am grateful to all. I also wish to extend a special word of
thanks to the authors and reviewers of the background papers as well as to
the moderators of the individual working groups: Georgina Mace, Eduardo
Brondizio, John Byrne, and Sharad Lele. The rapporteurs of the individual
working groups—Leticia Merino-Pérez and Esther Mwangi, Xumei Bai, Sun-
Jin Yun, and Amber Wutich—deserve special recognition, for to draft a report
during the Forum and finalize it afterward is no simple matter. Finally, I wish to
extend my appreciation to Thomas Sikor and Sharad Lele, the proposers of this
23rd Ernst Strüngmann Forum, as well as to the volume editors: Sharad Lele,
Eduardo Brondizio, John Byrne, Georgina Mace, and Joan Martinez-Alier.

A communication process of this nature relies on institutional stability and
an environment that encourages free thought. The Ernst Strüngmann Forum is
made possible through the generous support of the Ernst Strüngmann Foundation, established by Dr. Andreas and Dr. Thomas Strüngmann in honor of
their father. In addition, the following valuable partnerships are gratefully

acknowledged: the Scientific Advisory Board of the Ernst Strüngmann Forum, which ensures the scientific independence of the Forum; the German Science Foundation, for its supplemental financial support; and the Frankfurt Institute for Advanced Studies, which shares its intellectual setting with us.

Long-held views are never easy to put aside. Yet, when this is achieved, when we begin to probe the very limits to our understanding, existing gaps in knowledge start to become visible. Formulating strategies thereafter to address such gaps can be a most invigorating activity. On behalf of everyone involved, I hope this volume will spur further discourse and promote the use of more inclusive framings of environmentalism to enable socially relevant, impactful research and practice.

Julia R. Lupp, Director, Ernst Strüngmann Forum
Frankfurt Institute for Advanced Studies (FIAS)
Ruth-Moufang-Str. 1, 60438 Frankfurt am Main, Germany
https://esforum.de/

List of Contributors

Bai, Xuemei Fenner School of Environment and Society, College of Science, Australian National University, Canberra, Australia

Baker, Lucy Science Policy Research Unit (SPRU), University of Sussex, Falmer Campus, Brighton, BN1 9RH, U.K.

Baviskar, Amita Institute of Economic Growth, Delhi University Enclave, Delhi 110007, India

Boelens, Rutgerd Water Resources Management, Wageningen University, Center of Latin American Studies, University of Amsterdam, The Netherlands

Bond, Patrick School of Governance, University of the Witwatersrand, Johannesburg, South Africa; University of KwaZulu-Natal Centre for Civil Society, Durban, South Africa; and Leverhulme Centre for the Study of Value, University of Manchester, U.K.

Brondizio, Eduardo S. Department of Anthropology, Indiana University, Bloomington, IN 47405-710, U.S.A.

Bullard, Robert D. School of Public Affairs, Texas Southern University, Houston, TX 77004, U.S.A.

Byrne, John Center for Energy and Environmental Policy, University of Delaware, Newark, DE 19716-7301; and Foundation for Renewable Energy and Environment, New York, NY 10111, U.S.A.

Cardenas, Juan-Camilo Department of Economics, Universidad de los Andes, Bogotá, Columbia

Domínguez Guzmán, Carolina University of Amsterdam, Anthropology, Amsterdam, The Netherlands

Edwards, Gareth A. S. School of International Development, University of East Anglia, Norfolk NR4 7TJ, U.K.

Fantini, Emanuele IHE Delft Institute for Water Education, Department of Integrated Water Systems and Governance, Delft, The Netherlands

Fischedick, Manfred Wuppertal Institute for Climate, Environment and Energy, 42103 Wuppertal, Germany

Grimm, Nancy B. School of Life Sciences and Julie Ann Wrigley Global Institute of Sustainability, Arizona State University, Tempe, Arizona 85287-4501, U.S.A.

Hermwille, Lukas Wuppertal Institute for Climate, Environment and Energy, 42103 Wuppertal, Germany; and Institute for Environmental Studies (IVM), Vrije Universiteit Amsterdam, 1081 HV Amsterdam, The Netherlands

Kaufmann, Götz Environmental Policy Research Centre (FFU), Freie Universität Berlin, 14195 Berlin, Germany

Lehmann, Ina University of Bremen, artec Sustainability Research Center, 28359 Bremen, Germany

Lele, Sharachchandra Centre for Environment and Development, Ashoka Trust for Research in Ecology and the Environment (ATREE), Bangalore 560064, India

Lora-Wainwright, Anna School of Geography and China Centre, Oxford University, Oxford OX1 3UD, U.K.

Luhmann, Hans-Jochen Wuppertal Institute for Climate, Environment and Energy, 42103 Wuppertal, Germany

Lund, Peter D. School of Science, Aalto University, 00076 Aalto, Espoo, Finland

Mace, Georgina M. Department of Genetics, Evolution and Environment, University College London, London WC1E 6BT, U.K.

Martinez-Alier, Joan Department of Economics and Economic History, ICTA, Universitat Autònoma de Barcelona, Barcelona, Spain

Merino-Pérez, Leticia Instituto de Investigaciones Sociales, Universidad Nacional Autònoma de México, Ciudad Universitaria; UNAM Seminario on Society, Mexico City, c.p. 06140, Mexico

Minang, Peter A. Landscapes Governance Theme and Global Coordinator, ASB Partnership, World Agroforestry Centre, ICRAF, Gigiri, 00100 Nairobi, Kenya

Mwangi, Esther Forests and Governance Research, Center for International Forestry Research, Nairobi, Kenya

Özkaynak, Begüm Department of Economics, Boğaziçi University, 34342 Bebek, Istanbul, Turkey

Pahl-Wostl, Claudia Institut für Umweltsystemforschung, Universität Osnabrück, 49076 Osnabrück, Germany

Pascual, Unai Basque Centre for Climate Change (BC3) and Basque Foundation for Science, Ikerbasque, Leioa 48940, Spain

Rap, Edwin International Water Management Institute, Cairo, Egypt

Rauschmayer, Felix Department of Environmental Politics, Helmholtz Centre for Environmental Research, 04318 Leipzig, Germany

Redford, Kent H. Archipelago Consulting, Portland, ME 04112; and Department of Environmental Studies, University of New England, Biddeford, ME 04005, U.S.A.

Reyes-García, Victoria ICREA, Barcelona, and Environmental Science and Technology Institute (ICTA), Universitat Autònoma de Barcelona, Cerdanyola del Vallès, Barcelona, Spain

Schindler, Seth   Global Development Institute, University of Manchester, Arthur Lewis Building, Manchester M13 9PL, U.K.

Schleyer, Christian   Section of International Agricultural Policy and Environmental Governance, University of Kassel, 37213 Witzenhausen, Germany

Smit, Hermen   Institute for Water Education, Department of Integrated Water Systems and Governance, IHE Delft, The Netherlands

Stelzer, Franziska   Wuppertal Institute for Climate, Environment and Energy, 42103 Wuppertal, Germany

Suhardiman, Diana   Governance and Gender Research Group, IWMI, Vientiane, Lao PDR

Tallis, Heather   Office of the Chief Scientist, The Nature Conservancy, Santa Cruz, CA 95065, U.S.A.

Taminiau, Job   Foundation for Renewable Energy and Environment, New York, NY 10111, U.S.A.

Vallentin, Daniel   Wuppertal Institute for Climate, Environment and Energy, 42103 Wuppertal, Germany

van der Zaag, Pieter   Department of Integrated Water Systems and Governance, IHE Delft Institute for Water Education; and Water Resources Section, Delft University of Technology, Delft, The Netherlands

Wutich, Amber   School of Human Evolution and Social Change, Center for Global Health, Arizona State University, Tempe, AZ 85287-2402, U.S.A.

Yang, Fuqiang   Natural Resources Defense Council, China Program, Beijing, 100026, P. R. China

Yun, Sun-Jin   Graduate School of Environmental Studies, Seoul National University, Seoul 08826, Republic of Korea

Zwarteveen, Margreet   Department of Integrated Water Systems and Governance, IHE Delft Institute for Water Education, Delft; and Department of Human Geography, Planning, and International Development Studies, University of Amsterdam, 1018 WV Amsterdam, The Netherlands

1

Framing the Environment

Sharachchandra Lele, Eduardo S. Brondizio,
John Byrne, Georgina M. Mace, and Joan Martinez-Alier

Background

Among the many societal problems thrown up during a tumultuous twentieth century, it would be fair to say that "environmental problems" have been salient, and this salience has only grown as we entered the twenty-first century. Pockets of local pollution that popped up in the 1950s and 1960s, such as DDT, which led to thinning egg shells or methylmercury poisoning of fish and people in Minamata, Japan, were the harbingers of the larger and more dispersed crisis to follow—a crisis that has encompassed all aspects of human and nonhuman life, from deforestation and soil erosion to groundwater depletion and river basin closure in many river basins, from urban air pollution in Los Angeles to acid rain in Germany, and from dam-related displacement in China or India to Chernobyl- and Fukushima-type nuclear disasters. Cutting across all these locations, climate change, induced primarily by the burning of fossil fuels, is considered to be the "mother of all environmental problems," not for its own sake, but for the way it introduces stress and uncertainties into this already precarious socioenvironmental situation.

Concern about these problems, popularly labeled as "environmentalism,"[1] has triggered a large body of research and activism. If one uses the presence of terms such as "environment" or "sustainability" in the media or the large number of environmental pronouncements, policies, laws, agreements, and programs enacted at the local, national, and international levels since the 1970s as indicators, one would think that environmental concerns have been mainstreamed. Indeed, some successes are incontrovertible (e.g., the elimination of lead in gasoline or the phasing out of ozone-reducing chlorofluorocarbons). But systemic change is a far cry, and ideas about pathways forward are sharply divisive. We still get the sense that society is hurtling at an ever faster pace

---

[1] Although all "-isms" have an activist or ideological connotation, we simply mean here any research or action that recognizes some biophysical limits to and linkages between human actions and well-being in the broadest possible sense.

toward a world depleted of biodiversity, wracked by cataclysmic climate change, and facing a wide array of regional environmental crises due to novel hazards, resource scarcity, pollution, and an ever more "risk society."

What are the obstacles to making progress? Certainly a large part of the problem lies outside the spheres of environmental research or activism—in the deeply ingrained individual affinity to enjoy the fruits of the Industrial Revolution and its aftermath, while externalizing its downside onto future generations, the Global South, or nonhuman living beings. The problem is also embedded in the societal structures that facilitate these unjust, unsustainable, and arguably regressive forms of "development," "well-being," or the ideals of modernity disseminated since World War II.[2] Now there are even efforts to begin resource exploitation in the deep sea and to explore possibilities on other planets or asteroids, so insatiable is the human demand for scarce resources. Environmentalism poses a fundamental challenge to these ideas of development as well as the methods by which we try to achieve it. Thus it is not surprising that it engenders significant, if not virulent, opposition.

Tensions and misunderstandings among environmentalists also contribute to limiting progress on the ground. Some key examples are:

- biological conservation versus rights of indigenous communities (or "tigers versus tribals" as it is referred to in India) (Seminar 2005),
- climate sustainability versus climate equity positions (Dubash 2009),
- conflicts over mega-dam projects that pit nature-as-resource versus nature-as-life perspectives (Whitehead 2007).

Mirroring these conflicts in the activist world are bitter academic debates over the instrumental values of ecosystem services and the intrinsic value of biodiversity, over economic models of climate mitigation or the treatment of uncertainty in climate change mitigation policies, and over the role of population growth versus global consumption in tropical deforestation (Lambin et al. 2001) or environmental degradation, more generally. The intense and almost never-ending debate over whether "sustainable development" is a reasonable characterization of societal goals or a sellout to the status quo is a reflection of these tensions (Colby 1989; Lele 1991, 2013).

Within environmental thinking and research, these tensions originate from the different ways in which environmental problems are "framed." These problem framings differ on at least two dimensions: the values they prioritize and the explanatory theories they use, and therefore on the futures they envision. First, environmental research, like all applied research, is necessarily laden with values (Lele and Norgaard 1996; Jones et al. 1999). Environmental

---

[2] To give just one example, it is well known that "conventional economic accounting is false: it forgets the physical and biological aspects of the economy, it forgets the value of unpaid domestic and voluntary work, and it does not really measure the welfare and happiness of the population" (Martinez-Alier 2008). Yet, decision makers continue to use gross domestic product as the first measure of a country's health.

changes—whether in biodiversity, river flows, or forest cover—become "problems" only because some group of people in society cares about them. In other words, environmental change is simply a process: it is human interests and values that attribute negative or positive "value" to such change. Similarly, any goals that are set, such as sustainable development, as well as the criteria and indicators that will be used to measure progress toward them, are value-laden. The values that are included or prioritized determine which environmental processes are seen as problems, in what sense and context, and shape the solutions. When values are not shared widely or are not inclusive enough, the value framing becomes a major arena of debate and contestation, often hampering the achievement of what may ultimately be a common good.

Second, socioenvironmental research and action require an understanding of why human beings act in ways that lead to environmental degradation (in whatever sense of the term). Our understanding of social (and socioenvironmental) systems, however, is incomplete and, in spite of significant efforts, fragmented. The social sciences offer multiple, but often mutually incompatible perspectives, theories, and explanations for environmental change. More often than not, research on environmental problems appears to have been appropriated by academic communities in ways that reinforces epistemological territories as if sufficient in themselves to explain these complex problems (Brondizio et al. 2016). So we have environmental economics, environmental anthropology, cultural ecology, human ecology, environmental sociology, political ecology, environmental values, and environmental ethics, all of which contribute to advance understanding of such issues, but often limit the construction of an integrated understanding of environmental problems.

Objectives of the Forum

To move beyond the current fragmentation of ideas and approaches, environmental research and thinking require a multidimensional framing that transcends the divides between different ways of valuing the environment and understanding its condition. To achieve this requires a self-reflective exploration of how we, as researchers, study and mobilize evidence about environmental problems. This exploration was the unifying goal behind this Ernst Strüngmann Forum, which aimed (a) to understand how different framings of environmental problems are driven by differences in normative and theoretical positions and (b) to explore ways in which more inclusive framings might enable more societally relevant and impactful research and more concerted action/practice. Researchers from across the world gathered in Frankfurt, Germany, to discuss and debate these propositions in four sectoral or thematic areas:

- forests and other high-diversity ecosystems,
- urban environments,

- energy and climate change, and
- water.

This book is the outcome of those discussions.

In this introductory chapter, we outline the concept of framing, which was central to our deliberations at the Forum, and discuss in some detail two dimensions of framing environmental problems: the normative and the descriptive. The normative ideas of sustainability, diversity, and justice are central themes in the environmental discourse, and we provide an overview of the ways in which they have evolved as well as the nuances and linkages that have emerged. The descriptive (and analytical) dimensions of framing (i.e., the multiple perspectives on explaining and then proposing solutions to environmental problems) are then summarized in brief. A more tangible engagement with these and other dimensions of the framing of environmental problems emerges in the subsequent chapters, which are organized along the four themes mentioned above. These chapters are introduced in the penultimate section, followed by a summary of the key insights from the Forum.

Framing

In a highly cited article, Robert Entman (1993:52) provided a succinct definition of framing:

> [t]o frame is to select some aspects of a perceived reality and make them more salient in a communicating text, in such a way as to promote a particular problem definition, causal interpretation, moral evaluation, and/or treatment recommendation for the item described. Typically frames diagnose, evaluate, and prescribe...

The example he gives to illustrate this concept is quite pithy:

> An example is the "cold war" frame...[that] highlighted certain foreign events— say, civil wars [in third world countries]—as problems, identified their source (communist rebels), offered moral judgments (atheistic aggression), and commended particular solutions (U.S. support for the other side).

Cognitive scientists point out that all thinking and talking involves using structures—consciously or unconsciously—that provide meaning and predict relationships (Lakoff 2010). They also invoke specific emotions, and often simplistic stereotypes. A single word (e.g., whales, forests) that is closely associated with a frame can trigger a set of emotions and ideas, including stereotypes. From an activist perspective, therefore, the question is not whether framing can be avoided, but rather whose frame is activated in the brains of the public (Lakoff 2010). In political communication, the "selection" that Entman refers to can be very deliberate and even manipulative, as certain causes or outcomes may be blocked out and others emphasized so as to garner support for particular (often narrow) policies or actions.

In academia or research, frames emerge less deliberately and are deployed less manipulatively, being more the product of internal "sense making" (Fiss and Hirsch 2005; Oughton and Bracken 2009); that is, ways of structuring a complex real-world situation so that one can understand and grapple with it. Using Entman's definition and applying it to environmental research, we see that framing an environmental problem happens by

- identifying a *phenomenon* (e.g., say tropical deforestation),
- evaluating it, implicitly or explicitly; that is, indicating *why* it is a problem in a societal sense[3] (e.g., because it results in loss of biodiversity, which is the heritage of humankind),
- identifying possible *causes* (e.g., expansion of cattle ranching), and
- eventually offering *solutions* (e.g., promoting agroforestry as an alternative).

Given, however, that the task of research is actually to uncover these links (in this case, between deforestation and biodiversity loss, between cattle ranching and deforestation, or between different solutions and their impact on cattle ranching and forests), one might be tempted to assume that research does not involve framing, or at least that it is accompanied with a certain amount of reflexivity—an awareness that one is using a particular frame that both values, bounds, and simplifies a problem in particular ways. Many researchers consider "objectivity" to be a necessary feature of the scientific method. Yet while subjectivity can be minimized, the influence of the researcher cannot be completely removed and "frameless" research is impossible.[4]

Additionally, all applied research relates to societal goals and is thus necessarily value loaded. All research also involves making choices about scale and scope, variables to include, the functional form of their interaction, and method of data collection and analyses (Lele and Norgaard 2005). Disciplines and subdisciplines crystallize these practices into spaces where most of these choices about what to study and how to study it are taken for granted, leaving a narrow but comfortable space within which conventional research then continues (Oughton and Bracken 2009). In doing so, choices about problem scope and framing may be rendered less visible or, alternatively, self-evident, making reflection and questioning difficult (Spangenberg 2011). To the extent that some subdisciplines have emerged that take an explicitly normative label, such as conservation biology (Soule 1985) or sustainability science (Kates et al. 2001), there appears to be some willingness to make the normative concerns explicit. This is a step forward, but, as we discuss below, these framings may still not include other environmental concerns.

---

[3] That is, a problem as something societally undesirable rather than a problem as a puzzle (as in a mathematical problem).

[4] The widespread use of the drivers-pressures-state-impact-response (DPSIR) frame for environmental problems, for example, has been shown to implicitly favor some discursive positions over others (Svarstad et al. 2008).

Our attempt is not to suggest that these choices, and therefore frames, can be done away with. We would, however, like to see an increased awareness of them and, if possible, greater inclusiveness in framing socioenvironmental research. As a first step, it would be worth exploring the central ideas or central tendencies in such research when it comes to both the normative and the descriptive dimensions of environmental problem framing. These are, of course, not the only dimensions involved in framing an environmental problem: framing also involves important choices about epistemology, methods, handling of uncertainty, and so on (Leach et al. 2010). Neither are the normative and descriptive, or these other dimensions, entirely separable. Nevertheless, for brevity, we have focused on these two main dimensions.

Why Care about the Environment?

Environmentalism does not have a single origin, either historical or geographical (Guha 2000; Guha and Martinez-Alier 1997). Not surprisingly, therefore, it also does not have a unitary value framework. Indeed it may be more appropriate to talk of environmentalisms. If DDT and methylmercury were of concern because of the threat they posed to animal and human health, the concerns about charismatic species such as whales, pandas, or tigers have different ethical bases, and the destruction of marine fisheries due to overfishing or the vulnerability of the urban poor to environmental events are of concern from yet other perspectives. What constitutes the underlying values or ethical or moral arguments in such cases has been the subject matter of much discussion in popular and academic writing (Dietz et al. 2005).

From our reading of the discourse, the dominant sets of values underpinning environmentalist positions appear to fall into three broad categories: *sustainability, justice, and diversity*.[5] These are broad labels, each subsuming a range of concepts and terms. Both this "subsuming" as well as individual terms are highly contested. Taken together, however, they appear to capture most environmental concerns in one way or another and, at the same time, there is enough difference to make the categories worthwhile:

- *Sustainability*: Having originated from a specific meaning in forestry that dates back to the eighteenth century, this term has now become a catch-all phrase (Dixon and Fallon 1989) to the point where it is used to denote any form of pro-environment behavior (Thiele 2013). It is useful, however, to consider its original usage: maintaining something over time. Overfishing today will make fish unavailable tomorrow, and thus sustainability in the context of fisheries has intuitive appeal, as does sustainability in forest management. The major question has been whether the intertemporal trajectory can and should be in the form of something resembling an equilibrium or, given a highly dynamic and

---

[5]   This matches the three environmentalisms identified by Guha and Martinez-Alier (1997).

changing world, in the form of a system bouncing back from shock and stress; that is, resilience (Leach et al. 2010). For some analysts, resilience is a more robust concept than sustainability, especially since it can also incorporate growth and not just stability. Both sustainability and resilience may, in turn, depend on adaptability, whereby a function is maintained in some way despite changing circumstances. Nevertheless, the concern driving the search for sustainability/resilience/adaptability is clearly an intertemporal one: wanting to have tomorrow (by and large) that which you have today. Depending on the time horizon of concern, this can be a rather selfish concern (for one's own future) or one that is more altruistic (concern for future generations). Difficult moral issues can arise in choosing between a known need of people that exist today and a potentially greater (but unknowable) need for future people who have not yet been born.

- *Justice,* equity, fairness, and related concepts have a longer intellectual history than sustainability. They may be invoked in "purely" social contexts, such as the injustice of racial discrimination, but even here there are links to material processes, such as when such racial discrimination deprives some persons from access to land or water or resources essential for life and livelihood (Mohai et al. 2009). Injustice may also be the direct outcome of environmental actions, whether it is the release of pollutants into a river that affects downstream water users or the pumping of groundwater by some that deprives others of that resource.[6] Often, the social and biophysical dimensions are overlaid: people suffering air pollution in U.S. cities have often been people of color, and people displaced by dams have often been marginalized ethnic groups (Bullard and Johnson 2000). The ideas of justice, equity, and fairness as applied in environmental justice are, however, complex and multifaceted, even as the latter continues to expand globally as an approach to socioenvironmental issues (Agyeman et al. 2016). Distributional justice focuses on outcomes, whereas procedural justice and recognition justice address ways in which decisions are taken and who is involved (Schlosberg 2009). In the environmental context, justice has also been expanded to include intergenerational justice (Weiss 1990), thereby overlapping with the concern for sustainability, and fairness to nonhuman species, thereby overlapping with the concern for biodiversity.
- *Diversity*: The concept of *biodiversity* currently captures the core of naturalists' concerns for the environment, subsuming earlier formulations such as wilderness or wildlife (but see Soule and Noss 1998). Here

---

[6] "Biophysical injustice" could be a term to distinguish injustice caused purely by the environmental location of the pollutee vis-à-vis the polluter from "environmental injustice," which is currently used to refer to situations where these locations are the outcome of the social disadvantage of the pollutee, such as the siting of polluting industries in poor African-American or Latino neighborhoods in U.S. cities.

the goal is to maintain the variety of life on Earth, which in common parlance is usually translated to the local and global number of species. While this measure continues to be the main focus, it has also become clear that diversity is also necessary above and below the species level. For example, genetic diversity within species can buffer species from environmental changes and adds to the variety of valued attributes and functions. Above the level of species, ecosystems differ in composition of species, functions, and attributes. The Convention on Biological Diversity, adopted in 1992, recognizes these three levels (genes, species, and ecosystems) explicitly and is framed in terms of the connection between this diversity and the material and nonmaterial values that societies derive from the environment. This also represents a subtle shift in the discourse from biodiversity as the ultimate goal, to biodiversity as the provider of multiple goals (Chan et al. 2016; Mace 2014). Simultaneously, diversity has been formulated in more social terms—diversity of languages, ethnicities, knowledge systems and ontologies, and institutions or more generally cultural diversity—and this is seen as good in itself (UNESCO 2002). In addition, there is agro-biodiversity—the diversity of crops and livestock—at the biocultural interface (Maffi and Woodley 2012). We believe that these three forms of diversity—biodiversity, cultural diversity and agrobiodiversity—are mutually reinforcing, and so the idea of biocultural diversity has found policy support. While this has allowed indigenous and local communities to reclaim rights to land and resources, and to repair historical social injustices, it has also created homogeneous expectations that local cultures are the guarantors and the producers of biological diversity, often disregarding their marginal social and economic conditions (Kohler and Brondizio 2017) and thus important justice dimensions of diversity.

The above provides only a cursory overview of the depth and breadth of thinking and debate in each of these dimensions of environmental concern. Taken together, we believe that these three overarching concepts capture most, if not all, of the reasons why environmentalists care about the environment. Although there is some conceptual or operational overlap[7] between these concerns, it is, however, clear that they are still quite distinct: championing one does not ensure progress on the other. In fact, there can be trade-offs: creating pristine "wilderness" areas will definitely impinge on the livelihoods of forest-dwelling communities; an exclusive focus on reductions in greenhouse gas emissions will impose unfair burdens or constraints on those who have

---

[7] Conceptually, for example, when sustainability is articulated as a form of equity. Operationally, for example, when it is argued that lesser disparity in sharing a resource is more likely to ensure collective action that is required to prevent resource degradation. Or when it is claimed that conserving the tiger will also sustain the flow of rivers for downstream water users, because tiger conservation requires forest conservation.

been least responsible for climate change (Brondizio and Le Tourneau 2016). Since an inclusive or "cross-cultural" environmental ethic (as espoused by Guha 1997) is rare, tensions and fragmentation along the dimensions described above are common within the environmental movement, and are mirrored or refracted in environmental research in complex ways.[8]

Central Tendencies in Explanatory Theories

Environmentally damaging behavior, as viewed from these different lenses, can also be explained variously, ranging from theories that focus on political and economic processes to those concerned with various dimensions of human behavior. These theories, embraced by different social science subdisciplines, vary in scope, scale, and level of analysis, types of causality, and level of determinism (VanWey et al. 2005). They invoke, inter alia, individual agency and societal structures, sociocultural and environmental determinants, values and attitudes, population and consumption, technological change, and institutional arrangements (see, e.g., Robbins et al. 2011; Moran and Brondizio 2013). It would be impossible to do justice to these theories in the course of this chapter, but it would be fair to say that one of the major divides concerns structural versus agency-based explanations. Structural explanations privilege processes and conditions that drive and constrain individual actors, whereas agency-based explanations assume that individual actors have enough freedom to be considered as the drivers of change. In an environmental context, these divergent perspectives are exemplified by political ecology and neoclassical environmental economics, respectively. This and other divides, such as differences in language and terminology and different notions of evidence, constitute significant barriers to building more comprehensive explanations for environmental problems (e.g., Lele et al. 2002).

During the past two decades, however, conceptual and analytical "frameworks" (as opposed to theories) have emerged as metatheoretical tools aimed at uniting "pieces of a puzzle" and serving as vehicles for collaboration around complex and cross-scale socioenvironmental problems (e.g., Ostrom 2009). Such frameworks provide a common structure and language to support the analysis of a given phenomenon and/or problem. They identify relationships and directionality between components of a phenomenon without necessarily imposing a predefined causality between them. These frameworks can be organized at different levels of generality, from showing broad components and relationships that underlie a phenomenon (e.g., land use and cover change) to describing more specific processes (e.g., land-use intensification). Productive

---

[8] Tensions are not restricted to defining the "environmental goal." Sustainability and equity concern the temporal and spatial distribution of "human well-being," which is itself a value-laden concept. Making trade-offs between different goals requires additional choices about which process should be used to resolve these tensions. Thus, differences among environmentalists exist on these dimensions as well.

collaborations have emerged through the deployment of such frameworks (Binder et al. 2013) and, in the best cases, they have brought theories into the conversation rather than into false competition.

This is not to say that fundamental tensions have been resolved. However, the potential contribution of different theoretical tools to different problems at different scales is slowly being recognized. For instance, world system theory may best explain the unequal impacts of expanding extractive commodities, whereas collective action theory may explain the way people can overcome a commons dilemma. As the chapters that follow illustrate, there is increasing recognition that socioenvironmental problems are multidimensional, political, and value-laden; they are shaped by context and scale, and are subject to multiple framings. In other words, more than disputing the value of specific theories for their own sake, the focus needs to be on acknowledging the limitations of these theories, finding how they interrelate, and whether there are possible leveraging points of complementarity.[9] As such, we seem to be progressively moving toward subjecting multiple theories to a problem, rather than multiple problems to one theory or theoretical orientation. It is noticeable that the chapters converge in highlighting the importance of how a given social-environmental issue is "problematized," rather than starting with the selection of particular theories or specific conceptual frameworks. In other words, the authors ask what do we learn, who gains, and who loses when different theoretical, epistemological, and/or sociopolitical perspectives are used to address socioenvironmental problems. This is an important step toward bridging justice, diversity, and sustainability framings of environmentalism.

Overview of Chapters

As mentioned, four thematic areas were chosen to focus discussion at the Forum. Each section contains chapters that provide background to the theme as well as a synthesis of the discussions that took place during the Forum. Here we wish to highlight key aspects of these chapters.

Forests and Other High-Biodiversity Areas

The term forests today invokes ideas of naturalness, biodiversity, and various other environmental benefits with which high biodiversity areas are generally associated—ideas that drive conservation action. But what is it that we are trying to conserve, and is conservation even the best way to describe the goal? Speaking squarely to this question, Kent Redford and Georgina Mace (Chapter 2) focus on traditional biodiversity conservation and describe some recent debates in international conservation organizations and among academic

---

[9] For instance, political ecologists are asking how collaborations with commons or resilience theorists might be possible (Turner 2014, 2016).

conservation biologists. Despite the apparent simplicity of the idea that biodiversity conservation represents a concern to maintain the overall diversity of life on Earth, there are many different perspectives on what this means, how to measure it, at what scale, and using which kinds of values. Some of the most profound differences have arisen around the assumed or desired relationships between people and other species, how important these relationships are compared to purely biological or physical measures of diversity, and whose values are being prioritized. While long-running, these debates show little sign of convergence, and new issues are now emerging, such as debates about monetary valuation, new technologies such as genetic modification, and national versus international rights and responsibilities.

In Chapter 3, Peter Minang addresses the conceptual linkages between values and incentives in the context of forests, specifically focusing on three types of values: assigned, relational, and held values. Using the framing of ecosystem services and, in particular the payments for ecosystem services and reducing emissions from deforestation and degradation schemes, he shows how these different values interact in complex ways, affecting behaviors and choices as well as the outcomes of such schemes. Minang highlights especially the nonalignment of financial/economic and local/cultural values. Understanding the value-incentive relationship is shown to be important to avoid unanticipated and often perverse outcomes of apparently well-intentioned plans and policies. This requires sound knowledge of the context and may need to include multiple incentives or mixtures of incentives.

In the synthesis chapter, Leticia Merino-Pérez et al. analyze the diversity dimension of environmentalism. Although they use forests as a starting point, their discourse could apply to many other systems (e.g., coral reefs or tropical freshwater lakes), where the variety of species is a defining feature and diversity is valued in its own right. These areas are the traditional domain of conservation biology. The large, international conservation organizations have focused much of their work here, as these areas represent hotspots of both diversity and threat; substantial projects have been funded over many years, with some very successful and well-known outcomes. However, these areas are also where tensions between conservation and local people's rights have become increasingly evident, and where issues of justice and equity have grown over time at local and national scales. Merino-Pérez et al. review a suite of conservation initiatives across a range of geographic and political contexts, covering a variety of different values and objectives. The case studies presented highlight the wide diversity of values that underpin different framings of outcomes for forest systems, as well as the disparate governance mechanisms that are in place. Increasing attention has been directed to involving a wider range of stakeholders, especially local and indigenous communities, in the face of the evidence that some early conservation successes have stalled or will founder. The sustainability of many of these initiatives is also variable, especially as novel pressures and threats are encountered that were not originally anticipated when the initiatives were developed. Some novel pressures derive from global forces, such as climate change, international security, and migration, while others have to do with local factors such as land tenure and rights. All are affected by changing values over time, and new ways of working with or prioritizing different groups and their interests are needed. Often, these issues were not considered when the initiatives were initially planned and implemented, and in many cases, the project managers and governing institutions are not well set up to deal with them.

Urban Environments

Conventionally, urban environmental problems have been considered synonymous with air or water pollution and their associated challenges. The two background papers in this section, however, highlight other kinds of issues and, in the process, demonstrate how broad "environmental" framings can be.

In Chapter 5, Amita Baviskar presents an eloquent account of how social position and political-economic power influence the framing of urban environmental problems and priorities in unequal urban spaces in India. Through the lens of two neighbors—an upper middle-class family living in a comfortable high-rise apartment and their housemaid whose family occupies a shack next door—Baviskar shows how economic and sociopolitical power structures can define what is considered an environmental problem, and thus a priority, in complex and fast-changing urban areas. Drastically different lived experiences coexist next to each other. Deplorable sanitation conditions coexist with luxury, as much as manicured green spaces coexist with garbage dumps. This story of conviviality and distance, inequality and interdependence encapsulates the reality of cities across the Global South. It also addresses the way urban environmentalisms can be mobilized to the interest of different social groups, without necessarily addressing its contradictions and discrepancies.

Nancy Grimm and Seth Schindler (Chapter 6) use a social-ecological-technological system framing (SETS) to discuss the nature of cities as well as the nature in cities. They provide an instrumental approach to examine the potential integration of "green" and "gray" infrastructures as solutions to urban environmental problems. In doing so, they pinpoint the need to address deficiencies in urban services (e.g., sanitation), which particularly affect the Global South. Specifically, this is found in many fast-growing urban areas in Latin America, Africa, and Asia, where the absence of basic services and environmental degradation disproportionately burdens the urban poor. Grimm and Schindler provide an excellent overview of trends and patterns in global urbanization, raising questions about the social and environmental implications of highly concentrated settlements which, on the one hand, draw resources from vast areas around the globe and yet, on the other, represent the most vulnerable spaces to global environmental change. In approaching urban environments from the perspective of SETS, their aim is to avoid separating "nature" in urban spaces from built infrastructure. An urban SETS represents an ecosystem where nature, built infrastructure, and social conditions are coproduced by humans and nonhumans, biophysical endowments and the built environment. In making a case for the interdependence of the social, ecological, and technological components in urban spaces, they call for new approaches to urban design and planning in cities of the Global North and South.

In synthesizing the discussions, Xuemei Bai and colleagues provide in Chapter 7 a comprehensive review of multiple approaches to urban environmentalisms. They highlight how different types of concerns—from those related to species diversity to ecosystem services to the distant environmental impacts of cities—have influenced different types of framings and problem definitions. To examine these connections, the authors review five prominent framings applied to urban environmental issues and explore their relationship to persisting dualisms mobilized in such discussions: urban–rural, Global North versus Global South, brown–green agendas, and private versus common property rights. The five framings reviewed are (a) cities as SETS, (b) urban metabolism, (c) complex urban environments, (d) environmental justice, and (e) cities as solutions. Bai et al. show that urban environmental issues cannot be considered without attention to their regional and global connections. They issue a call for collaboration in the development of integrated conceptual framings and new analytical tools for reimagining urban futures. They make a case for the role of diverse urban constituents in bringing about desirable changes. Finally, recognizing that sustainable, diverse, and just urban futures require transformative change, they highlight the challenges associated with promoting plural environmental framings.

Energy and Climate Change

Since 1820, dramatic increases in per capita energy use have been matched by an eightfold increase in per capita income and a corresponding eightfold increase in per capita CO_2 emissions, the principal greenhouse gas responsible for climate change. While energy use, economic growth, and climate change are causally related, it is also true that their interrelationship has so far yielded socially unequal results. Different framings of the energy–climate problem emphasize different normative dimensions: unequal access to energy, long-term unsustainability of the global economy, threats to biodiversity under runaway climate change, or highly unequal impacts of even the current climatic changes. Analytically, the discourse is often polarized in terms of top-down governance versus bottom-up voluntarism and economic instruments versus more radical measures.

The two background chapters by Patrick Bond (Chapter 8) and Manfred Fischedick et al. (Chapter 9) take the Paris 2015 climate agreement as a starting point. Both chapters note that the governance "architecture" post-Paris seems fragmented and lacking vigorous enforcement mechanisms, but they offer different insights into how this may be triggering a rethinking of environmentalism in the energy–climate arena. Bond traces the positions taken by four individual South African environmentalists and a larger set of environmental organizations that have all been highly critical of the inadequate progress in international climate negotiations, but disagree over the implications of the Paris accord. At the heart of the divergence are different perspectives on the role of markets and of technology. The reformist approach accepts that change can only be incremental and achieved through market instruments and technological fixes. The radical approach rejects market-based solutions, such as cap-and-trade, and looks for stronger controls on corporate pollution complemented by local action. Bond suggests that a middle ground can, however, be found in concepts such as natural capital accounting, which uses some of the language of economics without going further down the path of complete monetization. Instead of treating capitalism as inevitable or just rejecting capitalism, Bond argues for an ecosocialist approach that uses multiple levels of mobilization and situates science to engage constructively with the challenge of reorganizing production and distribution.

For Fischedick et al. (Chapter 9), the recent emergence of "polycentric" social action has created more ambitious policy commitments than the earlier Kyoto Protocol structure, which had modest and weakly enforceable targets. The aspirations of decentralized or "bottom-up" efforts such as "100% renewable energy cities" and "carbon-free" mobility planning are collectively far surpassing the models championed by international treaties. Moreover, these efforts do not primarily rely on market mechanisms for their implementation. Instead, they use local and regional planning vehicles and civil society campaigns to contest carbon-intensive development and to mobilize communities to adopt much deeper energy conservation actions and accelerate renewable energy adoption more quickly than past national and international efforts. Indeed, these community-scale approaches deliver "governance by diffusion": multiple strategies are pursued and with each iteration a nonlinear process of action and innovation ensues. Finally, explicit inclusion of "climate justice" demands has been shown in the polycentric policy architecture to be crucial to obtaining diverse stakeholder support.

With this background, the synthesis chapter by Sun-Jin Yun et al. (Chapter 10) offers a detailed typology of framings of the energy–climate debate. This typology is presented as a means to distinguish clearly the aims, assumptions, and values of participants in the debate. The authors encourage an understanding of the conflicts between the framings as the basis upon which social change, or its hindrance, can be expected. They draw specific attention to the increasingly problematic status of market-based arguments and policies. Having been unable to realize sufficient political support to produce meaningful change after 20 years of the use of these arguments in international climate negotiations, and given their muted incorporation of climate justice concerns, Yun et al. suggest that participants in the debate are now focusing more intently on what are identified as analysis-focused framings and postmarket economy framings. Regarding the former, groups such as the Intergovernmental Panel on Climate Change could prepare justice-based transformation pathway analyses and climate life-cycle studies, while nongovernmental organizations could engage in critical policy analyses and action research to support the search for transformative change.

Postmarket economy framings are seen as focusing our attention on efforts that engage in political action rather than market-based policies to secure change. Examples are the efforts by a partnership of local government, citizen organizations, research groups, and advocacy movements in Seoul (South Korea) to reduce energy demand sufficiently to justify closure of coal and nuclear power plants; and a campaign to identify "unburnable fossil fuel reserves" as a means to require a shift toward sustainable energy options. The idea is to build political and economic support for "starving" the carbon energy regime by social means. Yun et al. recognize that transformative change is not, currently, favored by most political and economic leaders. The secession of the United States from the Paris Agreement stands as the most obvious example. Still, understanding framings and their conflicts, and looking for bridging concepts, is essential to addressing our mounting energy–climate conundrum, whether in the sphere of research or action.

Water

Water is as essential to human life and livelihood as energy. The particular characteristics of water—its mobility, bulkiness, cyclical nature, non-substitutability, and multiple uses (Savenije 2002)—make it one of the most contentious environmental issues. The academic literature on water is replete with instances where a disconnect results from alternative framings. Margreet Zwarteveen et al. (Chapter 11) explore one such disconnect: the differences in ways of knowing (modern versus traditional), in knowledge itself (universal versus particular), in the means of decision making (expert versus democratic), and the linkages that connect these issues to solving water problems. Taking a social constructivist position, the authors examine the water accounting approach (or frame) and argue that it is the product of layering (and mixing) particular values (e.g., efficiency or productivity) with certain readings of the waterscape (e.g., remote-sensing data interpreted in particular ways) to produce detailed explanations and policy recommendations (e.g., promoting drip irrigation in agriculture). They argue that while the water accounting approach is not "wrong," it is incomplete (because it misses out on other reasons for overuse), is insensitive to other concerns (such as equity), and often inaccurate (as remote sensing is plagued with inaccuracies that are only revealed by extensive ground-based work) (e.g., Heller et al. 2012). In addition, it can get easily aligned with a particular set of powerful actors that focus exclusively on technical and economic efficiency. In response, they call for toning down the ambitions and claims made by proponents of water accounting and combining it with other ways of understanding the values embedded in and the drivers of water use.

In the synthesis chapter (Chapter 12), Amber Wutich et al. engage in an in-depth discussion of how water questions are framed in different ways. Examining a range of frames common in the literature (e.g., integrated water resource management, water as a common-pool resource, the hydrosocial cycle), they describe the intellectual history of each frame, the values it emphasizes, the explanations it favors, and the assumptions it makes in the process. Further, they highlight points of overlap and tension between these different frames and outline some innovative ways to create more inclusive frames. They also ask whether more "inclusive" frames are always possible and/or desirable, and discuss the challenges and constraints connected with such inclusive framing. More integrated and inclusive framing of research may not emerge from academia but may need to be demanded by socioenvironmental movements.

Concluding Remarks

This Forum was convened with the idea of understanding the "internal" barriers to progress in environmentalism; that is, differences in the way that people concerned with environmental problems, particularly environmental researchers, think about these problems. Participants used the concept of problem "framing" as a tool to explore these differences in an effort to examine the potential for expanding the problem frames and the resulting challenges. To ground these explorations, we used the context of four sectors or thematic areas, each of which has a fairly distinct literature and set of environmental debates. Admittedly, much of the learning is individual, indirect, and hard to capture in words. A few common threads, however, did emerge that are worth summarizing in brief.

First, there is no question that different researchers bring very different perspectives to environmental problems: the normative dimension (why it is a "problem") and descriptive dimension (why the problem "occurs") are intertwined in complex ways, which makes mutual intelligibility and dialogue very challenging. Among academics, however, differences extend beyond the normative and descriptive dimensions into differences over method, over what constitutes evidence, or questions of "knowability" of the world. Among activists, differences may also be driven by strategic choices in a particular context.

Second, any discussion on sustainability, equity, and diversity is incomplete without a consideration of the fourth dimension—human well-being itself— that is sought to be sustained over time or distributed equally within society or modified to include the presence of wilderness or nature in it. To cast environmentalism as sustainability-ism or environmental justice-ism or diversity conservation-ism is to limit the idea. What is really needed is to rethink what constitutes societal well-being and how we might achieve it.

Third, the concept of "framing" does help unpack implicit normative and descriptive positions that are being taken in the analysis of an environmental problem, but such unpacking requires patience, reflexivity, and openness. Even then, there may be no easy "bridges" between different framings, because of the strongly intertwined nature of concerns, assumptions, methods, and so on. Some of the thematic groups concluded that the best case scenario might be increased self-awareness, or at least a partial integration of a few elements to increase understanding. There was also the perception that explicitly front-paging all values may sometimes reduce the chances of making an impact on the ground, because all stakeholders may not immediately be amenable to explicitly multidimensional approaches.

Fourth, academic structures, and the incentives or disincentives they create, generally reinforce centrifugal tendencies, aiding the mutual un-intelligibility of perspectives. Over the past decade or so, several attempts have been made to create space for a "different" science, such as a "sustainability science," or inclusive frameworks, such as the social-ecological systems framework. Much will depend, however, on how the structures and incentives within academia are changed to support such centripetal or integrative efforts.

Fifth, it seems likely that the push for integration will come from the crucible of action, and so the test of "adequacy" of integration will come from praxis. However, this requires reflective praxis, because community mobilization or resistance can be as limiting as dry intellectualism. In that context, the "bridging" across academics, practitioners, and activists seems as crucial as the bridges within academia.

Finally, it is clear that much of the "bridging" happens internally in unknowable ways: the process, in a sense, is the outcome. Multiple, continuous, and more diverse forums of this kind will enable more cross-disciplinary and cross-perspective dialogue within the environmental research community as well as between researchers, practitioners, and activists. This is necessary to bring about a better, more self-reflective understanding of environmentalism(s).

References

Agyeman, J., D. Schlosberg, L. Craven, and C. Matthews. 2016. Trends and Directions in Environmental Justice: From Inequity to Everyday Life, Community, and Just Sustainabilities. *Annu. Rev. Environ. Resour.* **41**:321–340.

Binder, C., J. Hinkel, P. Bots, and C. Pahl-Wostl. 2013. Comparison of Frameworks for Analyzing Social-Ecological Systems. *Ecol. Soc.* **18**:26.

Brondizio, E., and F.-M. Le Tourneau. 2016. Environmental Governance for All. *Science* **352**:1272–1273.

Brondizio, E. S., K. O'Brien, X. Bai, et al. 2016. Re-Conceptualizing the Anthropocene: A Call for Collaboration. *Global Environ. Change* **39**:318–327.

Bullard, R. D., and G. S. Johnson. 2000. Environmentalism and Public Policy: Environmental Justice: Grassroots Activism and Its Impact on Public Policy Decision Making. *J. Soc. Iss.* **56**:555–578.

Chan, K. M. A., P. Balvanera, K. Benessaiah, et al. 2016. Why Protect Nature? Rethinking Values and the Environment. *PNAS* **113**:1462–1465.

Colby, M. E. 1989. The Evolution of Paradigms of Environmental Management in Development, SPR Discussion Paper No. 1. Strategic Planning and Review Department, The World Bank. http://documents.worldbank.org/curated/en/160721468764740558/pdf/multi-page.pdf. (accessed Nov. 1, 2017).

Dietz, T., A. Fitzgerald, and R. Shwom. 2005. Environmental Values. *Annu. Rev. Environ. Resour.* **30**:335–372.

Dixon, J. A., and L. Fallon. 1989. The Concept of Sustainability: Origins, Extensions, and Usefulness for Policy. *Soc. Nat. Resour.* **2**:73–84.

Dubash, N. K. 2009. Environmentalism in the Age of Climate Change. *Seminar* **601**:63–66.

Entman, R. M. 1993. Framing: Toward Clarification of a Fractured Paradigm. *J. Commun.* **43(4)**:51–58.

Fiss, P. C., and P. M. Hirsch. 2005. The Discourse of Globalization: Framing and Sense Making of an Emerging Concept. *Am. Sociol. Rev.* **70**:29–52.

Guha, R. 1997. Towards a Cross-Cultural Environmental Ethic. In: Varieties of Environmentalism: Essays North and South, ed. R. Guha and J. Martinez-Alier, pp. 77–91. London: Earthscan.

———. 2000. Environmentalism: A Global History. New York: Longman.

Guha, R., and J. Martinez-Alier, eds. 1997. Varieties of Environmentalism: Essays North and South. London: Earthscan.

Heller, E., J. M. Rhemtulla, S. Lele, et al. 2012. Mapping Crop Types, Irrigated Areas, and Cropping Intensities in Heterogeneous Landscapes of Southern India Using Multi-Temporal Medium-Resolution Imagery: Implications for Assessing Water Use in Agriculture. *Photogramm. Eng. Remote Sensing* **78**:815–827.

Jones, P. C., J. Q. Merritt, and C. Palmer. 1999. Critical Thinking and Interdisciplinarity in Environmental Higher Education: The Case for Epistemological and Values Awareness. *J. Geogr. High. Ed.* **23**:349–357.

Kates, R. W., W. C. Clark, R. Corell, et al. 2001. Sustainability Science. *Science* **292**:641 – 642.

Kohler, F., and E. S. Brondizio. 2017. Considering the Needs of Indigenous and Local Populations in Conservation Programs. *Conserv. Biol.* **31**:245–251.

Lakoff, G. 2010. Why It Matters How We Frame the Environment. *Environ. Comm.* **4**:70–81.

Lambin, E. F., B. L. Turner, H. J. Geist, et al. 2001. The Causes of Land-Use and Land-Cover Change: Moving Beyond the Myths. *Global Environ. Change* **11**:261–269.

Leach, M., I. Scoones, and A. Stirling. 2010. Dynamic Sustainabilities: Linking Technology, Environment and Social Justice. London: Earthscan.

Lele, S. 1991. Sustainable Development: A Critical Review. *World Dev.* **19**:607–621.

———. 2013. Rethinking Sustainable Development. *Curr. History* **112**:311–316.

Lele, S., G. Kadekodi, and B. Agrawal, eds. 2002. Interdisciplinarity in Environmental Research: Concepts, Barriers and Possibilities. New Delhi: Indian Society for Ecological Economics.

Lele, S., and R. B. Norgaard. 1996. Sustainability and the Scientist's Burden. *Conserv. Biol.* **10**:354–365.

———. 2005. Practicing Interdisciplinarity. *Bioscience* **55**:967–975.

Mace, G. M. 2014. Whose Conservation? *Science* **345**:1558–1560.

Maffi, L., and E. Woodley. 2012. Biocultural Diversity Conservation: A Global Sourcebook. London: Routledge.

Martinez-Alier, J. 2008. Languages of Valuation. *Econ. Polit. Wkly.* **43(48)**:28–32.

Mohai, P., D. Pellow, and J. T. Roberts. 2009. Environmental Justice. *Annu. Rev. Environ. Resour.* **34**:405–430.

Moran, E. F., and E. S. Brondizio. 2013. Introduction to Human-Environment Interactions Research. In: Human-Environment Interactions, ed. E. S. Brondizio and E. F. Moran, pp. 1–24. Dordrecht: Springer Scientific.

Ostrom, E. 2009. A General Framework for Analyzing Sustainability of Social-Ecological Systems. *Science* **325**:419–422.

Oughton, E., and L. Bracken. 2009. Interdisciplinary Research: Framing and Reframing. *Area* **41**:385–394.

Robbins, P., J. Hintz, and S. A. Moore. 2010. Environment and Society: A Critical Introduction, vol. 13. Chichester: Wiley-Blackwell.

Savenije, H. H. 2002. Why Water Is Not an Ordinary Economic Good, or Why the Girl Is Special. *Phys. Chem. Earth* **27**:741–744.

Schlosberg, D. 2009. Defining Environmental Justice: Theories, Movements, and Nature. Oxford: Oxford Univ. Press.

Seminar. 2005. Forests and Tribals, Special Issue 552. http://www.india-seminar.com/semsearch.htm. (accessed Nov. 1, 2017).

Soule, M. E. 1985. What Is Conservation Biology? *Bioscience* **35**:727–734.

Soule, M. E., and R. Noss. 1998. Rewilding and Biodiversity: Complementary Goals for Continental Conservation. *Wild Earth* **8(3)**:18–28.

Spangenberg, J. H. 2011. Sustainability Science: A Review, an Analysis and Some Empirical Lessons. *Environ. Conserv.* **38**:275–287.

Svarstad, H., L. K. Petersen, D. Rothman, H. Siepel, and F. Wätzold. 2008. Discursive Biases of the Environmental Research Framework Dpsir. *Land Use Policy* **25**:116–125.

Thiele, L. P. 2013. Sustainability. Chichester: John Wiley & Sons.

Turner, M. D. 2014. Political Ecology I: An Alliance with Resilience? *Prog. Hum. Geogr.* **38**:616–623.

———. 2016. Political Ecology III: The Commons and Commoning. *Prog. Hum. Geogr.* **1**:1–8.

UNESCO. 2002. Universal Declaration on Cultural Diversity. UNESCO. http://unesdoc.unesco.org/images/0012/001271/127160m.pdf. (accessed Nov. 1, 2017).

VanWey, L. K., E. Ostrom, and V. Meretsky. 2005. Theories Underlying the Study of Human–Environment Interactions. In: Seeing the Forest and the Trees: Human–Environment Interactions in Forest Ecosystems, ed. E. F. Moran and E. Ostrom, pp. 26–53. Cambridge, MA: MIT Press.

Weiss, E. B. 1990. In Fairness to Future Generations. *Environ. Sci. Pol. Sust. Develop.* **32**:6–31.

Whitehead, J. 2007. Sunken Voices: Adivasis, Neo-Gandhian Environmentalism and State: Civil Society Relations in the Narmada Valley 1998–2001. *Anthropologica* **49**:231–243.

Forests and Other High-Diversity Ecosystems

2

Conserving and Contesting Biodiversity in the Homogocene

Kent H. Redford and Georgina M. Mace

Abstract

Discussions of environmentalism frequently become considerations of biodiversity and its conservation. Arguably, the defining feature of our planet is the extent and diversity of life on Earth, and there is increasing recognition that in addition to representing a loss of culturally valued elements, the ongoing loss of the diversity of life will prejudice human development in a multitude of ways. However, the framing of the problem, the approaches to defining and achieving change, and even the very definition of the term "biodiversity" are vague and malleable. One consequence is that the conservation of life on Earth is often at odds with other environmental and economic growth priorities, and this can be further confounded by different values among different stakeholders. This chapter reviews the background to conserving and contesting biodiversity especially from the perspective of conservation and with reference to high-diversity areas such as tropical forests.

Introduction

Over the last centuries, and accelerating since World War II, there has been a simplification of human and natural systems in the pursuit of productivity and efficiency for human use and consumption. Such global activities have resulted in erosion of site-specific biological diversity, agrobiodiversity, linguistic diversity, and cultural diversity, earning the current century its name as the "Homogocene" (Rosenzweig 2001). Local diversity of all these types has largely been eroded due to overlapping causes—the global spread, intensification, and homogenization of industry, agriculture, and culture (Redford and Brosius 2006). This chapter uses the framing of "homogenization" as it is a powerful result of loss of diversity of all types.

In this essay, we focus on the biological component of diversity and argue that although biodiversity is thought of as a single thing, the term has multiple meanings which differ in technical and value-based ways. In the broader context of this volume, considering the different ways in which environmental problems are "framed" around different underlying values and explanatory theories, biodiversity considerations have a central role.

Biological diversity, or biodiversity in all its forms, is being eroded by economic and cultural globalization, extinction, and non-native species. Being lost everywhere are unique and locally distinctive assemblages of species and their ecological interactions. Globally, most dimensions of biodiversity are decreasing, and there are many who believe that humanity has caused the sixth great extinction (Ceballos et al. 2015). This term has, over the last few decades, become the rallying cry for many people and organizations.

Many have called for greater support for local, national, and global efforts to conserve biodiversity, such as may be achieved through the Intergovernmental Science–Policy Platform on Biodiversity and Ecosystem Services (IPBES). This broad-based support is often predicated on the understanding that biodiversity is something good, and its conservation is therefore desired. In a colloquial fashion, biodiversity is favored in opposition to such clear hallmarks of modern human impact: shopping malls, urban sprawl, vast monocultures, oil palm plantations, and ocean life overexploitation. Biodiversity is often thought of more for what it is not—the human-dominated world—rather than for what it is—a bricolage. Often undefined, but commonly extolled, biodiversity has become something easy to love and yet hard for which to be held accountable.

"High-diversity" areas, such as tropical forests or coral reefs, provide a particularly good focus from which to examine some core conservation ideals about richness, intactness, native-ness, wildness, and endemism, and how these ideals have played out with local versus international interventions. The values people hold for such places vary among different actors (e.g., conservation NGOs, international aid donors, local communities) and over time. Understanding and reconciling these differences is a crucial step in avoiding further homogenization of cultural, agricultural, and biological diversity.

We begin with a review of definitions and uses of the terms biodiversity and conservation, principally from the framework of natural science and conservation NGOs. We highlight some major areas where there are differences in understanding, interpretation, and underlying values, and how these affect attempts to conserve life on Earth. We believe that to rethink or rebuild environmentalism, it is essential to consider not only justice, sustainability, and diversity, but also to look carefully at the underlying biological diversity that has powered a significant part of human progress.

Defining Biodiversity

Biodiversity has replaced nature as the object of interest for the conservation community with tens of millions of dollars spent to conserve it, organizations founded to save it, and global conventions put into place to regulate its management. In many conservation discussions, the term "biodiversity" is taken for granted; the assumption is made that when using it, everyone is talking about the same thing. Yet, biodiversity is a fairly new term and is often not defined in the same way by different people, or not defined at all.

Norse (1990) summarized the early history of the term, locating its roots in the late 1950s in the work of Hutchinson and MacArthur (this account is summarized and updated from Sanderson and Redford 1997 as well as Takacs 1996; see these references for a full list of citations). In the 1970s, the richness of species was called "natural diversity" by The Nature Conservancy while others described "genetic diversity." In 1980, Thomas Lovejoy used the term "biological diversity" without defining it, and the 1980 Annual Report of the U.S. Council on Environmental Quality also used a definition of biological diversity that included the concepts of genetic diversity and species richness.

Despite the lack of a specific definition, the term was picked up by the U.S. Government, which convened a "Strategy Conference on Biological Diversity," and in 1983 it became the goal of legislation passed by the U.S. Congress. By the mid-1980s, the first full definitions of the term were published by Burley (1984) and Norse et al. (1986). In 1988, E. O. Wilson edited the book *Biodiversity* based on a U.S. National Academy of Sciences meeting entitled "The National Forum on BioDiversity." This meeting focused on the value of biodiversity with talks from development experts, economists, and ethicists joining natural scientists in outlining what became known as the biodiversity crisis (Wilson 1988).

Article 2 from the Convention on Biological Diversity[1] provides a formal definition:

> "Biological diversity" means the variability among living organisms from all sources including, inter alia, terrestrial, marine, and other aquatic ecosystems, and the ecological complexes of which they are part; this includes diversity within species, between species and of ecosystems.

Biological diversity is usually interpreted as occurring at three major levels (Redford and Richter 1999)—genes, species, and ecosystems—though some practitioners include populations, communities, ecosystems, and biomes as well. The specific ways of measuring biodiversity vary by different practitioners (see Mace 2014a) but often include the following:

- Diversity of the genetic component refers to the variability within a species, as measured by the variation in genes within a particular species, subspecies, or population.
- Diversity of the species component refers to the variety of living species and their component populations at the local, regional, or global scale.
- Diversity of the ecosystem component refers to a group of diverse organisms, guilds, and patch types occurring in the same environment or

---

[1]   https://www.cbd.int/convention/articles/default.shtml?a=cbd-02 (accessed April 24, 2017).

area, and strongly interacting through trophic, spatial biotic, and abiotic relationships.

In practice, any effort at biodiversity measurement faces enormous problems due to gaps and biases in the information available. Probably less than 10% of all the species on Earth have been described and named, and those that are known are strongly biased toward vertebrates, terrestrial, and temperate areas. Different disciplines favor different measures of biodiversity. Ecologists tend to think about biodiversity in terms of the forms and functions of organisms in a place, especially in a community or an ecosystem, because it is the structuring of varieties in space and time that leads to functions and dynamics that they seek to understand. Similarly, evolutionary biologists think about the dynamics, but with an increasing focus on the historical or inherited variation, and therefore the genetic and phylogenetic attributes. Conservation biologists are sometimes concerned with function and process, but often also with preservation of species or genetic diversity, seeking efficient and achievable solutions to the allocation of limited resources. For nature conservationists and wildlife managers, biodiversity often simply means the maintenance of wild habitats and species (Mace 2014b). In other disciplines, the concept of biodiversity often lacks the notion of diversity; for example, in economics, biodiversity is generally understood simply to mean species, natural resources, or forests (Kontoleon et al. 2007). Many people use the term biodiversity in one of two ways: either as a general word to refer to "all life on Earth" or as a measure of the number of species—species richness.

The ecosystem component of biodiversity has received significantly less attention and the genetic component hardly any at all. The focus on diversity within defined areas (such as hotspots) has also been a persistent theme despite the commonly held view that it is global biodiversity that is being discussed.

It is in the high-diversity areas of the world, especially those undergoing rapid economic development, where the lack of a common understanding of the multiple roles of biodiversity most often becomes an obstacle to planning and policy implementation. Here we consider the framing of the issue of biodiversity conservation from both the perspective of conservation science and conservation practice.

Conservation Science and the Measurement of Biodiversity

The measurement of diversity in ecological communities has a long and rich history in ecological and evolutionary science that is rather weakly linked to the conservation and policy activities described above. A suite of metrics has been developed for summarizing different dimensions of variability, over space and time, and across different hierarchical levels in the classification of species and of ecosystems. There are several monographs dedicated to biodiversity measurement in theory and in practice (Gaston 1996; Magurran 2004). Recognizing the difficulty that this lack of standardization poses for policy making, there has been a recent effort to identify a set of "essential biodiversity variables"; these are intended to constitute a more manageable set of metrics for policy makers, yet they represent the most important patterns in a range of policy-relevant contexts (Pereira et al. 2013). However, even this essential set contains six classes of metrics and over 25 categories of measurement (Brummitt et al. 2016). Without doubt, this complexity is an obstacle to the establishment of goals and targets, but it is also important to recognize that there is no single simple measure of biodiversity, especially given the very wide range of values, purposes, and contexts to which science and policy may be applied.

The ecological science metrics focus strongly on species richness as well as abundance. Abundance is important because many ecological processes are more affected by biomass than by diversity alone (Diaz et al. 2007). These measures vary over time and space, and recent reviews have focused on patterns of change in local diversity over time, changes in local diversity across the landscape, and combinations of these (McGill et al. 2015; Newbold et al. 2015), as well as changes to global diversity (Ceballos et al. 2015; Dirzo et al. 2014). These studies show how local (or small-scale) biodiversity change may be very different in both extent and nature from global (or large-scale) biodiversity change. Local diversity loss is variable but often smaller than global diversity loss, because local losses may be at least partially compensated for by non-native species migrating in, and generalist, wide-ranging species replacing local specialists. In some cases, this effect actually leads to no loss locally or perhaps even small increases (Sax and Gaines 2003). This may result in substantial changes to local ecological communities that may not be represented by metrics that count species but ignore species identity. These compositional changes driven by land-use change and intensification may be very profound (Newbold et al. 2015) and may have important consequences locally as well as globally, especially considering the potential consequences for ecological functions.

In practice, metrics used for biodiversity assessment in conservation do include other attributes of species. Especially important here is the state of the species assemblage in an area relative to some reference state, often pre-disturbance by industrialized humans. Measures of intactness (lack of disturbance), native-ness (species native to the area), and endemism (species that are only found in the local area) are thus all commonly prioritized in conservation planning. Levels of extinction risk are often important modifiers, especially in plans for protection and restoration with priority given to species closer to a risk of extinction.

In recent decades, with rapid improvements in the availability of both species and landscape occurrence data as well as remote-sensed observations and the analytical capability of Geographical Information System (GIS) tools (Jetz et al. 2012), there has been a proliferation of analyses of priority places and systems that consider ecological processes and patterns (Pressey et al. 2007) as well as future changes. However, whether these large-scale approaches embrace the full suite of locally and functionally relevant biodiversity components is unclear. In addition, though little recognized, GIS is not an entirely objective technology. Its use can entail significant assumptions about biodiversity distribution, in general, and human modified systems, in particular (Putz and Redford 2009). Local human needs and wants may be at odds with global or regional perspectives, and the biodiversity relevance for development is often contested, especially with respect to use and values.

Biodiversity and Conservation Values and Approaches

Biodiversity is often glossed as "the variety and variability of life"—a broad definition that makes the term of relevance to a very wide range of stakeholders.

Agricultural scientists and others concerned about the loss of crop and livestock breeds became advocates for biodiversity as well as the importance of agrobiodiversity (Jackson et al. 2007). Ethnobiologists working with agriculturalists growing traditional landraces joined the biodiversity bandwagon (Nazarea 2006), as did pharmaceutical companies prospecting for new drugs in wild species. Zoos, seeking new support for their traditional breeding of endangered species, joined indigenous and traditional peoples who positioned themselves as keepers of biodiversity.

When the possibility of a global treaty began to be discussed, all these and more interest groups lobbied to have their interests included. An early (1991) draft of the Convention on Biological Diversity (CBD) reflected this range of interests, as contained in the statement:

> Human cultural diversity could also be considered part of biodiversity....Cultural diversity is manifested by diversity in languages, religious beliefs, land management practices, art, music, social structure, crop selection, diet, and any number of other attributes of human society.

Though not kept in the final text, this plethora of interests and interest groups remains an important legacy of the original enthusiasm for the broad nature of the concept and the lack of an operational definition. When working within the international political system, it has proved impossible to resist the inclusion of the positions of divergent stakeholders—a fact that continues to make biodiversity difficult to measure.

Early support for the newly emerging term of biodiversity came from a wide range of stakeholders, but most influential were a handful of U.S. and British academics and conservationists, in particular E. O. Wilson, Peter Raven, Norman Myers, and Thomas Lovejoy. What these people had in common was a deep affinity for species. Led by Wilson and Raven, taxonomists themselves, and united by a common love of tropical forests and deep concern about their destruction, biodiversity rapidly became cast as the number of species in an area—for which tropical forests were particularly notable. Myers took these interpretations and built the concept of conservation "hotspots" where global attention should be focused. Promulgated by Conservation International and funded by the MacArthur Foundation and Global Environment Facility, the hotspots approach became a global movement, influencing billions of dollars in spending toward "biodiversity hotspots" which were really areas of high, and threatened, species richness. By focusing attention on hotspots, one prominent goal for conservation became to reduce the rate of species extinction.

Hotspots were not, however, adopted by most other conservation organizations because of different underlying values. All priority-setting exercises are based on values, and the value-based nature of priority setting is important to tease out because it explains differences between organizational priorities, such as the difference between the ecoregional approach and the hotspot approach (Redford et al. 2003). Values underlying hotspots include (a) preventing extinction as the highest priority conservation action and (b) the total number of species saved is more important than what those species are. On the other hand, the ecoregional approach is based on the value of representation: it is important to preserve biodiversity within its natural distribution everywhere it occurs, from the tundra, to savannas, to tropical forest.

Such differing value positions have been combined with a lack of clear agreement on the role of human activity and diverse knowledge types in creating and/or maintaining biodiversity. Diverging views about what biodiversity is most important is reflected in the use of the concept of biocultural diversity within the new (2012) IPBES. Biocultural diversity refers to human cultural diversity linked to biological diversity through use, tradition, or practice. This includes forest types resulting from long-term human practices, traditional grazing practices, crop varieties intercrossing with wild relatives, and rotational agriculture. IPBES parties (countries) working together have developed a conceptual framing of the linkages between people and nature which reflect a wider set of knowledge and value systems than earlier efforts, such as the Millennium Assessment, which were more straightforward products of Western scientific methods and approaches (Diaz et al. 2015). As such, we are set to see implementation of global biodiversity conservation that returns to the earlier interpretation of biodiversity as including human activities.

Part of the legacy of this pattern of inclusiveness from the 1990s to the present day, and one little discussed, is the plethora of values represented by all those declaring their interest in biodiversity (Pascual et al. 2017). Unlike other international environment issues such as climate change or desertification, the precise objects of interest and targets for action in biodiversity conservation are broad and vague. Different values are embraced, often implicitly, and increasingly explicitly. Values are defined as trans-situational goals that serve as guiding principles in the life of a person or group (Schwartz 2011) and are used to contrast the foundational goals of groups involved in an issue, clarify the basis of conflict among stakeholders, and more generally provide for the understanding and prediction of human behavior (Manfredo et al. 2016). As such, the global conservation community does not necessarily have the same values as local conservation groups, indigenous people, national development officials, international aid donors, or multinational businesses. Yet given the vague ways in which biodiversity is used, these different groups can all seem to be in harmony with one another's values with no apparent trade-offs. It is only when specific actions are proposed that the veneer of biodiversity as all things to all people is torn, reflecting the need to have stakeholder values laid out early in all negotiating arenas and to consider the existence of trade-offs and the need to negotiate them explicitly. Biodiversity is seen by many as a subject whose study is pursued by scientists working in universities or conservation NGOs with tools like remote sensing, habitat modeling, radio tracking, and priority setting. As scientists, most of this group of stakeholders is not explicit about the values that underpin their work, often denying that their work is value-based, seeming to believe in the positivist view that science is objective and value free.

Yet, conservation biology is "inescapably normative" (Barry and Oelschlaeger 1996), and values are an important part of its study. There are other types of values that underpin work on biodiversity including social, economic, and cultural values. Decisions and positions that are argued on the basis of evidence may often be in disagreement due to lack of acknowledgment of divergent values.

The Focus of Conservation

Though often used to modify biodiversity, the word conservation has a much longer and more complicated history than biodiversity itself. To some, conservation is equivalent to preservation—keeping away from human exploitation. To others, conservation is equivalent to sustainable use—"conserving" the resource. Mace (2014b) outlined four framings of modal positions on conservation since the 1960s:

1. Nature for itself with an emphasis on species, wilderness, and protected areas
2. Nature despite people with an emphasis on extinction, threats and threatened species, habitat loss, pollution, and overexploitation
3. Nature for people with its emphasis on ecosystems, ecosystem approaches, ecosystem services, and economic values
4. People and nature with its emphasis on environmental change, resilience, adaptability, and socioecological systems

These four framings are not exclusive; they intertwine and overlap with some manifestations of each framing found at all times. Through time, the tension between biocentric values and anthropocentric values has been woven into the fabric of conservation, surfacing at times and submerging at others.

In previous decades, the single strongest axis of tension in the biodiversity community was between preservation and use. The preservation camp has driven the world's focus on protected areas and has turned into the world's largest coordinated single land-use effort and been a critical tool in conservation's tool chest. The coverage of protected areas globally has increased, especially over the last decade, and is arguably the single greatest success of the conservation movement. Yet, the focus of the protected-area community is on how detrimental human activities have been to biodiversity with a simple response to ameliorate these activities: separate key biodiversity from use and change detrimental use patterns. The call for more protected areas persists and reached its apogee in E. O. Wilson's quixotic recent call for half of the Earth's lands and seas to be set aside as protected (Wilson 2016).

A very different view comes from those who view biodiversity as essential for sustainable use and the betterment of human kind. For example:

- The CBD states that "biodiversity is the basis of agriculture."[2]
- The United Nations Environment Programme (UNEP) states that "biodiversity provides the basis for ecosystems and the services they provide, upon which all people fundamentally depend" (UNEP 2007).
- The United States Agency for International Development states that "conserving the diversity of life on Earth is fundamental to human well-being."[3]

In all of these examples, biodiversity is assigned worth in that it helps humans—the anthropocentric value. Following this view, one approach is to estimate the total economic value of ecosystems with biodiversity intact to those that have been converted or otherwise simplified for agriculture or industrial use (Balmford et al. 2002; Costanza et al. 2014). This approach has been applied globally, nationally, and locally and generally leads to the conclusion that unconverted areas can have high total economic values but that crucially these values lie outside standard market mechanisms and so cannot be realized under current market-based economic systems. This market failure of public goods and services is pervasive, and a key reason why market mechanisms cannot deliver successful conservation, at least in the absence of effective regulation.

The tension most evident in the last several years in the United States has been (once again) between biocentric and anthropocentric approaches, with the self-styled "new conservation" advocates maintaining that a sharp turn toward human-centric conservation is essential because biocentric conservation has failed (Hunter et al. 2014). This distinction is much less clear in the European context, though there are rising calls for re-wilding in Europe—a

---

[2]   https://www.cbd.int/ibd/2008/basis (accessed April 24, 2017).

[3]   https://www.usaid.gov/biodiversity (accessed April 24, 2017).

marked biocentric approach. As is clear from this brief overview, there is no such thing as a single approach to "conservation"; practitioners display divergent and overlapping sets of values and norms that change over time.

Different Stakeholders Differentially Calculate the Effects of Direct and Indirect Human Action on Biodiversity

Most people would agree on the global trends showing losses in biodiversity—as measured in all its components and attributes. However, different stakeholders have different perspectives on this loss. To some, the clearing of a forest for an oil palm plantation is a triumph in regional development but to others it is the loss of prime habitat for orangutans and other tropical rainforest plants and animals. To illustrate this diversity, we briefly describe how five different stakeholder groups might think about biodiversity use and loss. Not all members of each group will hold the same views, but we use a modal perspective to emphasize the differences between groups. There have also been fads in funding that have changed values and politics.

First, to those committed to biodiversity conservation, there are generally considered to be five major threats to biodiversity:

1. Habitat loss and degradation
2. Introductions of invasive alien species
3. Overexploitation of natural resources
4. Pollution and diseases
5. Human-induced climate change

To conservationists, there are only downsides to most major human activity, as indicated by the losses of genetic variation, species, and ecosystems. Protected areas are the prime tool being deployed globally to minimize such losses (Ferraro and Pressey 2015).

Large, global NGOs, like World Wildlife Fund, The Nature Conservancy, and BirdLife International, largely share a focus on certain groups of threatened species and distinctive, diverse habitats. The values underpinning these organizations are largely overlapping and not necessarily the same, as local or national conservation organizations may have more of an emphasis on the needs of local populations (Redford et al. 2003). While there are a few clear areas of overlap in interests, such as wildlife tourism), there are more often conflicting interests. In truth, there is significant variation within the conservation community, partially because there is no single definition of "conservation," with differences mostly arrayed around whether conservation is for the sake of biodiversity itself or for sustainable use by humans.

Second, for indigenous or traditional groups there is a long cultural tradition of interaction between biodiversity and culture. Culturally embedded rules may govern the management of certain genetic resources, species, and ecosystems that are important for subsistence or for sociocultural reasons. There is often no difference between use and conservation, and "biodiversity" can be defined as including human beings (Redford and Mansour 1996). Strong political positions have been taken by some of these groups and their advocates who claim that biodiversity is conserved, or in some cases even created, by such groups. Evidence assessing such claims is mixed, but the power of the argument has proved strategically effective. Recent work done under IPBES has reached a consensus framework that includes a wide variety of views (Diaz et al. 2015). Additionally, the establishment of protected areas by conservation advocates has in some cases displaced human communities and/or resource uses causing many to claim that protected areas are bad for indigenous and traditional peoples (cf. Hutton et al. 2005).

Many indigenous and traditional groups are experiencing strong pressure from externally driven forces focused on markets for species or converting native ecosystems to commercial plantations. Many have also been displaced from their lands. As a result, and despite the strong rhetoric referred to above, there have been alliances by such groups and conservationists to establish protected areas that serve to inhibit negative development and secure land rights (Redford and Painter 2006).

Third, national development officials responsible for increasing economic activity and decreasing poverty often view biodiversity as either a resource to be exploited through activities like lumbering and fishing or converted to significant use like agriculture and mining. There are oft-cited examples where biodiversity itself can be used for economic progress as in ecotourism, though this is often more vaunted than proven.

Fourth, urban dwellers are often disconnected from the immediate natural world of biodiversity and conservation is not seen as relevant to their lives. Built infrastructure buffers them, decreasing interaction with, and often appreciation for, the natural world. However, there is a broadly increasing appreciation of the need for cities to be more active in ensuring supplies of fresh water that has caused a rise in connectivity between the city and the watershed on which it relies. In some cases this has resulted in cities paying to conserve watersheds. There is also growing interest in the public health benefits of urban green spaces. Urban biodiversity is becoming a focus for city planners that is often quite disconnected from biological and conservation objectives, and is influenced by green-ness, including green and blue infrastructures on roofs, walls, and in waterways.

Fifth, bilateral and multilateral aid for biodiversity from wealthy countries has varied according to country and current fashion. There have been large investments in programs that explicitly tied sustainable human development and conservation, and others that have been directed exclusively at protected areas, and yet others that funded human well-being programs with an expectation that they would generate biodiversity benefits. Though there is no single pattern, in general, this group of stakeholders views biodiversity through the lens of human development, a position summed up in the March 2016 posting from the Center for International Forestry Research: How forests and trees contribute to the global development agenda.[4] In sum, biodiversity conservation is not a single entity with a single constituency but a name for a broad set of beliefs, policies, and practices based on underlying values. When questioning a given conservation intervention or policy, key questions to answer early on include:

- What are the underlying values?
- How do these influence the desired purpose of biodiversity?
- Which components and attributes are of interest?
- Over what time period?
- At what scale (local, regional, global)?
- What loss will be tolerated and who will feel this loss?

Major Contemporary Tensions in Biodiversity Conservation

A set of issues in conservation and biodiversity is currently drawing significant attention and funding. It is worth highlighting these issues because they serve as heuristics that help shed light on a set of tensions underlying the practice and illustrate many of the points made above. They may also become, or already are, part of the way biodiversity conservation is defined. Below we provide only sketches of the complicated issues, values, and science that underlie each of these pairings:

- *Access and benefit sharing*: CBD is not only designed to conserve biodiversity but also to ensure access to and benefits from the use of biodiversity, particularly to local/indigenous peoples. These twin objectives sometimes work in concert with one another but at other times are in opposition. Their pairing in the Convention is further evidence of the social nature of conservation.
- *Biocultural diversity*: As discussed above, the practice of conservation sits uncomfortably astride the arguments about the role of human activity in creating biodiversity. Positions on this issue vary with the historical patterns of human use and the target component or attribute of biodiversity. For example, if genetic diversity of crops is the target, as is the case of the Potato Park in the Peruvian Andes, then ongoing human farming is necessary. Or if grassland biodiversity is the target in Southern Europe, then continued grazing by domestic species is also required. This differs from many settings where human activity must be restricted to maintain desired biodiversity, as is the case with Asian elephant conservation.

---

[4] http://us7.campaign-archive2.com/?u=68cb62552ce24ab3c280248d7&id=14d18d74b6&e=9 30f0acdf2 (accessed April 24, 2017).

- *Biodiversity and poverty*: One of the dominant forcible pairings in the last decade has been between biodiversity conservation and poverty alleviation (Roe et al. 2013). The complicated ways in which biodiversity is defined and deployed are matched by the complications inherent in defining and measuring poverty. Despite this, major funders have created funding streams based on assumptions such as that poor people were mostly found where biodiversity conservation is a priority or that alleviating poverty would result in poverty alleviation. Neither of these has proven to be true across the board.
- *Payment for ecosystem services*: Another popular trend in the last decade has been payment for ecosystem services, based on the assumption that if properly priced in the marketplace, those goods and services of use to humans that were produced by "nature" could be conserved. The most common manifestation is in urban water funds where clean water from a neighboring watershed is ensured through payments to conserve vegetation in the watershed. Though working reasonably well for water, it is not clear if "natural" biodiversity is necessary for clean water, if the model applies to many other services, or if it works where there are no "services" at all.
- *Urban nature*: Recent work is showing that urban parks may play important roles in public and mental health for urban dwellers. Though some evidence shows that more diversity in these green spaces is better, it is not clear that a handful of exotic trees and a monospecific sward of grass (aka lawn) might not serve equally well. So, although the results are promising for green space conservation, it may be less promising for biodiversity conservation.
- *Synthetic biology*: The rise of synthetic biology—the ability to engineer genomes to cause organisms to produce goods and services for humans—is still in its early stages. These technologies offer the possibility of dramatically changing the relationship between humans and biodiversity since the genetic code itself can become domesticated for human purposes (Redford et al. 2013).

Conclusion

In 2001 Rosenzweig (attributing to Gordon Orians) laid out the concept of the Homogocene before the rise in popularity of a term that largely overwrote it, the Anthropocene (Rosenzweig 2001). Both terms describe Earth as it has become impacted by broadscale and pervasive human actions. The former term describes the result whereas the latter, the main actor. Agriculture, industry, fishing, hunting, urbanization, mining, commerce, and attendant climate change have combined to thrust humankind into the spotlight as the dominant ecological and evolutionary actor on our planet.

The rise of nature conservation has been a response to the threat and loss of local diversity—biodiversity conservation is only its most recent manifestation. One of the reasons that biodiversity has met with such widespread and immediate use is that the term "nature" was no longer considered an acceptable target for conservation efforts, though in the last couple of years, through the work of IPBES, it is coming back into fashion (Diaz et al. 2015). The other term in widespread use, "wilderness," had encountered strong opposition for its lack of relevance to more populated parts of the world and its tacitly antihuman perspective. As a new term, biodiversity has no baggage and if left vaguely defined as "all life" could be all things to all people. Who could be against conserving all life?

But despite its pretensions to the contrary, biodiversity is not a term with a universally agreed-upon definition. Rather it is a value proposition: diversity is good and should be maintained. As such, the definition shifts like a skin over the underlying social values, and those stakeholders whose values are taken into consideration. Lack of appreciation for this living, value-based use of the term biodiversity underlies frustrated critiques like that of Maier (2012). Politics is the public contestation of values and in that regard, biodiversity conservation is politics (Sanderson and Redford 1997).

Thinking of biodiversity conservation as inextricably linked to a living political discourse allows us to ask why it doesn't include clean water, urban living, and soils; why there is virtually no attention to environmental justice in the biodiversity conservation world; and how, or whether, agriculture and culture should be included in biodiversity conservation efforts. The challenge is to acknowledge the worth of these other initiatives and to support those who champion their persistence, without diluting the vital job of ensuring the persistence of the truly voiceless—the rest of life on Earth.

As such, important questions remain as to the operational definition of biodiversity and why it is viewed so separately from other human concerns such as clean water, urban living, climate, and soils. One explanation is that biodiversity conservation projects are sometimes seen as a "nice-to-have" rather than as "essential-to-have," as is the case with water and soils, for example. There are large, influential, and well-funded NGOs operating at national and global levels to secure conservation priorities and targets, to an extent unmatched by other environmental concerns. These two factors may often put biodiversity conservation at odds with other environmental issues in development projects. Instead of being central to them, biodiversity can become an awkward addition with contested and hard to estimate values. This is not a good outcome because there is plenty of evidence that securing local and global biodiversity, at least in some forms and configurations, is critical to sustainable development and underpins many of the other more straightforward environmental resources.

The fluid definition of biodiversity has also allowed a climate of "win-win" solutions where human uses are claimed to be achieved while simultaneously conserving biodiversity. Once such arena is payment for ecosystem services, but careful examination (Howe et al. 2014; Sikor 2013) reveals that most declared "win-wins" are in fact trade-offs with values held publically and often lost in exchange for privately held values.

While the values attached to biodiversity and its conservation are more diverse than these other environmental priorities, many of the issues are similar. In particular, the considerations of local versus global, present versus future, public versus private, and monetary versus intrinsic are similar. The lack of clarity over the term simply adds another layer of confusion to what is already a complicated and interacting set of issues.

References

Balmford, A., A. Bruner, P. Cooper, et al. 2002. Economic Reasons for Conserving Wild Nature. *Science* **297**:950–953.

Barry, D., and M. Oelschlaeger. 1996. A Science for Survival: Values and Conservation Biology. *Conserv. Biol.* **10**:905–911.

Brummitt, N., E. C. Regan, L. V. Weatherdon, et al. 2016. Taking Stock of Nature: Essential Biodiversity Variables Explained *Biol. Conserv.* **213**:252–255.

Burley, F. W. 1984. The Conservation of Biological Diversity: A Report on United States Government Activities in International Wildlife Resources Conservation with Recommendations for Expanding U.S. Efforts. Washington, D.C.: World Resources Institute.

Ceballos, G., P. R. Ehrlich, A. D. Barnosky, R. M. Pringle, and T. M. Palmer. 2015. Accelerated Modern Human–Induced Species Losses: Entering the Sixth Mass Extinction. *Sci. Adv.* **1**:e1400253.

Costanza, R., R. de Groot, P. Sutton, et al. 2014. Changes in the Global Value of Ecosystem Services. *Global Environ. Change* **26**:152–158.

Diaz, S., S. Demissew, C. Joly, et al. 2015. The IPBES Conceptual Framework: Connecting Nature and People. *Curr. Opin. Environ. Sustain.* **14**:1–16.

Diaz, S., S. Lavorel, F. de Bello, et al. 2007. Incorporating Plant Functional Diversity Effects in Ecosystem Service Assessments. *PNAS* **104**:20684–20689.

Dirzo, R., H. S. Young, M. Galetti, et al. 2014. Defaunation in the Anthropocene. *Science* **345**:401–406.

Ferraro, P. J., and R. L. Pressey. 2015. Measuring the Difference Made by Conservation Initiatives: Protected Areas and Their Environmental and Social Impacts. *Phil. Trans. R. Soc. B.* **370**:20140270.

Gaston, K. J. 1996. Biodiversity: A Biology of Numbers and Difference. Oxford: Blackwell Science Press.

Howe, C., H. Suich, B. Vira, and G. M. Mace. 2014. Creating Win-Wins from Trade-Offs? Ecosystem Services for Human Well-Being: A Meta-Analysis of Ecosystem Services Trade-Offs and Synergies in the Real World. *Global Environ. Change* **28**:263–275.

Hunter, M. L., K. H. Redford, and D. B. Lindenmayer. 2014. The Complementary Niches of Anthropocentric and Biocentric Conservationists. *Conser. Biol.* **28**:641–645.

Hutton, J., W. M. Adams, and J. C. Murombedzi. 2005. Back to the Barriers? Changing Narratives in Biodiversity Conservation. *Forum Develop. Stud.* **32**:341–370.

Jackson, L. E., U. Pascual, and T. Hodgkin. 2007. Utilizing and Conserving Agrobiodiversity in Agricultural Landscapes. *Agric. Ecosyst. Environ.* **121**:196–210.

Jetz, W., J. M. McPherson, and R. P. Guralnick. 2012. Integrating Biodiversity Distribution Knowledge: Toward a Global Map of Life. *Trends Ecol. Evol.* **27**:151–159.

Kontoleon, A., U. Pascual, and T. Swanson. 2007. Biodiversity Economics. Cambridge: Cambridge Univ. Press.

Mace, G. M. 2014a. Biodiversity: Its Meanings, Roles, and Status. In: Nature in the Balance: The Economics of Biodiversity, ed. D. Helm and C. Hepburn, pp. 1–22. Oxford: Oxford Univ. Press.

————. 2014b. Whose Conservation Is It? *Science* **345**:1558–1560.

Magurran, A. E. 2004. Measuring Biological Diversity. Malden, MA: Blackwell.

Maier, D. S. 2012. What's So Good About Biodiversity? A Call for Better Reasoning About Nature's Value. New York: Springer.

Manfredo, M. J., J. T. Bruskotter, T. L. Teel, et al. 2016. Why Social Values Cannot Be Changed for the Sake of Conservation. *Conserv. Biol.* **31**:772–780.

McGill, B. J., M. Dornelas, N. J. Gotelli, and A. E. Magurran. 2015. Fifteen Forms of Biodiversity Trend in the Anthropocene. *Trends Ecol. Evol.* **30**:104–113.

Nazarea, V. D. 2006. Local Knowledge and Memory in Biodiversity Conservation. *Annu. Rev. Anthropol.* **35**:317–335.

Newbold, T., L. N. Hudson, S. L. Hill, et al. 2015. Global Effects of Land Use on Local Terrestrial Biodiversity. *Nature* **520**:45–50.

Norse, E. 1990. Threats to Biological Diversity in the United States. Washington, D.C.: Office of Policy Planning and Evaluation, U.S. Environmental Protection Agency.

Norse, E. A., K. L. Rosenbaum, D. S. Wilcove, et al. 1986. Conserving Biological Diversity in Our National Forests. Washington, D.C.: The Wilderness Society.

Pascual, U., P. Balvanera, S. Díaz, et al. 2017. Valuing Nature's Contributions to People: The IPBES Approach. *Curr. Opin. Environ. Sustain.* **26–27**:7–16.

Pereira, H. M., S. Ferrier, and M. Walters. 2013. Essential Biodiversity Variables. *Science* **339**:277–278.

Pressey, R. L., M. Cabeza, and M. E. Watts. 2007. Conservation Planning in a Changing World. *Trends Ecol. Evol.* **22**:583–592.

Putz, F. E., and K. H. Redford. 2009. The Importance of Defining Forest: Tropical Forest Degradation, Deforestation, Long-Term Phase Shifts, and Further Transitions. *Biotropica* **42**:10–20.

Redford, K. H., W. Adams, and G. M. Mace. 2013. Synthetic Biology and Conservation of Nature: Wicked Problems and Wicked Solutions. *PLoS Biol.* **11**:e1001530.

Redford, K. H., and J. P. Brosius. 2006. Diversity and Homogenization in the End Game. *Global Environ. Change* **16**:317–319.

Redford, K. H., P. Coppolillo, E. W. Sanderson, et al. 2003. Mapping the Conservation Landscape. *Conserv. Biol.* **17**:116–131.

Redford, K. H., and J. A. Mansour, eds. 1996. Traditional Peoples and Biodiversity Conservation in Large Tropical Landscapes. Arlington: The Nature Conservancy/ America Verde Press.

Redford, K. H., and M. Painter. 2006. Natural Alliances between Conservationists and Indigenous Peoples. Working Paper No. 25. New York: Wildlife Conservation Society.

Redford, K. H., and B. Richter. 1999. Conservation of Biodiversity in a World of Use. *Conserv. Biol.* **13**:1246–1256.

Roe, D., E. Y. Mohammed, I. Possas, and A. Giuliani. 2013. Linking Biodiversity Conservation and Poverty Reduction: De-Polarizing the Conservation-Poverty Debate. *Conserv. Letters* **6**:162–171.

Rosenzweig, M. L. 2001. The Four Questions: What Does the Introduction of Exotic Species Do to Diversity? *Evol. Ecol. Res.* **3**:361–367.

Sanderson, S. E., and K. H. Redford. 1997. Biodiversity Politics and the Contest for Ownership of the World's Biota. In: Last Stand, ed. R. Kramer et al., pp. 115–132. New York: Oxford Univ. Press.

Sax, D. F., and S. D. Gaines. 2003. Species Diversity: From Global Decreases to Local Increases. *Trends Ecol. Evol.* **18**:561–566.

Schwartz, S. H. 2011. Values: Cultural and Individual. In: Fundamental Questions in Cross-Cultural Psychology, ed. F. J. R. van de Vijver et al., pp. 463–493. Cambridge: Cambridge Univ. Press.

Sikor, T. 2013. The Justices and Injustices of Ecosystem Services. Abingdon: Routledge.

Takacs, D. 1996. The Idea of Biodiversity: Philosophies of Paradise. Baltimore: Johns Hopkins Univ. Press.

UNEP. 2007. Global Environment Outlook 4: Environment for Development. New York: United Nations Environment Programme.

Wilson, E. O., ed. 1988. Biodiversity. Washington, D.C.: National Academy of Sciences Press.

———. 2016. Half-Earth: Our Planet's Fight for Life. New York: W. W. Norton.

3

Values, Incentives, and Environmentalism in Ecosystem Services

Peter A. Minang

Abstract

Values have been widely considered important in environmentalism. Similarly, incentives have been widely deployed as mechanisms for influencing environmentalism. Yet little attention has been given to understanding the relationships between values and incentives and how that understanding can best serve environmentalism. This chapter explores connections between values and incentives in the context of ecosystem services in high-diversity tropical forest systems. It highlights the potential linkages between held, assigned, and relational values and multiple incentive types through two main decision-making framings: rational choice and bounded rationality. While economic and financial incentives are largely linked to assigned values, nonfinancial and noneconomic incentives are more linked to relational and held values. In reality though, values influence each other; as a result, the complex processes through which values influence each other and influence behavior become important. Four main value-related implications emerge for designing and implementing incentives that can change environmentally significant behavior in the context of ecosystem services in tropical forests: (a) the need for multiple incentives or mixes of incentives in recognition of diversity in values needed to enable sustainability, (b) mitigating crowding out in incentive structures, (c) the rise of theory of place, and (d) the need for further empirical research to better understand the interactions between values and incentives in the realm of environmentalism with implications for ensuring diversity, justice, and sustainability.

Introduction

The Millennium Ecosystem Assessment (2005) established the tremendous importance and global reliance on Earth's ecosystems for services such as food, fiber, water, fuel, disease management, climate regulation, and many others. It also stressed that most of these services are being severely degraded and exploited unsustainably, largely through human activity. For example, tropical forests represent a direct source of food, fuel, and fiber for more than 1.2 billion people globally (Agrawal 2007), yet tropical forests have shrunk significantly over the years at an average of about 13 million ha per year. Between 1990 and 2011 the world lost in total 135,494,000 ha of forests (74,927,000 ha in the Amazon, 35,769,000 ha in Southeast Asia, and 5,271,000 ha in the Congo). During this period, rainforest loss represented 85.59% of the world's total forest loss (FAO STAT). As a result, there have been growing calls for action to stem the unsustainable exploitation and degradation of ecosystem services (Millennium Ecosystem Assessment 2005).

Understanding environmentalism is an essential part of any effort to influence anthropogenic causes of degradation in ecosystem services. Generally, environmentalism refers to ideas about how humans interact with the natural environment and how the environment should be protected through human effort (Thomas Sikor, unpublished). Behaviorally, environmentalism can be defined as the propensity to take actions with pro-environmental intent (Stern 2000). More specifically, understanding what motivates and informs human choices and actions toward the natural environment is crucial. In addition, values have long been cited as an important underlying motive for human actions on the environment (Stern 2000; Jones et al. 2016). On the other hand, incentives have been widely used as a key intervention to influence human behavior, in a bid to enhance ecosystem services (van Noordwijk et al. 2012). Little attention, however, has been paid to understanding how values relate to incentives and how this can be deployed to improve design and implement more effective and efficient incentive mechanisms. This chapter hopes to contribute to a better understanding of the linkages between values and incentives in ecosystem services management and is structured as follows. I begin with a brief presentation of both values and incentives. Thereafter, decision-making frames of rational choice and bounded rationality are used to establish linkages between values and incentives, and the interactive processes between values and how values influence behavior are explored. Specific ways in which incentives design and implementation could benefit from such an understanding are discussed, and avenues for further research are highlighted.

Values in Environmentalism

According to the Oxford English Dictionary, "values" refers to a person's principles or standards of behavior; one's judgment of what is important in life; or simply, the regard, worth, or usefulness of something. Schwartz (1994) defines a value as "a belief pertaining to desirable end states or modes of conduct that transcends specific situations, guides selection or evaluation of behavior, people and events, and is ordered by importance relative to other values to form a system of value priorities." While values can be acquired during the formative years, they are shaped throughout life, but largely remain stable in later stages of life. They are said to be the bedrock that shapes attitudes, norms, and behaviors (Stern 2000; Jones et al. 2016). Values, therefore, guide decision making and a sense of what is right. Values tend to be mostly individual; however, they can also be collective (i.e., community values) especially in disciplines such as anthropology and sociology. In common-pool or community-managed resources, it might be most appropriate to think more in terms of community or commonly held values as opposed to individual values only.

In a review of human values in understanding and managing socioecological systems, Jones et al. (2016) identify three distinct types of values: held, assigned, and relational. *Held values* refer to ideals of what is desirable, how things ought to be, and how one should interact with the world (Brown 1984; Bengston 1994). *Assigned values* are those attached, for example, to various ecosystem goods, services, and places. They represent the expressed relative importance or worth of an object to an individual or group in a given context (Brown 1984). *Relational values* emerge from the relationships between people and nature and are associated to preferences and hence feelings (Brown 1984; Chan et al. 2016). These three types of values are related. Assigned values are largely shaped by held values, whereas relational values explain the relationships between held and assigned values (Brown 1984; Jones et al. 2016).

The environmental literature identifies other types of values including intrinsic values (those which protect nature for nature's sake) and instrumental values (those which protect nature for humans' sake (Stern 2000; Chan et al. 2016). Kellert (1996) elaborated a typology that includes ten nature-related values: aesthetic, dominionistic, ecologistic-scientific, humanistic, moralistic, naturalistic, negativistic, spiritual, symbolic, and utilitarian. This diversity of value concepts reflects the deep considerations that shape and motivate decisions and actions of individuals as they interact with nature. While values are important, the ways and processes through which values influence and change behavior are even more important in terms of how decision making can be influenced or shaped (Guagnano et al. 1995; Stern 2000; Wunder 2005; van Noordwijk et al. 2012).

Incentives in Environmentalism

Incentives have long been established to be an important element in environmental performance, including restoration of ecosystem services in tropical forests (Wunder 2005; van Noordwijk et al. 2012). Incentives are "anything that can motivate an agent to take a particular course of action" or "any policy, program, institution or economic instrument that motivates conservation and management of forest ecosystems" (Casey et al. 2006:18). They can be broadly listed as being fiscal (e.g., taxes, tariffs, subsidies), economic (e.g.,

low interest loans, compensation for certain investments, conditional payments, premiums), reputational (e.g., name, fame, and shame awards), and/or
administrative (e.g., privileged access, land or tree rights, shorter processing
times) in nature (Rademaekers et al. 2012). Heimlich et al. (1998) discuss five
categories of incentives:

1. Involuntary regulatory disincentives
2. Voluntary, nonregulatory economic incentives
3. Institutional innovations that encourage market, legal, and planning authorities to enhance resource conservation
4. Facilitative incentives, including administrative and technical
   assistance

Several examples of incentives in ecosystem services management exist:
prohibition of use, tenure and property rights, taxes and penalties, subsidies,
quotas, permits, etc. fall into the regulatory category; payments for ecosystem
services (PES), rewards for ecosystem services, certification programs, mar-
keting labels, etc. are in the voluntary category (Casey et al. 2006; FAO 2015).
Below, I present two examples of incentive schemes for ecosystem services
from the forest sector: PES and Reducing Emissions from Deforestation and
Forest Degradation (REDD+). I have chosen PES and REDD+ because they
are by far the most widely used in natural resources management in developing
countries, and I use them to ground this discussion in reality.

Payments for Ecosystem Services

PES is an innovative set of instruments used to protect and conserve ecosystem
and environmental services that has tried to move beyond generic instruments
(e.g., protected areas, community conservation, integrated conservation and
development). The most widely cited definition of PES is "a voluntary, conditional transaction where at least one buyer pays at least one seller for maintaining or adopting sustainable land management practices that favor the provision
of well-defined environmental services" (Wunder 2005:3). Following close to
a decade of practice and research, Wunder (2015:241) suggests an expanded
definition of PES as "voluntary transactions between services users and service
providers that are conditional on agreed rules of natural resource management
for generating off-site services." This new definition takes into account critics
of bias toward monetary rewards, exclusions of nonmarket transactions, and
other weaknesses found in previous definitions. Conditionality emerges as the
primary feature of the current definition. Monitoring, reporting and verifica-
tion, specific standards, participation, transparency, safeguards and pro-poor
conditions are emerging as principal conditions in the suite of conditionalities
in the PES arena.

Flows of financial capital remain the basic vehicle through which buyers can
express their appreciation for environmental services in the most widely used

PES mechanisms (van Noordwijk et al. 2012). In the United States and Latin America, where privately owned forests abound, payments or transfers have largely gone to individuals (Casey et al. 2006; Sierra and Russman 2006). In Africa, where communal ownership of forests dominates, transfers of financial resources have been to communities as well as to individuals: the CAMPFIRE Programme on Wildlife in Zimbabwe or the Nhambita Community Carbon Project in Mozambique (Frost and Bond 2007). Transfers may also be viewed as flexible mechanisms through which stakeholders affected by changes in land use can try to influence actors that change land use on a day-to-day basis (van Noordwijk et al. 2012). Land-use proxies have thus been traditionally used as indicators for measuring progress in ecosystem services, as they are more direct and cheaper to measure. Other examples of practical PES schemes include the Reward for Upland Poor Ecosystem Services, water biodiversity, and carbon in Singkarak, Indonesia; Payment for Forest Ecosystem Services in Vietnam at national level; and the Noel Kempff Mercado Carbon Action Project in Bolivia.

Reducing Emissions from Deforestation and Forest Degradation

Incentives and policy change to reduce emissions from deforestation and forest degradation, REDD+, has been promoted as an approach to address climate change and achieve other sustainable development benefits. REDD+ is an evolving concept currently under negotiation within the United Nations Framework Convention on Climate Change (UNFCCC), in which countries can elect and engage in the reduction of emissions from forests against an agreed baseline or reference level. Economic incentives, market and/or fund-based, are to be provided once reported emission reduction has been verified. According to the UNFCCC, emission reductions can consist of reducing emissions from deforestation, reducing emissions from degradation, conservation of forest carbon stocks, sustainable management of forests, and enhancement of carbon stocks. REDD+ is also expected to generate sustainable development co-benefits such as biodiversity, water, and poverty reduction.

Several countries have engaged actions to develop the necessary technical and institutional capacity to implement any mechanism as recommended and supported by the UNFCCC (§70–73, UNFCCC Decision 1/CP16). Actions aimed at developing technical and institutional capacity in developing countries are referred to as REDD+ readiness. While accounting and accountability for emission reduction will be primarily at the national scale, change in behavior and practice will have to reach all forest areas of a country. Where subnational implementation structures can be designed within existing institutions and policies, a major break with business as usual is needed to shift from enhancing to reducing emissions.

There are currently no payments for REDD+ as described within the UNFCCC. However, there are several REDD+ projects within the voluntary market for which payments are being made, such as the Kasigau Corridor REDD+ Project in Kenya (Cerbu et al. 2011; Bernard et al. 2014). Most of these are subnational level activities and are verified and certified through either the Climate Community and Biodiversity Alliance and the Voluntary Carbon Standards. Several of these REDD+ projects are built on integrated conservation and development projects, hence they are highly linked to biodiversity and protected area management (Minang and van Noordwijk 2013).

Like all incentives, PES and REDD+ are designed to influence choices or behaviors of individuals and communities in human environment relations (environmentalism). In most instances, they are designed to build on any behavior or choice that can significantly enhance environmental services. Hence, decision-making processes through which choices are made or behaviors changed are important dimensions to environmentalism.

Linking Values and Incentives: Decision and Behavioral Science Framings

Thaler and Sunstein (2008) distinguish two types of "species": econs and humans. Econ, or *Homo economicus* (the economic man), is one who thinks most rationally, efficiently calculating the details of all options before making rational choices, as imagined by economic theory. Humans, *H. sapiens*, are real people who fall well below the efficient, analytical, and predictive decision-making rules espoused by economic theory. Humans tend to be predictable, show more "feelings," and tend to be stable in choices following long held values and beliefs. These two dimensions of being perhaps explains the two main framings that have shaped decision science in recent times; namely rational choice and bounded rationality, respectively.

Rational Choice

Rationale choice as a paradigm has dominated decision-making thinking for the last decades. It refers to a situation where decisions are made to maximize net benefits from investments in time as well as resources (land and/or nature). Accordingly, aggregate social behavior is the result of the behavior of individual actors, each of whom makes a preferred choice after taking into account the costs and benefits, probabilities of events, and all other necessary information, following a logical process (Scott 2000). An important feature of rational choice is that it assumes that almost complete and perfect information is available to the decision maker. Rational choice considers the individual to be essentially *H. economicus*: this person will balance costs against benefits to arrive at actions that will maximize personal advantage. This thinking has undoubtedly shaped how economics is applied to natural resources and by extension economic incentives for forest resources management. Rational choice theory can thus be directly associated with assigned values.

In large part, economic and financial incentives that currently make up the bulk of incentives in ecosystem services in tropical countries have been driven by bounded rationality thinking. A good part of the PES literature has addressed the determination of the right level of reward, compensation, or payment needed to change a particular course of action in the environment or a given land use (van Noordwijk et al. 2012). A number of studies have discussed the opportunity cost of REDD+ as a possible minimum requirement to enable any deforestation-related PES scheme (White and Minang 2011). The literature on valuation of ecosystem services (de Groot et al. 2012) also demonstrates that economic and financial incentives for ecosystem services are largely based on assigned values. One of the most reported cases of PES—the New York drinking water company and the Catskills catchment—illustrates the key role of rational choice and assigned values in the logic of PES. Instead of spending between 8–10 billion USD on water purification installations, New York City negotiated with local governments in the Catskills catchment area, and subsequently enacted water-friendly land-use restrictions, thereby enabling protection of the watershed and reducing costs dramatically (Appleton 2002). Seymour et al. (2012) argued that assigned values are better predictors of behavior than held values in the natural resource management context. Figure 3.1 illustrates the linkages between financial incentives and underlying values.

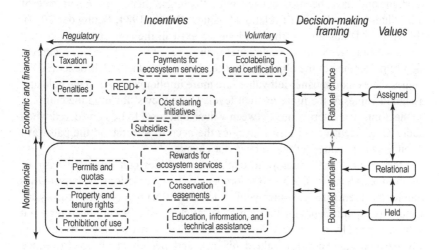

Figure 3.1 Overview of the dynamic and complex relationships between incentives and values in environmentalism, as mediated by decision-making framings.

Bounded Rationality

Within behavioral economics, bounded rationality has emerged as a main-stream perspective on how individuals make decisions when information is limited, time for making such decisions is short, the problems being addressed are complex and intractable, and the cognitive abilities of individuals are finite. Under this scenario, individuals employ a "satisficing" heuristic to reach the best possible decision, rather than search endlessly for the "optimal" solution (as in rational choice). Three "bounds" or deviations from the rational choice of standard economic theory have been recognized (Jolls et al. 1998):

1. Bounded rationality is incomplete information that interacts with limits to human cognition and leads to judgment errors or objectively poor decision making.
2. Bounded willpower involves taking actions that individuals know to be in conflict with their long-term interests.
3. Bounded self-interest is the willingness to sacrifice individual interests for those of others.

A number of judgment errors and biases typically enter into decision making in very predictable ways, for both individuals and groups, and there is a tendency to rely on mental short cuts, "rules of thumb," or heuristics. Hence it is conceivable to "nudge" behavior and affect decision making (Jolls et al. 1998).

Nonfinancial and regulatory incentives can be directly linked to bounded rationality thinking and therefore tied to more relational and held values. Relational values emanate from relationships between humans and the environment and have been associated with "feelings," hence the emergence of felt values as a subclass of relational values (Brown 1984; Schroeder 2013). Brown (1984) has argued that held values exist in the conceptual realm and influence judgments (in the relational realm), resulting in behavioral expression of preference in the object realm (i.e., assigned values). Schroeder (2013) argues the exact opposite: felt values are more implicit and can shape held and assigned values at the more explicit level. These views illustrate the complex connections and interactions between values that need to be considered when addressing incentives. It also illustrates the need to understand the pathways through which values influence behavior change.

Stern et al. (1999) attempt to explain how values constitute the basis of environmentalism (defined as the propensity to take actions with pro-environmental intent) using values-beliefs-norms (VBN) theory. VBN theory attempts to explain behavior through a causal chain of five variables: personal values, especially altruistic values; beliefs in terms of ecological worldview, adverse consequences for valued objects, and perceived ability to reduce threats; and pro-environmental personal norms or a sense of obligation to take pro-environmental actions. The argument is that these variables act directly on one another, in the order presented, to shape a series of pro-environmental behaviors ranging from public (e.g., activism) to private. Further, VBN theory argues that values underlie and influence beliefs and norms, which leads to various kinds of behaviors. Several studies have found that altruistic values are stronger among people with pro-environmental behavior (Stern 2000). VBN theory illustrates complex processes behind the behavior being targeted by incentives and the relationships to one or more values.

This discussion illustrates how the logic of certain incentives can be linked to certain values more than others in theory. However, in practice, incentives might, more often than not, be indirectly connected to more than one value. For a conceptual map of how a selected set of incentives might be linked to values, see Figure 3.1.

Implications for Praxis and Research

To understand the linkages between values and incentives in the design and implementation of incentive schemes, multiple incentives (or mixtures of incentives) are required. In addition, for incentive structures, crowding out needs to be mitigated and the rise of theory of place in the deployment of incentives needs to be studied and supported by empirical research.

The Need for Multiple Incentives: Recognizing Diversity for Sustainability

The interdependencies observed between held, assigned, and relational values (see above) suggests that the dominance of single financial or economic incentives is questionable, as currently applied in tropical forestry cases of REDD+ or PES in many countries/places. A diverse set of actors, with diverse values and interests in tropical forests, impose trade-offs and create synergies that must be negotiated if sustainability is to be achieved. This points to the need for multiple instruments and incentives in order to be successful (Minang and van Noordwijk 2013). With his typology of ten nature-related values, Kellert (1996) argues that an individual or community might actually relate to or hold several nature-related values simultaneously, thus making the case for deploying multiple complementary incentives in any given place (see also Schroeder 2013).

Consider, for example, pan-Tropical research on increasing productivity along tropical forest margins in Brazil, Indonesia, Cameroon, Thailand and Peru. This work has shown that subsidies, enhanced financing, and other incentives for intensifying agricultural production along forest margins, as a means of reducing encroachment into forests, was necessary but insufficient as long as demand for agricultural commodities remained inflexible (Palm et al. 2005; Minang and van Noordwijk 2013). Given the deeply instrumental or assigned values of the concerned actors, farms are likely to be established to meet the ever-increasing global demand. Therefore, regulatory mechanisms driven by intrinsic public value (e.g., the creation of protected areas as well as incentives for enforcement) are needed to stop the expansion of agriculture. Despite negotiations and the combination of instruments, agriculture continues to grow. This has precipitated arguments on sustainable or ecological intensification and the need for behavior and value changes in agriculture-forest landscape management—a debate that requires resolution if tropical forestry ecosystems are to be maintained (Pretty et al. 2011; Tittonell 2014).

Economic or financial valuations often represent only a single scale of values—often monetary in nature. Such a scale would not apply, for example, to the quantification of the spiritual, symbolic, or life-saving medicinal value of sacred forests in Africa. To these communities, forests are a part of life. They provide an inextricable connection to nature that can be associated to the biophylia hypothesis (Caston 2013), in which the community is dependent on the forest/nature for their very being. A purely monetary valuation or financial incentive is unlikely to be successful in this environment unless accompanied perhaps by property rights-based regulatory incentives. Understanding the diverse set of values in a given community can help establish the right kind of mix of incentives needed to enable environmentalism.

Bemelmans-Videc et al. (1998) articulated a framework for policy incentives and evaluations: where carrots, sticks, and sermons served as mechanisms to construct a mix of incentives capable of simultaneously addressing multiple value-driven behaviors. Carrots could refer to PES-type or economic incentives to address assigned values. Sticks could be dis-incentives in the form of regulations, rules, punishments, and prohibitions to discourage instrumental or utilitarian behaviors. Sermons could represent more training, education, capacity building and information provision type incentives aimed at reinforcing or changing held intrinsic prosocial and pro-environmental behavior (Figure 3.1). Applying this to the tropical landscape mentioned above, subsidies for intensification (carrots) could coexist with the development and enforcement of protected areas (sticks) and the promotion of sustainable intensification (sermons).

Mitigating Crowding Out: Justice for Pro-Environmentalism

In the realm of multiple incentives, understanding the diversity and depth of values can help mitigate potential crowding out of intrinsic pro-environmental behavior in the use of external incentives within ecosystem services. Increasingly, classic financial or economic incentives are being challenged by experiences in environmentalism from the rural developing world (Martinez-Alier 2002), in terms of how they interface with social motivations. A number of these studies have found that monetary payments can undermine prosocial behavior or altruistic/intrinsic motivation at critical points, especially when they are withdrawn (Cardenas et al. 2000; Reeson and Tisdell 2008).

Consider, for example, forest carbon and REDD+ payments, which are allocated by project according to a set budget and timeframe. Once the project ends and payments are phased out, it has been found that the actions previously supported are also withdrawn. Imagine REDD+ payments being received by forest countries that have managed sacred forests in Africa for many years out of intrinsic and cultural values. Similarly, there is a risk that fewer countries would seek to ensure that the 17% of national territory remain as protected areas (per the Convention on Biological Diversity Aichi target 11) after REDD+ payments for conservation come to an end. Therefore, considerations of rules and conditionalities (e.g., caps for specific incentives) to address potential for crowding out might be necessary.

The body of knowledge on how incentives crowd out prosocial behavior is largely outside the realm of natural resources management, and ecosystem services in particular (van Noordwijk et al. 2012). Thus there is a pressing need for research in this area, if incentives for ecosystem services are to achieve the desired impact. This research needs to address trade-offs, synergies, and potential tipping points through modeling and other approaches (Gneezy et al. 2011). Without deep understanding of values and value interactions within incentives, it will be impossible to mitigate crowding out. Finally, addressing the free-rider problem in ecosystem services that results from incentives also requires a good understanding of local values, with some potential for influencing justice in environmentalism.

The Rise of Theory of Place

Inherent in every incentive design and implementation is a theory of change: what is the expected impact on human environment relations and how will it unfold? All too often, this is not clearly articulated in terms of the context (i.e., complex institutional landscape, political economy, norms, policies, etc.) within which the incentive will take place and the interactions and implications that can be expected from the incentives (van Noordwijk et al. 2015). The latter has been referred to as theory of place.

Stern et al. (1999) emphasized the importance of context and place in VBN theory by pointing out the central role of ecological worldviews, the perceived ability to take pro-environmental action as key contextual factors that modify values and shape environmental behavior. Guagnano et al. (1995) put context and environment as paramount components in understanding behavior through attitude-behavior-context (ABC theory). ABC theory attempts to increase understanding of behavior as a function of the organism and its environment. It argues that behavior (B) is an interactive product of personal-sphere attitudinal variables (A) and contextual factors (C). In a curbside recycling study, Guagnano et al. (1995) found that attitude and behavior are strongly associated when context is neutral, but almost nonexistent when contextual forces are strong, because context largely determines behavior (see also Stern 2000).

Values in natural resources management tend to be place specific. Assigned values, in particular, pertain mainly to given objects or services from a given place (Seymour et al. 2012). Relational and held values have a strong influence on assigned values and are themselves shaped by interactions between multiple variables, such as social processes, interpersonal influences, government policies, private sector, and market influences (Brown 1984; Stern 2000; Chan et al. 2016; Jones et al. 2016). This body of evidence suggests that incentive design and implementation as well as environmentalism actions would benefit from a better understanding of theories of place, in much the same way theories of change are articulated in the management of ecosystem services.

Need for Further Research

For natural resources and ecosystem services, the need for further exploration into the mechanisms and relationships between value, behavior, and incentives cannot be overemphasized. This work can borrow from the economic policy and technology innovations arena, where more understanding exists. Although some work is beginning to emerge (van Noordwijk et al. 2012; Jones et al. 2016), there is a dearth of empirical and experimental work as a whole. Future research needs to pay attention to this to advance environmentalism.

Summary

In this chapter, I have explored the linkages between values and incentives in environmentalism and, in particular, ecosystem services. To enhance the design and implementation of incentives for ecosystem services, it is critical that we understand the potential influences of values on incentive mechanisms (and vice versa). Here, I have focused the discussion around the three main values—held, assigned, and relational values—and highlighted other useful nature-related values (e.g., intrinsic, instrumental, dominionistic, scientific, etc.). Using PES and REDD+ as examples to ground the discussion, linkages between values and incentives were explored through two main decision-making framings: rational choice and bounded rationality. Conceptual linkages between values and incentives in environmentalism are summarized in Figure 3.1. While economic and financial incentives are largely linked to assigned values, nonfinancial and non-economic incentives are linked primarily to relational and held values. In reality, though, values influence each other as well as the behavior that results, and thus understanding the complex processes becomes extremely important. VBN and ABC theories in environmental behavior science can be used to deepen our understanding of the interactions between values and incentives. To change environmentally significant behavior in the context of ecosystem services, four areas emerge pertinent to the design and implementation of effective incentives:

1. To respond to the diversity of values, multiple incentives or mixes of incentives are needed to enable sustainability.
2. Crowding out needs to be mitigated in incentive structures.
3. The rise of theory of place requires further study in the context of incentive deployment.
4. Further empirical research is needed to enable a better understanding of the links between values and incentives.

Meeting these challenges will positively impact diversity, justice, and sustainability in tropical forest environmentalism.

References

Agrawal, A. 2007. Forests, Governance, and Sustainability: Common Property Theory and Its Contributions. *Int. J. Commons* **1**:111–136.

Appleton, A. F. 2002. How New York City Used an Ecosystems Services Strategy Carried out through an Urban-Rural Partnership to Preserve the Pristine Quality of Its Drinking Water and Save Billions of Dollars. In: The Katoomba Conference. Tokyo: Forest Trends.

Bemelmans-Videc, M. L., R. C. Rist, and E. Vedung, eds. 1998. Carrots, Sticks and Sermons: Policy Instruments and Their Evaluation. New Brunswick, NJ: Transaction.

Bengston, D. N. 1994. Changing Forest Values and Ecosystem Management. *Soc. Nat. Resour.* **7**:515–533.

Bernard, F., P. A. Minang, B. Adkins, and J. T. Freund. 2014. REDD+ Projects and National-Level Readiness Processes: A Case Study from Kenya. *Clim. Pol.* **14**:788–800.

Brown, T. C. 1984. The Concept of Value in Resource Allocation. *Land Econ.* **60**:231–246.

Cardenas, J. C., J. Stranlund, and C. Willis. 2000. Local Environmental Control and Institutional Crowding-Out. *World Dev.* **28**:1719–1733.

Casey, F., S. Vikemann, C. Hummon, and B. Taylor. 2006. Incentives for Biodiversity Conservation: An Ecological and Economic Assessment. Defenders of Wildlife. http://www.defenders.org/publications/incentives_for_biodiversity_conservation.pdf. (accessed Jan. 20, 2017).

Caston, D. 2013. Biocultural Stewardship: A Framework for Engaging Indigenous Cultures. *Minding Nature* **6**:22–32.

Cerbu, G. A., B. M. Swallow, and D. Y. Thompson. 2011. Locating REDD: A Global Survey and Analysis of REDD Readiness and Demonstration Activities. *Environ. Sci. Pol.* **14**:168–180.

Chan, K. M. A., P. Balvanera, K. Benessaiah, et al. 2016. Opinion: Why Protect Nature? Rethinking Values and the Environment. *PNAS* **113**:1462–1465.

de Groot, R., L. Brander, S. van der Ploeg, et al. 2012. Global Estimates of the Value of Ecosystems and Their Services in Monetary Units. *Ecosys. Serv.* **1**:50–61.

FAO. 2015. Incentives for Ecosystem Services in Agriculture. Food and Agriculture Organization of the United Nations. http://www.fao.org/in-action/incentives-for-ecosystem-services/en/. (accessed July 27, 2017).

Frost, F., and I. Bond. 2007. The Campfire Programme in Zimbabwe: Payments for Wildlife Services. *Ecol. Econ.* **65**:776–787.

Gneezy, U., S. Meier, and P. Rey-Biel. 2011. When and Why Incentives (Don't) Work to Modify Behavior. *J. Econ. Perspect.* **25**:191–210.

Guagnano, G. A., P. C. Stern, and T. Dietz. 1995. Influences on Attitude-Behavior Relationships: A Natural Experiment with Curnside Recycling. *Environ. Behav.* 27:699–718.

Heimlich, R., K. Wiebe, R. Claassen, D. Gadsby, and R. House. 1998. Wetlands and Agriculture: Private Interests and Public Benefits: Agricultural Economic Report No. (AER-765). Resource Economics Division Economic Research Service U.S. Dept. of Agriculture. Washington, D.C.: GPO.

Jolls, C., C. R. Sunstein, and R. Thaler. 1998. A Behavioral Approach to Law and Economics. *Stanford Law Rev.* 50:1471–1548.

Jones, N. A., S. Shaw, H. Ross, K. Witt, and B. Pinner. 2016. The Study of Human Values in Understanding and Managing Social-Ecological Systems. *Ecol. Soc.* 21(1):15

Kellert, S. R. 1996. The Value of Life: Biological Diversity and Human Society. Washington, D.C.: Island Press.

Martinez-Alier, J. 2002. The Environmentalism of the Poor: A Study of Ecological Conflicts and Valuation. Cheltenham: Edward Edgar.

Millennium Ecosystem Assessment. 2005. Ecosystems and Human Well-Being: Synthesis. Washington, D.C.: Island Press.

Minang, P. A., and M. van Noordwijk. 2013. Design Challenges for Achieving Reduced Emissions from Deforestation and Forest Degradation through Conservation: Leveraging Multiple Paradigms at the Tropical Forest Margins. *Land Use Policy* 31:61–70.

Palm, C., S. Vosti, P. Sanchez, and P. Ericksen. 2005. Slash and Burn Agriculture: The Search for Alternatives. New York: Columbia Univ. Press.

Pretty, J., C. Toulmin, and S. Williams. 2011. Sustainable Intensification in African Agriculture. *Int. J. Agric. Sustain.* 9:5–24.

Rademaekers, K., R. Williams, R. Ellis, et al. 2012. Study on Incentives Driving Improvement of Environmental Performance of Companies. Rotterdam: ECORYS.

Reeson, A. F., and J. G. Tisdell. 2008. Institutions, Motivations and Public Goods: An Experimental Test of Motivational Crowding *J. Econ. Behav. Organ.* 68:271–281.

Schroeder, H. 2013. Sensing Value in Place. In: Place-Based Conservation: Perspectives from the Social Sciences, ed. W. P. Stewart et al., pp. 73–87. Dordrecht: Springer Science + Business Media B.V.

Schwartz, S. H. 1994. Are There Universal Aspects in the Structure and Contents of Human Values? *J. Soc. Iss.* 50:19–45.

Scott, J. 2000. Rational Choice Theory. In: Understanding Contemporary Society: Theories of the Present, ed. G. Browning et al., pp. 126–138. London: Sage Publications.

Seymour, E., A. Curtis, D. Pannell, C. Allan, and A. Roberts. 2012. Understanding the Role of Assigned Values in Natural Resource Management *Australas. J. Env. Manag.* 17:142–153.

Sierra, R., and E. Russman. 2006. On the Efficiency of Environmental Service Payments: A Forest Conservation Assessment in the Osa Peninsula, Costa Rica. *Ecol. Econ.* 59:131–141.

Stern, P. C. 2000. New Environmental Theories: Toward a Coherent Theory of Environmentally Significant Behavior. *J. Soc. Iss.* 56:407–424.

Stern, P. C., T. Dietz, T. D. Abel, G. A. Guagnano, and L. Kalof. 1999. A Value-Belief-Norm Theory of Support for Social Movements: The Case of Environmentalism. *Hum. Ecol. Rev.* 6:81–97.

Thaler, R., and C. R. Sunstein. 2008. Nudge: Improving Decisions About Health, Wealth, and Happiness London: Penguin Books.

Tittonell, P. 2014. Ecological Intensification of Agriculture: Sustainable by Nature. *Curr. Opin. Environ. Sustain.* **8**:53–61.

van Noordwijk, M., B. Leimona, R. Jindal, et al. 2012. Payments for Environmental Services: Evolution toward Efficient and Fair Incentives for Multifunctional Landscapes. *Annu. Rev. Environ. Resour.* **37**:389–420.

van Noordwijk, M., P. A. Minang, O. E. Freeman, C. Mbow, and J. de Leeuw. 2015. The Future of Landscape Approaches: Interacting Theories of Place and Change. In: Climate-Smart Landscapes: Multifunctionality in Practice, ed. M. van Noordwijk et al., pp. 375–388. Nairobi: World Agroforestry Centre (ICRAF).

White, D., and P. A. Minang, eds. 2011. Estimating the Opportunity Costs of REDD+: A Training Manual. Washington, D.C.: The World Bank.

Wunder, S. 2005. Payments for Environmental Services: Some Nuts and Bolts. Jakarta: CIFOR.

———. 2015. Revisiting the Concept of Payments for Environmental Services. *Ecol. Econ.* **117**:234–243.

4

Frameworks for Biodiversity and Forests

Leticia Merino-Pérez, Esther Mwangi,
Georgina M. Mace, Ina Lehmann, Peter A. Minang,
Unai Pascual, Kent H. Redford, and Victoria Reyes-García

Abstract

This work reviews diverse definitions of biodiversity and forests used in different discourses as well as the most common conceptual frameworks that influence the understanding of the dynamics of forests and biodiversity. It presents the ways in which different frameworks (conservation biology, ecological economics, environmental policy, and collective action—institutional analysis theory) address issues of sustainability, diversity, and justice, themes commonly used as analytical dimensions and evaluative criteria of policies and programs aiming to avoid and/or revert socioenvironmental deterioration. It reflects on how these frameworks are driven by differences in normative and theoretical positions, and how these positions influence actions and outcomes. Examples are presented of programs that have conservation and sustainability goals in forest and other high-diversity systems. These cases illustrate how diverse framings and values approach issues of justice and governance and influence conservation and sustainable management programs.

To minimize conflicts and achieve more balanced actions and outcomes, it finds that value systems present in discourses and policies be recognized and that dialogue among them be enhanced. This is important not only for interdisciplinary work, but for dialogues aimed at integrating the questions, concerns, and tools of different frameworks to construct more holistic, inclusive, and effective approaches to socioecological realities.

Group photos (top left to bottom right) Leticia Merino-Pérez, Esther Mwangi, Georgina Mace, Peter Minang, Georgina Mace and Kent Redford, Unai Pascual and Victoria Reyes-García, Esther Mwangi, Leticia Merino-Pérez and Peter Minang, Unai Pascual and Victoria Reyes-García, Georgina Mace and Kent Redford, Ina Lehmann, Xuemei Bai and Ina Lehmann, Peter Minang and Gareth Edwards, Unai Pascual and Victoria Reyes-García, Kent Redford and Leticia Merino-Pérez, Unai Pascual, Ina Lehmann, Kent Redford, Esther Mwangi, Victoria Reyes-García, Georgina Mace

Introduction

The relationship between power and knowledge pervades the production of discourse and practice, including environmental knowledge and policies. The power–knowledge balance influences the construction of framings and frameworks[1] and shapes the definition of those discourses and practices regarded as legitimate. From the perspectives of various environmental framings and diverse academic fields, biodiversity and forests are regarded as key contemporary themes and intertwined issues. The ways in which they are perceived and the salience of their inherent dimensions and aspects varies according to the diverse representations of environment and society as well as to the optics of the diverse conceptual frameworks that are used to research and understand the processes under analysis.

While recognizing the relations among framings and frameworks in diverse realms of environmentalism, the systemic analysis of some of the most prominent contemporary environmental framings (Dauvergne 2016; Escobar 2008; Martinez-Alier 2011) and "environmentalities" (Agrawal 2006)[2] is outside the reach of this work. Our aim is more limited as we focus on the ways in which different environmental frameworks treat and regard biodiversity and forests, the questions and problems they raise, the concepts and models used to explain phenomena, and the prescriptions derived from these diverse frames.

Our main proposals are:

- Forests and other high-biodiversity regions are valued (and therefore sought to be conserved) for different reasons and values: intrinsic, instrumental (for a variety of benefits they provide at different scales), and relational.
- Definitions of "forest" (or "good forest"), "deforestation," and especially "degradation" vary because the definers hold different values.
- Although conservation may be the dominant discourse for forests, discourses of sustainable use and of justice have also emerged significantly over the last few decades.
- Different disciplinary perspectives within academia are not only influenced by different values, they also involve different assumptions

---

[1] We understand framings as a set of concepts and perspectives based on how individuals, groups, and societies organize, perceive, and communicate about reality. It involves a social construction of social phenomena. Framing selects certain aspects of an issue and makes them more prominent so as to elicit certain interpretations and evaluations of an issue (Goffman 1974). In very general terms, a theoretical framework is regarded as a set of concepts, criteria, proposals, and assumptions that are relationally organized, together with their definitions and reference to relevant theory. Frameworks seek to explain the meaning, nature, and challenges associated with a phenomenon, often experienced but unexplained in social life, so that this knowledge and understanding enable more informed and effective actions.

[2] The concept of "environmentality," as proposed by Agrawal (2006), refers to environmental policies that are dependent on the nature of constituting elements such as knowledge, politics, institutions, and identities.

about human behavior and hence about how forests and biodiversity should be governed.

We recognize that real-world "conservation" interventions vary in the amount of attention given to concerns for sustainable use and justice. They also vary in the assumptions made about human behavior. Although academic perspectives and implementation approaches may share common ground, implementation efforts must confront the diversity of values among stakeholders and face the need to become more inclusive over time.

We begin with a discussion of definitions and values attached to the concepts of biodiversity and forests. We address the primary challenges for biodiversity and forest sustainability and relate the definitions and policies to different value systems. Thereafter we review, from our perspective, the most relevant contemporary frameworks used in the understanding of problems related to biodiversity and forests, policy design, and social action. Using diverse cases of programs with conservation and/or sustainable management goals, we analyze the values that have impacted their design and implementation, and the type of governance systems in place.

Biodiversity and Forests: Definitions and Values

It is generally agreed that diversity of life is a key feature of planet Earth that should be preserved for the continuity of the presence and evolution of life. The meanings of biodiversity and conservation tend to be taken for granted, but in reality there are multiple understandings which differ in conceptual and technical approaches, value bases, and policy implications (Redford and Mace, this volume). The term "biodiversity" is a successor to the broad and polysemic notion of "nature,"[3] in particular the post-1960 view of nature opposed to the idea of nature largely prevalent in the Western world as something that humans have to dominate. The term "biodiversity" was first used in the 1950s in American and British academic circles. Some years later it was adopted by international NGOs and U.S. government agencies and became central in conservation discourses and policies. As in the case of "nature," the definition of the concept of biodiversity has proved complex to operationalize. The Convention on Biological Diversity (2014) defines biodiversity as "the variability among living organisms from all sources including, *inter alia*,

---

[3] Whitehead (1920/2007) proposes to distinguish different meanings of "nature": nature as the essence of a thing, nature as the "natural world" object of study of natural sciences, nature as opposed to artificial, nature as opposed to culture, nature as wilderness, and nature as expression of the divine. Whitehead (1920/2007) reviews the process of social construction of "nature," distinguishing three related meanings: nature as a cultural construction, the cultural management of the environment, and the relations between humans, animals, and plants. More recently the Inter-Governmental Panel for Ecosystems and Biodiversity positions "nature" as a central category for diagnosis and policy making (Diaz et al. 2015).

terrestrial, marine and other aquatic ecosystems and the ecological complexes of which they are part; this includes diversity within species, between species and of ecosystems." For conceptual, methodological, and policy purposes, this definition refers to variation of genes, species, and ecosystems occurring in local, regional, and global scales (see Redford and Mace, this volume).

Over the last half century, "conservation" has been understood in different ways:

- Conservation of nature, wilderness, and positions achieved mainly through protected areas
- Conservation oriented to prevent extinction, threats and threatened species, habitat loss, pollution, and overexploitation
- Through strategies such as logging vans and prohibition of extraction and commercialization of endangered species
- With an emphasis on ecosystems, ecosystem services, and economic values by promoting policies such as Payment for Ecosystem Services (PES), green taxes, and certification of sustainably produced goods
- In the context of environmental change, looking to build resilience and adaptability of socioecological systems, recurring to varied policy multiscale and multisectorial tools

Nevertheless, most nature conservationists and wildlife managers understand biodiversity as "all life on Earth" and/or as "species richness" (Redford and Mace, this volume). The concern for the loss of species, particularly in tropical forests, has been an important orientation of biodiversity conservation policies for a long time. From this perspective, the focus on "hotspots" (i.e., places with a high number of species and under significant threats) was regarded as a central problem and a priority for action (Vandermeer and Perfecto 2013). It remains a persistent feature of conservation policies. Other aspects, such as genetic and biocultural diversity, tend to receive much less attention. The complexity of this theme and the specialized knowledge it demands leads to the generalized perception that biodiversity is a field for scientists and conservation biology that is primarily science driven in spite of its highly normative approach.

Among the high-diversity ecosystems (e.g., deserts, grasslands, wetlands, and oceans), forests are particularly important as they host more than 80% of the terrestrial biological diversity (statistics from The World Bank). Thus, forest conservation is a main component of biodiversity conservation. The category of forests includes a wide variety of ecosystems. Different types of ecological communities with prevalence of perennial vegetation are currently classified as forests (humid tropical, dry tropical, temperate, boreal, and cloud forests). Vegetation in different conditions and successive stages (e.g., old growth forests, degraded forest, managed and secondary forests, and forest gardens) fall into this broad category. There are also various definitions of forest that incorporate factors such as tree density, tree height, land use, legal standing, and ecological function (Convention on Biological Diversity 2014; UNEP 2010). The most widely used definition is the one provided by the Food and Agriculture Organization (FAO). It defines forests as areas with more than 0.5 ha and more than 10% of tree forest canopy, and trees as plants capable of growing more than 5 m (FAO 2015).

Deforestation is broadly understood as the disappearance of forest vegetation and is mostly used as a binary concept. The concept of degradation is more difficult to assess as it is related to losses of various environmental values, which are more difficult to capture and measure. In 1990 the FAO estimated that the world had 4128 million ha of forest (31.6% of the global land area). By 2015, twenty-five years later, forest extension had decreased to 3999 million ha (30.6%) (FAO 2016). From 2005–2015, the global deforestation rate halved in relation to the previous decade. Deforestation is a complex process with losses in some regions and gains in others. Forest recovery took place in much of the developed countries during the last decades of the nineteenth century and the beginning of the twentieth century. During the last decades, regions in developing countries have experienced processes of forest regrowth (Hecht et al. 2013).

Forests have been and are contested landscapes that bring to the fore tensions in values systems, economic goals, and social movements—tensions of environmentalisms that reflect justice, diversity, and sustainability issues. According to Hecht (2014), ideologies about forests, imagined histories, iconography, institutional arrangements, and competing knowledge systems structure the understanding of the social nature of forests. Forests—long regarded as remote areas, frontiers of civilization, wasteland, and unproductive lands—are home to local groups that are often externally or weakly controlled by central colonial powers. For the last 30 years, forests have been central to the framing of environmentalism: the isolation of forest ecosystems and forest dwellers has been associated with conservation of natural areas and biodiversity. Forests have also played a fundamental role in the debate on climate change, mitigation, and policy making.

Different stakeholders hold diverse instrumental, intrinsic, and relational values of forests and biodiversity. For local communities, forests are key providers of food, wood fuel, fodder, and medicines (Vira et al. 2016),[4] and they are often a source of cultural identity. Forests provide important sources of revenue and raw materials for central authorities and corporations that largely control commercial extraction of timber[5] and minerals from the mountainous subsoil of forest regions (Boyer 2015; Putz and Redford 2010; Scott

---

[4]   The United Nations estimated that 300 million people live in forests and 1.6 million people's livelihood depends on forest resources (UN 2011).

[5]   Trade in forest products was estimated at $327 billion in 2004, most of this production comes from public forests under concessions to private firms, 80% of the world's forests are publicly owned (UN 2011).

1998). More recently, forests are valued as providers of various ecosystem services, including provisioning services already mentioned but also key supporting, regulating, and cultural services: soil formation, carbon sequestration, hydrological services, ecotourism, and so forth (Millennium Ecosystem Assessment 2005).

Problems arise when values held by different actors are not compatible when translated into actions toward nature that often diverge and conflict. Shifting cultivation is considered by some to be a sustainable management strategy that enables the presence of patches of vegetation in different successional stages; for others it is simply regarded as deforestation. Commercial forestry, which has largely been regarded as "rational" and "scientific," is associated with the homogenization of forest systems. Strict conservation through restrictive protected areas are considered by many to be the best way to preserve forests and the biodiversity they host, whereas for others forest conservation should be achieved in the context of forest land-use planning and biocultural landscapes, which combine conservation with various forest uses.

Biodiversity and Forests through the Lens of Different Conceptual Frameworks

Over the last three decades, the conceptualization of forests and biodiversity has been enriched by insights and analytical tools from the ecological as well as social and economic sciences. In this section, we briefly describe the most relevant contemporary theoretical frameworks that we find useful to emphasize aspects of real problems and explain the main causes of forest and biodiversity losses or conservation, and build proposals aiming to revert environmental deterioration. We are fully aware that our review is far from exhaustive; however we hope that it will spur a more comprehensive reading of socioenvironmental realities. Following Lele and Kurien (2011), we review the perspectives of conservation biology, ecological economics, political ecology, and collective action theory. For each of these frameworks we reflect on their thematic/conceptual focus, their understanding of the causes and drivers of forest and biodiversity losses, and the general proposed prescriptions in terms of technical management and/or governance and the values explicitly or implicitly held. Thereafter, we review the extent and the ways in which these frameworks treat the themes of justice as well as biological and sociocultural diversity, which are increasingly relevant for conservation and sustainability movements and policies.

Conservation Biology

Conservation biology is the oldest and most established approach used to define biological diversity, and the natural sciences (biology, ecology, genetics)

are its primary contributors. This framework distinguishes different components of biodiversity—ecosystems, species, and genes (Redford and Richter
1999; Redford and Mace, this volume)—that are considered and evaluated in
terms of composition (identity and variety of constituent elements), function
(evolutionary and ecological processes acting among elements), and structure (physical organization pattern of elements). Different methodological
and conceptual approaches are used to study biological diversity: Ecologists
tend to focus on forms and functions of organisms in a given place, aiming to
understand functions and dynamics of systems within communities and ecosystems. Evolutionary biologists focus on dynamics but with an emphasis on
historic or inherited variations; that is, on genetic and phylogenetic attributes.
Conservation biologists consider function and processes to be of primary concern for the preservation of species, genetic diversity, and to reach achievable solutions. Different metrics have been developed for the assessment of
biodiversity in conservation planning; generally, all include attributes such as
richness of species, intactness, native-ness, endemism, and risk of extinction.

Impacts of human activity (primarily implying redirection of matter, energy,
and flows) in one or more biodiversity components are generally unknown and
unappreciated (Redford et al. 2003). In terms of richness of species, the current
rate of diversity loss is estimated to be 1000 times higher than the (naturally
occurring) background extinction rate, and this is expected to continue to grow
in the future. Another classification of biological diversity is the distinction
between alpha diversity (the mean species diversity in sites or habitats at a local scale), beta diversity (the differentiation among those habitats), and gamma
diversity (the total species diversity in a landscape). Given the wide variety
of values, purposes, and contexts, there is no single simple measure of biodiversity. This complexity creates important challenges for the establishment of
policy goals and targets.

The most relevant values from this framing are intrinsic (inherent) "from
their uniqueness to their rights" (Pascual et al. 2016), making nature's conservation justifiable and valuable in itself. Forests are intrinsically valued as
high-diversity ecosystems whose preservation responds to the recognition of
the rights of nonhuman species and to the importance of maintaining the evolution of life for which diversity is an integral dimension.

Biodiversity and forest losses are regarded as mainly consequences of
"anthropogenic" behavior and activities. Social processes are conceptualized in
a very general, homogenizing way, without systemic reflection on their structure
and dynamics. "Anthropogenic presence" tends to be treated as a general
analytical variable with negative impacts, leaving aside the role of societies in
the protection and construction of natural landscapes and the complexity of the
interactions between nature and societies. This framework distinguishes direct
and indirect drivers of deterioration of natural systems. Direct drivers include

64 *L. Merino-Pérez et al.*

human activities[6] and natural processes that interact with natural landscapes in ways defined as negative (Salfsky et al. 2008). Indirect drivers refer to a set of socioeconomic conditions (e.g., population growth, poverty, development) related to environmental performance based on overgeneralized, often ideological assumptions (e.g., population growth and poverty unavoidably lead to natural destruction). Explanations of environmental degradation lack a systemic approach that ignores fundamental social phenomena related in multiple ways to ecosystem conditions (e.g., production systems, power, heterogeneity–inequality, livelihoods, property, rights, governance, economic externalities, institutions, and knowledge systems).

Based on the notion of pristine nature as being distinctive and opposed to society and culture, proposals derived from this concept tend to recommend the reduction, ideally the absence, of human activities (even human presence) in areas of interest of biodiversity conservation. To achieve this purpose, this approach relies mostly on centralized governance schemes such as highly restrictive protected areas controlled by central governments (Dellas and Pattberg 2009; Mascia et al. 2014). Around the developing world, particularly in countries with tropical forests, the implementation of these policies carries severe social costs, often ignoring the rights of indigenous peoples and local communities. This approach is mostly held by global conservation agencies and national governments in conjunction with national academic conservation groups, but very rarely by local stakeholders (Hecht et al. 2013).

As discussed above, this framework has traditionally been concerned with the conservation of biological diversity: cultural and social diversity are usually external to its conceptual limits. The same is true for the consideration of justice. Nevertheless, the work of some conservationists in countries of the Global South, which has the largest share of global biodiversity, has led them to the recognition of local needs and of the necessity to embed them within conservation goals. In this sense, the scope of conservation biology within certain academic circles has started to include the understanding of local meanings and uses of nature, as well as the relational values that sustain the links of local societies with their natural surroundings.

Ecological Economics

Typically, markets fail to address forests and other high-diversity ecosystems as sources of marketable goods and nonmarketable services (positive externalities) essential for human life (Angelsen and Kaimowitz 1999; Doupe 2015). Economics and knowledge of ecosystem services are the scientific disciplines that contribute predominantly to the framework of ecological economics.

---

[6] Mainly those that imply vegetation removal and pollution of ecosystems and natural resources (agriculture, cattle raising, urban development, etc.) and "natural" or socioenvironmental phenomena (e.g., climate change, forest fires and pests, ocean acidification).

Forest and biodiversity losses are considered to result from market failures: difficulties of markets to internalize the whole range of values provided by forests and biodiversity. These failures lead to the absence of incentives for the users and/or owners of natural resources to commit to the maintenance and protection of forests and other natural systems over time. Threats created by market failures are more sensible when preservation of natural systems implies high opportunity costs, defined as "the loss of potential gain from other alternatives when one alternative is chosen" (Buchanan 2008). This is often the case in contexts where actors have subsistence or developmental options that conflict with conservation of natural systems (Angelsen and Kaimowitz 1999; Geist and Lambin 2002; Muñoz-Piña et al. 2008).

Proposals influenced by ecological economics are directed to the creation of market-oriented conservation/sustainability tools and aim to "internalize" the value of environmental goods in the pricing of natural resources and services. These proposals and policies seek to create incentives for local users and landowners to commit to sustainable management and conservation measures. Because of the diversity of goods and services provided by forests as well as the global concerns associated with deforestation, most of these tools and practices have been applied primarily to forest ecosystems.

PES schemes represent one of the most common strategies derived from this approach. Their aim is to compensate landowners for the environmental services provided by their lands (mostly forests); in exchange, landowners must commit with management measures defined by the paying parties, which generally include the abandonment of production on the lands involved in the programs (Pagiola 2008; Wunder 2005, 2015).[7] Despite the emphasis given to the need for environmental service markets, these markets have been difficult to create. To date, many instances of PES rely on government subsidies, not on real markets. Other cases, mostly those working on carbon sequestration and climate change, have engaged in the creation of international markets for this service with the participation of international agencies and banks, NGOs, and corporations. Other difficulties of PES programs relate to their additionality, effective, and permanent impacts (Calvet-Mir et al. 2015).

Another mechanism oriented toward the creation of incentives for sustainable practices is the certification of forest products resulting from sustainable, nature-friendly production processes. These practices (e.g., the Forest Stewardship Council or Rainforest Alliance) are geared mostly to niche markets, where consumers agree to pay premium prices. For some forest products, certification can be difficult as the premium prices for sustainable forest producers in global markets are frequently hard to achieve (Molnar et al. 2003). Products destined for direct consumption (e.g., coffee, cacao) show more

---

[7] PES covers mostly hydrological "services" provided by forests and carbon storage in forest biomass; biodiversity is considered to a lesser extent.

favorable tendencies. Other schemes involve the disappearance of subsidies to unsustainable activities.

Challenges that result from the implementation of some PES programs include local "sovereignty" issues of communities taking part in these programs, where people's autonomy to decide about their lives and territories is increasingly challenged. Some interventions result in perverse incentives (e.g., paying people for what they are supposed to do, or paying them for doing nothing, crowding out prosocial behavior). These challenges may be related to the top-down, centralized nature of the policy design, implementation, and evaluation.

The values behind the environmental economics framework are predominantly instrumental and related to ecological efficiency: the maintenance of the flows of ecosystemic services and the provision of economic incentives for local users and/or landowners. Concerns about cultural diversity and justice generally fall outside the scope of this framework though recent critics expose the convenience and need to include considerations of equity into the reflection and practice of PES programs and certification initiatives (Calvet-Mir et al. 2015; Pascual 2010; Pascual et al. 2014). Stakeholders who sustain this approach are mostly international agencies working on sustainable development policies and national governments looking for alternative conservation tools for protected areas. Some NGOs and local businesses also take part in both PES and forest certification. In the search for viable policies to mitigate global climate change, this approach has gained in importance.

Political Ecology

Political ecology addresses the relationships between nature and society, and characterizes the human environment (as a social field) as a set of power relations involving confrontation, domination, and negotiation. The analytical framework of political ecology has been enriched by collaborations between political economists, ecologists, anthropologists, and historians. This transdisciplinary research field hosts different approaches that share central themes and concerns: questions about social marginality and unequal access to natural resources, the political causes and effects of resource allocation, the attention to the cultural, socioeconomic, and political contexts that shape human use and control of nature (Bryant and Bailey 1997). Blaikie and Brookfield (1987) suggest that political ecology "combines the concerns of ecology and a broadly defined political economy." Focus on contentions and struggles over land and natural resources, power asymmetries, and social inequalities are critical points of departure. Analysis of capital accumulation and political economy provide the overall framework for understanding dispossession and displacement of local communities by global forces of state and market (Bengt 2015).

Over the last two decades, the field of political ecology has expanded considerably to include detailed examinations of politics as well as recurring historical and ethnographic approaches. The adoption of a poststructuralist orientation[8] (Escobar 2008; Peet and Watts 1996) has led to the recognition of different "environmentalisms" (representations, discourses, and practices), which result from different cultural and social experiences, as well as social positions that hold different, often incompatible, world views (Agrawal 2006; Dauvergne 2016; Hecht 2014; Martinez-Alier 2011). Hegemonic environmentalism justifies the prevalent distribution of costs and gains of different actors, and tends to reproduce political and economic inequalities both in national contexts and at the global level.

Justice is a central concern, treated as the search for (a) equity, recognition, and fair procedures in socioenvironmental realities and (b) fairness in the distribution of environmental assets, gains, and costs of economic activities and consumption patterns. The proposals of some political ecologists give prominence to the recognition and devolution of rights over forest lands and access to natural resources to marginalized and disempowered actors (notably indigenous groups, local rural communities, and dwellers of poor urban neighborhoods), emphasizing the need for decentralization and empowerment of local governments and actors as a possible means to revert exclusion and environmental deterioration (Ribot 2009; Ribot et al. 2006; White and Martin 2002). Values embedded in this framework can be characterized as "relational," given the importance of the relations of local communities with nature in contexts of equity and justice.

Forests and other high-diversity ecosystems are regarded as places of significant human action. Complex institutions, ecologies, and economies have transformed these landscapes in the past and continue to shape them in the present (Hecht et al. 2013). These landscapes are often contested, subject to different and even conflicting meanings and claims, objects of struggle over appropriation among opposed stakeholders.

Environmental deterioration and environmental conflicts are seen as closely related (Boyer 2015; Merino 2004, 2016). The central causes are the absence of internalization of the enormous environmental impacts and costs of contemporary production processes (including high impact activities such as mining, fracking, and industrial agriculture); consumption patterns with high ecological footprint; and the political capacity of corporations, governments, and national elites to impose these costs on those with weaker political voices (i.e., vast numbers of people who live in developing countries as well as future generations) (Dauvergne 2010, 2016). Over the last decade, environmental injustice and deterioration have worsened under the global schemes of land grabbing and neo-extractivism. These harmful activities are enabled through

---

8   A new relation between agency and structures, and an attention to different knowledge systems and their influence in theoretical reading of socioenvironmental processes. Focus has expanded from rural issues to include environmental politics in urban settings and addresses contemporary questions: climate change, genetically modified organisms, food industries, pollution, city planning, and infrastructure development.

economic incentives: governments of developing countries rely on the fiscal revenues received from extractive transnational corporations working in their countries (Campodónico 2007).

A frequent critique to political ecology points out its rather limited focus on the analysis of conflicts and denouncement of environmental injustice. In addition, not enough attention is given to the construction of sustainable socioenvironmental strategies and policies, neglecting the comprehension of ecological dynamics and viable governance schemes engaged in fair and sustainable management.

Stakeholders that use this framework tend to be some international and national NGOs and groups advocating for human rights and traditional rights of indigenous people, federations of indigenous groups, local rural communities, as well as groups of those affected by the environmental impacts of industrial agriculture and extractive activities. Currently, an increasing amount of scholarship is studying local stakeholders' notions of environmental justice in conservation interventions (Sikor et al. 2014). This is an important way forward in understanding where problems of justice arise.

Collective Action and Institutional Analysis Theory

The framework based on collective action and institutional analysis theory is strongly influenced by contributions from political science, natural resource economics, and experimental economics. Its focus is on themes such as collective action (coordination and cooperation), property rights, governance, social capital, and institutions (understood as rules in use) involved in shared resources, as natural resources tend to be. This conceptual proposal gained broad international attention in the early 1990s, when Elinor Ostrom, the lead proponent in this field, responded to proposal ofset forth in the "*Tragedy of the Commons*" (Hardin 1968), which was widely accepted in the conservation and development fields. Hardin proposed that this tragedy was a universal and unavoidable destiny of common goods characterized, in his view, by unrestrained access to natural resources. One of Ostrom's main arguments held that Hardin confused community property/management and open access to natural resources effectively associated with the deterioration of the resource systems. Based on ample empirical evidence, including cases of community property/ management of pastures, rivers, irrigation, forests, and fisheries, Ostrom and colleagues demonstrated that when collective users were able to communicate, self-organize, and had control over the natural resources upon which they depend, they often created rules that enabled cooperation and sustained use of common resources (Ostrom 1991).

Ostrom's findings refuted the universality of the "*Tragedy of the Commons*" without denying its reality. She recognized that these tragedies were present in many cases and proposed that they were caused by the inability to collaborate around common purposes and shared resources: this inability to cooperate and coordinate around use and management of common environmental resources was the root of environmental deterioration. Adequate responses to these challenges of shared resources face dilemmas concerning collective action—this refers to a wide range of situations with conflict between individual short-term profits and collective long-term benefits (Cardenas 2009; Ostrom 2005). Lack of information and understanding of the resource system, absence of communication and trust, elite capture, and weak or null incentives to cooperate are all obstacles to collective action. Key conditions enabling collective users to self-organize successfully and sustain their *commons* include the existence of meaningful levels of autonomy, local participation in rule making, monitoring systems accountable to local users, shared understanding of the resource systems, and trust among group members.

Sustainable use of resources requires addressing appropriation and provision needs: the first refers to the sustainability of harvesting, and more generally use practices; the second to the investments (of work, time, knowledge or money) needed to maintain resource systems. Ostrom's well-known typology of goods permits an understanding of the types and levels of pressures deriving from the conditions of excludability and subtractability/rivalry. Private goods with high subtractability and high excludability face potential appropriation problems and provision needs, but no collective action challenges. Pressures for club goods tend to be low because rivalry is low and excludability is high. Public goods are those with low excludability and low subtractability. Even if, in principle, the use of these resources faces limited appropriation problems, their maintenance poses provision needs that are difficult to address as collective action dilemmas among often anonymous actors (the public). Finally, common-pool resources (CPRs)[9] have high rivalry and low excludability: they face important appropriation and provision challenges as well as collective action dilemmas that need to be resolved by communities and user groups whose members often share rights and duties (in regard to the commons) and frequently know each other. Because of these pressures and challenges, the need for agreements and rules is particularly important for the sustainability of CPRs.

The distinction between types of goods, property regimes (attentive to the nature of the resource holders), and property rights has a heuristic value for this framework. Property rights are regarded as "bundles of rights": use rights (withdrawal, access) and control rights (exclusion, management, and alienation) (Schlager and Ostrom 1992). From an institutional perspective, property rights are important for sustainability as they provide incentives for rights holders to commit to the use and maintenance of the resources based on long-term perspectives, and to participate in the construction of governance systems that are capable of responding to sustainability challenges (Dietz et

---

[9] Ostrom rarely referred to the term "commons," instead she used the more technical concepts of CPRs and public goods when referring to shared resources.

al. 2002). As the distribution of rights varies within the frames of property regimes (private, collective, or public) and different agents (owners or not) can
hold diverse property rights, no property regime constitutes an environmental
or social panacea. The distribution of property rights among different social
actors, rather than the property regime, is weighted more for collective action
and sustainability. The concentration of property rights and imposition of private, public, or even collective property regimes as panaceas may lead to mis-
use and deterioration of natural systems and resources. From the perspective
of this framework, forests and other high-diversity ecosystems (e.g., prairies,
wetlands, costal zones) are CPRs with important appropriation and provision
problems under public, private, and public property. The diverse distribution
of use and control rights involved create incentives and disincentives for different actors to engage in sustainability.[10]

One frequent criticism of this framework is that while focusing on the construction of collective action and governance, institutional analysis neglects
the themes of conflict, power, and inequality. In fact, Ostrom and colleagues
aimed to develop and follow comprehensive analytical frames such as the institutional analysis and development framework and the social-ecological systems framework, which attempts to integrate different theories and categories
(e.g., actors, contextual socioeconomic conditions, institutional arrangements,
governance systems, and interactions) (McGinnis and Ostrom 2014; Ostrom
2005; Poteete et al. 2009).

As in the case of political ecology, the proposals derived from collective
action and institutional analysis underline the importance of recognizing the
rights of local communities to use and manage natural resources. The analysis
of the relations and asymmetries among different actors with stakes in use
or conservation of the resources is considered fundamental for environmental
governance. Procedural justice—the fair access to political decision making—
is a prominent concern, as participation of local communities in rule making
has a prominent role in robust governance systems. Nevertheless, local control,
stewardship, or community property are not regarded or proposed as panaceas.
Collective action is costly to achieve, and sustainable management of natural
resources is a complex task that requires local governance capacities nested
in polycentric governance systems[11] as well as the use of different knowledge
systems (traditional and local knowledge as well as scientific understanding
and recommendations). The values assigned to nature by this framework can

---

[10] White and Martin (2002) document that the vast majority of the forests in the world are public
property under concessions to transnational corporations which have acquired use and management rights, and whose incentives are mostly oriented to maximize short-term profits. Traditional property rights over forests are denied in conditions of public property, particularly in
forests under concession. Local people often lose incentives to invest in provision measures
and to follow appropriation rules that limit their access to important means for livelihoods.

[11] Polycentricity is understood as governance systems based on multiple decision-making centers
operating at multiple scales (Aligicia and Tarko 2012).

be instrumental, scientific, and relational: individuals engage in cooperation for sustainable use of nature when they depend on natural goods, but it is also acknowledged that the sense of sacredness, belonging, identity, and knowledge that people develop in relation to nature and landscapes provides powerful incentives for conservation and uses based on long-term horizons (Berkes 1999). Finally, the relationships that people establish with territories and between themselves create a sense of community and identity.

Important challenges for conservation, sustainability, and justice signaled by research oriented within this framework include uncertainty over property rights as well as incomplete or fragmented property rights. This includes countries where communities lose their rights to minerals but have rights to lands, forest, and trees, or governments that hold rights over minerals and gas in the subsoil.

This framework has largely remained in academic circles. However, since 2000, a vast array of groups demanding democratization of environmental, urban, and knowledge governance in developed and developing countries have increasingly adopted the idea of the commons (Bollier and Helfrich 2012; Capra and Mattei 2015; CAPRi 2010; Hess 2008; Iaione 2013), as a result of social mobilization and often social creation.

Conservation Programs, Framings, and Values

To enhance our understanding of the values and frameworks present in different environmentalisms (influenced by values and frameworks), we selected and analyzed a set of eight initiatives from the field that represent different approaches to biodiversity conservation (Table 4.1). Our analysis has three objectives: (a) to understand what kinds of normative concerns drive these initiatives, (b) to explore ways in which interaction of different frameworks and different values interrelate in concrete cases, and (c) to analyze how these diverse conceptual and ethical framings relate with the main challenges faced by initiatives seeking conservation and/or sustainability.

Conservation biology is the dominant paradigm for the oldest initiatives as well as for those with conservation purposes created by externally driven interventions (Yellowstone, Monarch Butterfly Biosphere Reserve, and The Gulf of Mannar Biosphere Reserve). Intrinsic values related to the preservation of nature tend to be predominant in these cases. Local appreciation of natural systems involved in these cases depends largely on the livelihoods that local residents are able to obtain from them. Conflict between local and external values is common, and is enhanced in national contexts characterized by power and economic asymmetries, lack of communication, and trust. Conflicts are acute in cases where there is a high concentration of decision-making capacities in the hands of agencies external to local populations. The criticisms of protected areas often emanate from the perspective of political ecology and collective

72

Table 4.1 Description of the conservation initiatives.

Initiative	Property regime, location, and extension	Type of intervention or action, date established	Values	Governance schemes
Yellowstone National Park (Schullery 2004; UNESCO 2012)	Public lands 9000 km² in the Rocky Mountains, U.S.A.	Protected area, national park. Limited resource use, regulated tourism. Established in 1872.	*Intrinsic:* landscape conservation. *Relational:* public education and appreciation of nature. *Compatible local and external values.*	Centralized* decision making, centralized in the federal government.
Tsimane Indigenous Territory (Reyes-García et al. 2014)	Community property of 400,000 km² in Beni Dept., Bolivia	Communal indigenous lands. Devolution of resource management rights to indigenous people in 1991; limited actions by nonindigenous (colonists, logging companies).	*Relational:* sense of identity and belonging through the recognition of traditional rights.	Decentralized local indigenous institutions are legally recognized but lack enforcement capacities.
PES program, Mexico (Alix-García et al. 2005)	Community and private forests in Mexico: 4000 km²	PES program, run by the central government of Mexico. Payment to forest owners for the maintenance of forest cover. Established in 2004.	*Instrumental:* maintain flow of ecosystemic services (hydrological, carbon sequestration, conservation of biodiversity).	"Monocentric" decision making centralized in the federal (central) government.
Unión Zapoteco-Chinanteca (UZACHI) (Bray and Merino 2004; Chapela 2005)	Community property of 300 km² in Oaxaca, Mexico	Fight against a forest concession to an external firm. Sustained community forestry and conservation. Initiated in 1991. Participatory land-use planning, sustainable-certified forestry, community conservation and ecotourism. Also PES	*Instrumental:* create incentives for local sustainable forest management and land-management practices. *Relational:* enhance local stewardship and appreciation of nature. *Compatible local and external values.*	"Polycentric" central and local decision-making capacities, often conflicting.
Monarch Butterfly Biosphere Reserve (Merino and Hernández 2003; Tucker 2004; UNESCO 2008)	Community property in Central Mexico: 562.5 km²	Protected area, biosphere reserve. Established in 1996, extension more than doubled in 2000. PES (conservation) to some communities in 2001.	*Intrinsic:* conservation of the migration phenomena of the monarch butterfly through North America, by preserving its winter habitat. *Instrumental:* held by most community members within the biosphere reserve whose livelihoods depend on forest resources. *External and local values are often conflictive.*	"Monocentric" decision making centralized in the federal (central) government.

Table 4.1 (continued)

Initiative	Property regime, location, and extension	Type of intervention or action, date established	Values	Governance schemes
Gulf of Mannar Marine Biosphere Reserve (Kasim and Edwards 2013; Rajagopalan 2008)	Public property of 560 km² in the Gulf of Mannar, Tamil Nadu, India	Marine biosphere reserve, conservation and development, created in 1986. Alternative livelihood activities for local fishing communities, planned and funded by the GEF, UNDP and the governments of India and Tamil Nadu	*Intrinsic*: restore and preserve marine ecosystems. *Instrumental*: maintain ecosystem services through support of local livelihoods, which depend on marine resources. *Conflict between local needs and external values has occurred.*	"Monocentric" decision making centralized in the federal (central) government.
Kasigau Corridor REDD+ Project (Chomba et al. 2016)	Large private holdings, national parks and communal lands; 1685 km² in Kasigau, Coast, Kenya	REDD + project, conservation corridor. Payment to locals for reduced shifting cultivation and charcoal production. Began in 2008.	*Instrumental*: maintain provision of ecosystem services (mainly carbon sequestration). *External and local values are sometimes conflictive.*	"Monocentric" decision making centralized in the federal (central) government.
Blanket Bog Restoration (2015)	Private lands in the Uplands of the Pennines, Dartmoor, Exmoor, Lake District, York, Border Moors, Cheviots, and Forest of Bowland, U.K.	Habitat restoration for biodiversity (nationally significant habitat) and ecosystem services (carbon, water regulation). Negotiations with landowners and managers, water companies, nature conservation organizations, and soil managers to block drainage, reduce grazing, restore peat bog. Began in 2014.	*Intrinsic*: conservation of biodiversity. *Instrumental*: maintain or restore provision of ecosystem services. *Relational*: promote involvement of local landowners in restoration activities. *Negotiated local and external values.*	"Polycentric" local capacities for decision making.

*Aligicia and Tarko (2012) distinguish monocentric strongly centralized governance systems, decentralized systems where decision-making capacities are transferred from national to local agencies, and polycentric systems with multiple interconnected decision-making centers.

action theory, underlining aspects of frequent undemocratic decision-making processes, disempowerment, and alienation of local communities.

In seeking to achieve sustainability through policy schemes with lesser political and social costs than traditional protected areas, more recent initiatives (e.g., National Parks and Biosphere Reserves) have found an important orientation in ecological economics. This framework attempts to achieve a sustained flow of ecosystem services while taking into account the incentives, needs, and instrumental values held by local actors. In many cases, these initiatives have failed to introduce context-specific elements in the design and implementation of PES programs, leaving aside critical aspects of elite capture, inequality, and poverty (e.g., lack of access to land property rights) that challenge the results of these interventions (Rodríguez and Merino 2017). PES and particularly REDD+ programs often rely on governance schemes that tend to re-centralize control rights in central governments. Participation of local stakeholders and marginalized groups in PES/REDD+ programs is limited or null, thus increasing the risks for conflict as these groups' values are poorly taken into account. It is rare for PES and REDD+ programs to incorporate elements of frameworks different to ecological economics that may enable more context-sensitive environmental policy schemes.

The perspective of actors involved in the only intervention driven by local actors considered in this work (UZACHI) has widened over time to include different values from those originally adopted. This intervention, which initiated opposition to a forest concession in communal lands, brought about sustainable forest logging, forest certification, diversification of forest uses, conservation, and the adoption of intrinsic and relational values.

The wide national program of PES in Mexico, with more than a decade of experience, has yielded mixed results that vary in different contexts: When applied in lands under protected areas, it has helped to support social costs of restrictions imposed on local communities. In other places, it has complemented the revenues that local communities obtain from commercial forestry, thus contributing to local livelihoods and stewardship of natural resources. Still in other cases, it has conflicted with local livelihoods and affected the most vulnerable groups. A similar process seems to be taking place in the new REDD+ project in Kasigau Corridor, Kenya.

Over the years, most initiatives have widened the framings and values that guide their actions, incorporating incentives, values, and sometimes the participation of local stakeholders (e.g., visitors in the case of Yellowstone Park): The management of Yellowstone has actively incorporated environmental education to increase public relational values of the Park. The English initiative of the restoration of Blanket Bog promotes conservation and restoration with a strong base of local civic actors. In addition, UZACHI has incorporated elements of collective action and institutional analysis theory, conservation biology, and ecological economics in the participatory land-use planning, community by-laws, and conservation strategy. This development has been made possible through sustained community participation and the collaboration between national and international actors engaged in sustainable development programs.

The claims of the Tsimane people over their traditional lands and forests fall within the framework of political ecology. Nevertheless, the isolation of this group, which gained formal rights over an enormous territory, and the lack of governmental capacity to back up indigenous rights in the face of loggers abusing the forest and indigenous people, are key factors which threaten this initiative and restrain further socioenvironmental innovation.

The outcomes of these initiatives, in terms of ecological, economic, and social sustainability, varies: from the highly successful case of Yosemite National Park, to the fragile socioenvironmental conditions in the Tsimane Territory or the Monarch Butterfly Biosphere Reserve, which faces challenges of illegal logging, drug-trafficking, and crime. Some of the strongest challenges currently being faced derive from international, even global, expectations that conflict with local livelihoods and rights not considered when many of these initiatives were first implemented.

Conclusions

Frameworks are epistemic developments constructed and used to understand the world. Following Entman (1993), we assume that frameworks involve the selection of "some aspects of a perceived reality and make them more salient in a communicating text, in such a way as to promote a particular problem definition, casual interpretation, moral evaluation and/or treatment recommendation." Research questions and hypothesis, data collection and interpretation, as well as interventions and policy proposals are produced within the limits of particular frameworks. Disciplines provide paradigms, theories, and concepts that name, define, and explain certain dimensions of reality. Worldviews are always shaped by formalized, informal, explicit or implicit theories, understood as concepts related in systematic patterns. Theories spotlight phenomena viewed as valid and relevant, leaving aside other dimensions, receiving less importance or not even considered. This process of discursive delimitation and problem definition takes place within particular social and historical contexts and dynamics, where values and power relations play significant roles (Foucault 1969/2008; Piaget and Garcia 1982). Research methodologies and validation criteria are developed and established within the fields of particular disciplines, theories, and knowledge systems—modern Western science being just one of them (Foucault 1971; Poteete et al. 2009). In addition, theories are influenced directly or indirectly by the values and visions of different stakeholders, through various influences and mechanisms of prestige and funding (Fairhead et al. 2012), as well as those of the researchers themselves (Lele and Norgaard 2005).

Evaluation criteria utilized to evaluate the outcomes of conservation and sustainability policies and programs are also influenced from specific frameworks and framings. They are determined by academic traditions, disciplinary and theoretical perspectives, as well as values held by different stakeholders (Fairhead and Leach 1996; Foucault 1969/2008). Consequently, these criteria are generally limited, even subjective. Thus, researchers need to be conscious about the relativistic nature of their knowledge, not only of their strengths but also their limits, and their conceptual and policy implications (Fairhead and Leach 2006; Fairhead et al. 2012).

In academic and policy fields, we find different definitions of what biodiversity and forests are, a varying focus on socioenvironmental processes and understandings of what is being lost, and diverse ideas of what should be done to revert deterioration. Recognizing the different frameworks and values held by different stakeholders taking part in particular forest or biodiversity programs constitutes a first step in efforts to build more inclusive, potentially better accepted policies, minimizing conflict, and ultimately increasing the possibilities to achieve socioecological objectives.

Values are principles associated with a given worldview or cultural context, a preference someone has for something, and the importance of something for itself or for others (Pascual et al. 2016). The value of nature or biodiversity is a contested domain and a source of conflict over the way humans relate in and through nature (O'Neill et al. 2008). There is no conceptual agreement on the value of biodiversity. Values of biodiversity and nature (including forests) are commonly classified as intrinsic, instrumental, and relational. Thus, emphasis can be placed on the instrumental role of biodiversity to support a good quality of life through the capacity to provide material (e.g., food, fiber) and immaterial (e.g., recreation, mental health) benefits. Alternatively, emphasis can be on honoring Earth as sacred (Diaz et al. 2015). Sustainability, diversity, and justice are common values in the academic and policy approaches to forests and biodiversity. Such a wide spectrum of values is rarely taken into account in environmental decision making. Instead, a struggle over dominance regarding monistic worldviews over nature is a constant feature. This is often manifested in global conflicts over resources, a sense of environmental injustice, and unsustainable development (Pascual et al. 2016).

The struggles over worldviews and associated values and the resulting conflicts are a direct manifestation of environmental injustices perceived and felt by disempowered actors in society, such as those whose worldviews and values are dominated. Likewise, harmonization of social, environmental, and economic goals inherent in sustainability goals are hard to achieve if proper institutions are not designed and put in place to help resolve environmental conflicts and injustices over time. It is thus impossible to detach the issue of worldviews and associated values from institutions, understood as norms, rules, and strategies which determine the normative views on the appropriateness of policy interventions (Ostrom 1991).

Over the years, frameworks for biodiversity conservation have expanded substantially, shifting from diversity-based frameworks to integrating sustainability, including the human dimensions of well-being and livelihoods. This is in line with the realization—by way of contestation—that conservation can be undermined if local livelihoods, rights, and needs are not integrated into conservation and sustainable management programs. However, strategic issues such as participation, representation, and the distribution of benefits and burdens of conservation and sustainability are issues that deserve more discussion and recognition in policy making. If forests and other high-diversity ecosystems are to have a future, justice needs to be incorporated into biodiversity frameworks for a more holistic understanding of the social outcomes of conservation and, importantly, a more realistic design of actions for conserving biodiversity.

References

Agrawal, A. 2006. Environmentality: Technologies of Government and the Making of Subjects. Durham: Duke Univ. Press.

Aligicia, P., and V. Tarko. 2012. Polycentricity: From Polanyi to Ostrom and Beyond. *Governance* 25:237–262.

Alix-Garcia, J., A. de Janvry, E. Sadoulet, and J. M. Torres. 2005. An Assessment of Mexico's Payment for Environmental Services Program. Agricultural and Development Economics Division (ESA), Food and Agriculture Organization of the United Nations. ftp://ftp.fao.org/es/ESA/Roa/pdf/aug05-env_mexico.pdf. (accessed Dec. 15, 2016).

Angelsen, A., and D. Kaimowitz. 1999. Rethinking the Causes of Deforestation: Lessons from Economic Models. *World Bank Res. Obs.* 14:73–98.

Bengt, G. 2015. Political Ecology: Anthropological Perspectives. In: Karlsson International Encyclopedia of the Social and Behavioral Sciences, 2nd edition, ed. J. D. Wright, pp. 350–355. Oxford: Elsevier.

Berkes, F. 1999. Sacred Ecology: Traditional Knowledge and Resource Management. Philadelphia: Taylor and Francis.

Blaikie, P., and H. Brookfield. 1987. Land Degradation and Society. London: Methuen.

Blanket Bog Restoration. 2015. A Strategy for the Restoration of Blanket Bog in England: An Outcomes Approach. http://www.moorsforthefuture.org.uk/sites/default/files/UMG%20Restoration%20Strategy%20for%20England.pdf. (accessed Nov. 29, 2017).

Bollier, D., and S. Helfrich, eds. 2012. The Wealth of the Commons: A World Beyond Markets and State. Amherst: Levellers Press.

Boyer, C. 2015. Political Landscapes: Forest Conservation and Communities in Mexico. Durham: Duke Univ. Press.

Bray, D. B., and L. Merino. 2004. La Experiencia de Las Comunidades Forestales de México. Semarnat. Mexico: Instituto Nacional de Ecología.

Bryant, R. L., and S. Bailey. 1997. Third World Political Ecology. London: Routledge.

Buchanan, J. M. 2008. Opportunity Cost. In: The New Palgrave Dictionary of Economics Online, second edition, ed. S. N. Durlauf and L. E. Blume. Palgrave Macmillan. http://www.dictionaryofeconomics.com/article?id=pde2008_O000029. (accessed Dec. 15, 2016).

Calvet-Mir, L., E. Corbera, M. A., and J. Fisher. 2015. Payments for Ecosystem Services in the Tropics: A Closer Look at Effectiveness and Equity. *Curr. Opin. Environ. Sustain.* **14**:150–162.

Campodónico, H. 2007. Economía Peruana: la Dependencia en Los Precios de Las Materias Primas y la Fallida Reforma Tributaria. http://www.cristaldemira.com/descargas/HCS-Peru_Hoy_2007A.pdf. (accessed Dec. 15, 2016).

Capra, F., and U. Mattei. 2015. The Ecology of Law: Toward a Legal System in Tune with Nature and Community. Oakland: Berrett-Koheler.

CAPRi. 2010. Resources, Rights, and Cooperation: A Sourcebook on Property Rights and Collective Action for Sustainable Development. CGIAR Program on Collective Action and Property Rights (CAPRi). Washington, D.C.: Intl. Food Policy Research Institute.

Cardenas, J. C. 2009. Dilemas de Los Colectivo: Instituciones, Pobreza y Cooperación en el Manejo Local de Los Recursos de Uso Común. Bogota: Universidad de los Andes.

Chapela, F. J. 2005. Indigenous Community Forest Management in Sierra de Juarez, Oaxaca. In: The Community Forests of Mexico. Managing for Sustainable Landscapes, ed. D. B. Bray et al. Austin: Texas Univ. Press.

Chomba, S., J. Kariuki, J. F. Lund, and F. Sinclair. 2016. Roots of Inequity: How the Implementation of REDD+ Reinforces Past Injustices. *Land Use Policy* **50**:202–2013.

Convention on Biological Diversity. 2014. Definitions: Indicative Definitions Taken from the Report of the Ad Hoc Technical Expert Group on Forest "Biological Diversity." https://www.cbd.int/forest/definitions.shtml. (accessed July 31, 2017).

Dauvergne, P. 2010. The Shadows of Consumption: Consequences for the Global Environment. Cambridge, MA: MIT Press.

———. 2016. Environmentalism of the Rich. Cambridge, MA: MIT Press.

Dellas, E., and P. Pattberg. 2009. Assessing the Political Feasibility of Global Options to Reduce Biodiversity Loss. *Int. J. Biodiv. Sci. Ecosys. Serv. Manag.* **9**:347–363.

Diaz, S., S. Demissew, J. Carabias, et al. 2015. The IPBES Conceptual Framework: Connecting Nature and People. *Curr. Opin. Environ. Sustain.* **14**:1–16.

Dietz, T., N. Dolsak, E. Ostrom, and P. Stern. 2002. The Drama of the Commons. In: The Drama of the Commons, pp. 3–36. Washington, D.C.: National Academy Press.

Doupe, P. 2015. The Costs of Error in Setting Reference Rates for Reduced Deforestation. *Land Econ.* **91**:723–738.

Entman, R. M. 1993. Framing: Toward Clarification of a Fractured Paradigm. *J. Comm.* **43**:51.

Escobar, A. 2008. Territories of Difference, Place, Movements, Life, *Redes.* Durham: Duke Univ. Press.

Fairhead, J., and M. Leach. 1996. Misreading the African Landscape: Society and Ecology in a Forest-Savanna Mosaic. Cambridge: Cambridge Univ. Press.

———. 2006. Science, Society and Power. Cambridge: Cambridge Univ. Press.

Fairhead, J., M. Leach, and I. Scoones. 2012. Green Grabbing: A New Appropriation of Nature? *J. Peasant Stud.* **39**:237–261.

FAO. 2015. Incentives for Ecosystem Services in Agriculture. Food and Agriculture Organization of the United Nations. http://www.fao.org/in-action/incentives-for-ecosystem-services/en/. (accessed July 27, 2017).

———. 2016. State of the World's Forests 2016. Forests and Agriculture: Land-Use Challenges and Opportunities. Rome: Food and Agriculture Organization of the United Nations. http://www.fao.org/publications/sofo/2016/en/. (accessed Oct. 6, 2017)

Foucault, M. 1969/2008. L'archéologie Du Savoir. Paris: Gallimard.

————. 1971. L'ordre Du Discours. Leçon Inaugurale Au Collège de France Prononcée Le 2 Décembre 1970. Paris: Gallimard.

Geist, H., and E. Lambin. 2002. Proximate Causes and Underlying Driving Forces of Tropical Deforestation. *BioScience* **52**:143–150.

Goffman, E. 1974. Frame Analysis: An Essay on the Organization of Experience. Cambridge, MA: Harvard Univ. Press.

Hardin, G. 1968. The Tragedy of the Commons. *Science* **162**:1243–1248.

Hecht, S. 2014. Rethinking Social Lives and Forests Transitions: History, Ideologies, Institutions and the Matrix. In: The Social Lives of Forests. Past, Present and Future Woodland Resurgence, ed. S. Hecht et al. Chicago: Univ. of Chicago.

Hecht, S., K. Morrison, and C. Padoch, eds. 2013. The Social Lives of Forests. Past, Present and Future Woodland Resurgence. Chicago: Univ. of Chicago.

Hess, C. 2008. Mapping the New Commons. In: 12th Biennial Conf. of the International Association for the Study of the Commons, Cheltenham, July 14–18, 2008. http://surface.syr.edu/cgi/viewcontent.cgi?article=1023&context=sul. (accessed Dec. 15, 2016).

Iaione, C. 2013. La Città Come Bene Comune. *Aedon Revista di Arte e Diritto* **1**:31–40.

Kasim, M., and P. Edwards. 2013. Conservation and Sustainable Use of Gulf of Mannar's Biosphere Reserve's Coastal Biodiversity. Draft Report of the Terminal Evaluation Mission. United Nations Development Programme. https://www.the-gef.org/project/conservation-and-sustainable-use-gulf-mannar-biosphere-reserves-coastal-biodiversity. (accessed Dec. 15, 2016).

Lele, S., and A. Kurien. 2011. Interdisciplinary Analysis of the Environment: Insights from Tropical Forest Research. *Environ. Conserv.* **38**:211–233.

Lele, S., and R. B. Norgaard. 2005. Practicing Interdisciplinarity. *BioScience* **55**:967–975.

Martinez-Alier, J. 2011. El Ecologismo de Los Pobres: Conflictos Ambientales y Lenguajes de Valoración. Barcelona: Editorial Icaria.

Mascia, M. B., S. Pailler, R. Krithivasan, et al. 2014. Protected Area Downgrading, Downsizing, and Degazettement (PADDD) in Africa, Asia, and Latin America and the Caribbean, 1900–2010. *Biol. Conserv.* **169**:355–361.

McGinnis, M. D., and E. Ostrom. 2014. Social-Ecological System Framework: Initial Changes and Continuing Challenges. *Ecol. Soc.* **19**:30.

Merino, L. 2004. Conservación O Deterioro: Los Impactos de Las Políticas Públicas en Las Comunidades y en Los Usos de Los Bosques en México. Mexico City: Instituto Nacional de Ecología.

————. 2016. Rights, Pressures and Conservation in Forest Regions of Mexico. In: Environmental Governance in Latin America, ed. F. de Castro et al., pp. 234–256. Basingstoke: Palgrave McMillan.

Merino, L., and M. Hernández. 2003. Destrucción de Instituciones Comunitarias y Deterioro de Los Bosques en la Reserva de la Biosfera Mariposa Monarca, Michoacán, México. *Rev. Mex. Sociol.* **66**:261–309.

Millennium Ecosystem Assessment. 2005. Ecosystems and Human Well-Being: Synthesis. Washington, D.C.: Island Press.

Molnar, A., R. Butterfield, F. Chapela, et al. 2003. Forest Certification and Communities: Looking Forward to the Next Decade. Washington, D.C.: Forest Trends. http://www.cifor.org/publications/pdf_files/reports/forest_communities.pdf. (accessed Dec. 16, 2015).

Muñoz-Piña, C., A. Guevara, J. M. Torres, and J. Braña. 2008. Paying for the Hydrological Services of Mexico's Forests: Analysis, Negotiations and Results. *Ecol. Econ.* **65**:725–736.

L. Merino-Pérez et al.

O'Neill, J., A. Holland, and A. Light. 2008. Environmental Values. Abingdon: Routledge.

Ostrom, E. 1991. Governing the Commons: The Evolution of the Institutions for Collective Action. Cambridge: Cambridge Univ. Press.

———. 2005. Understanding Institutional Diversity. Princeton: Princeton Univ. Press.

Pagiola, S. 2008. Designing Payments for Environmental Services in Theory and Practice: An Overview of the Issues. *Ecol. Econ.* **65**:663–674.

Pascual, U. 2010. Exploring the Links between Equity and Efficiency in Payments for Environmental Services: A Conceptual Approach. *Ecol. Econ.* **69**:1237–1124.

Pascual, U., P. Balvanera, S. Diaz, et al. 2016. Valuing Nature's Contributions to People: The IPBES Approach. *Curr. Opin. Environ. Sustain.* **26/27**:7–16.

Pascual, U., J. Phelps, E. Garmendia, et al. 2014. Social Equity Matters in Payments for Ecosystem Services. *BioScience* **64**:1027–1036.

Peet, R., and M. Watts, eds. 1996. Liberation Ecologies: Environment, Development, Social Movements. London: Routledge.

Piaget, J., and R. Garcia. 1982. Psicogénesis e Historia de la Ciencia. Mexico: Siglo XXI.

Poteete, A. R., M. A. Janssen, and E. Ostrom. 2009. Working Together: Collective Action, the Commons, and Multiple Methods in Practice. Princeton: Princeton Univ. Press.

Putz, F. E., and K. H. Redford. 2010. The Importance of Defining Forest: Tropical Forest Degradation, Deforestation, Long-Term Phase Shifts, and Further Transitions. *Biotropica* **42**:10–20.

Rajagopalan, R. 2008. Marine Protected Areas in India. Chennai, India: International Collective in Support of Fishworkers. https://core.ac.uk/download/pdf/11017358.pdf. (accessed Dec. 16, 2016).

Redford, K. H., P. Coppolillo, E. W. Sanderson, et al. 2003. Mapping the Conservation Landscape. *Conserv. Biol.* **17**:116–131.

Redford, K. H., and B. Richter. 1999. Conservation of Biodiversity in a World of Use. *Conserv. Biol.* **13**:1246–1256.

Reyes-García, V., J. Paneque-Gálvez, P. Bottazzi, et al. 2014. Indigenous Land Reconfiguration and Fragmented Institutions: A Historical Political Ecology of Tsimané Land (Bolivian Amazon). *J. Rural Stud.* **34**:282–291.

Ribot, J. 2009. A Theory of Access. *Rural Sociol.* **68**:153–181.

Ribot, J., A. Agrawal, and A. M. Larson. 2006. Recentralizing While Decentralizing: How National Governments Reappropriate Forest Resources. *World Dev.* **34**:1864–1886.

Rodríguez, K., and L. Merino. 2017. Contextualizing Context in the Analysis of Payment for Ecosystem Services. *Ecosyst. Serv.* **23**:259–267.

Salfsky, N., D. Salzer, A. J. Stattersfield, et al. 2008. A Standard Lexicon for Biodiversity Conservation: Unified Classifications of Threats and Actions. *Conserv. Biol.* **22**:897–911.

Schlager, E., and E. Ostrom. 1992. Property-Rights Regimes and Natural Resources: A Conceptual Analysis. *Land Econ.* **68**:249–262.

Schullery, P. 2004. Searching for Yellowstone: Ecology and Wonder in the Last Wilderness. Helena: Montana Historical Society Press.

Scott, J. C. 1998. Seeing Like a State: How Certain Schemes to Improve the Human Condition Have Failed. New Haven, CT: Yale Univ. Press.

Sikor, T., A. Martin, J. Fisher, and J. He. 2014. Toward an Empirical Analysis of Justice in Ecosystem Governance. *Conserv. Lett.* **7**:524–532.

Tucker, C. M. 2004. Community Institutions and Forest Management in Mexico's Monarch Butterfly Reserve: Society and Natural Resources. *Soc. Nat. Resour.* **17**:569–587.

UN. 2011. Forests for People: A Fact Sheet. http://www.un.org/esa/forests/wp-content/uploads/bsk-pdf-manager/83_FACT_SHEET_FORESTSANDPEOPLE.PDF. (accessed Nov. 29, 2017).

UNEP. 2010. Annual Report. New York: United Nations Environment Programme. http://staging.unep.org/annualreport/2010/. (accessed Oct. 6, 2017).

UNESCO. 2008. Monarch Butterfly Biosphere Reserve: World Heritage List. http://whc.unesco.org/en/list/1290. (accessed Dec. 16, 2016).

———. 2012. World Heritage List: Yellowstone National Park. http://whc.unesco.org/en/list/28. (accessed Dec. 16, 2016).

Vandermeer, J., and I. Perfecto. 2013. Paradigms Lost: Tropical Conservation under Late Capitalism. In: The Social Lives of Forests. Past, Present and Future Woodland Resurgence, ed. S. Hecht et al., pp. 114–129. Chicago: Univ. of Chicago.

Vira, B., C. Wildburger, and S. Mansourian. 2016. Forests and Food: Addressing Hunger and Nutrition across Sustainable Landscapes. Cambridge: Open Books Publisher.

White, A., and A. Martin. 2002. Who Owns the World's Forests: Forest Tenure and Public Forests in Transition. Washington, D.C.: Forest Trends. http://www.cifor.org/publications/pdf_files/reports/tenurereport_whoowns.pdf. (accessed Dec. 16. 2016).

Whitehead, A. N. 1920/2007. The Concept of Nature. The Tarner Lectures Delivered in the Trinity College, 11/1919. Sophia Project, Philosophical Archives. Whitefish: Kessinger Publ.

Wunder, S. 2005. Payment for Environmental Services: Some Nuts and Bolts. Jakarta, Indonesia: Center for International Forestry Research. http://www.cifor.org/publications/pdf_files/OccPapers/OP-42.pdf. (accessed Dec. 16, 2016).

———. 2015. Revisiting the Concept of Payments for Environmental Services. *Ecol. Econ.* **117**:234–243.

Urban Environments

5

City Limits

Looking for Environment and Justice in the Urban Context

Amita Baviskar

Abstract

As cities have overtaken the countryside as habitat for most of humanity, their environmental politics have become all the more critical. However, the contours of urban environmental politics—their discursive frame and ultimate aims, their authorized cast of actors, and modes of action—often have little to do with ecology or justice. Why is this so? This chapter argues that the power to define and address an issue as an "environmental problem" is unequally distributed. Social location and cultural capital shape interpretive frameworks and capacities to act. Selective and superficial framings of environmental issues derive from urban inequality. Indeed, the urban environment poses a peculiar perceptual problem because it does not seem to be composed of commonly understood features of "nature." That is, the predominantly artifactual aspect of the urban environment complicates understandings of ecological issues based on the template of rural environments. Historically, urban environments have been managed in terms of securing spatial and social order. This logic continues to dictate environmental politics in the city, to the detriment of ecology and justice.

The City: Two Views

Anuj Gupta, 45, senior manager with a corporate firm in Gurgaon, a suburb of Delhi, stood at the window of his luxury apartment in Malibu Towne, looking down at the scene below, he recounted:

> This used to be all green. Trees and fields of wheat and yellow mustard in winter. That's why we moved here. Because of the fresh air and peace. But now it's a mess.

Pointing to the glass and steel skyscrapers that punctuate the distance, he continued:

It's all built up. I drive out of the gate of the colony and I'm stuck in traffic for hours. There's no order, no discipline. The air is so bad that my son has severe asthma. We have air purifiers installed at home and his school bus is air-conditioned but he has to carry his inhaler with him all the time. Half the kids in his class do the same.

Turning back, as a maid served us glasses of lemonade, he gestured at the cool, marble-floored room:

This apartment, too. You'd think this was worth it but we have no water. We pay a fortune for private tankers to fill our reservoir and we buy filtered water for drinking and cooking from another supplier. We're supposed to have 24-hour electricity but the power supply is so bad that our generators work overtime. The maintenance charges are through the roof. The only thing I can say is that at least it's safe. Out of the colony gates, it's another story, but inside we're all right. The security guards have strict instructions to check everyone who comes in; we have CCTVs (closed-circuit television) and intercom. All the maids and drivers have ID cards issued by our RWA (Residents' Welfare Association) and verified by the police. It's OK here, safetywise. We don't go out much, only to work, school or the shopping malls. The malls here are good, at least there's that: nice stores, lots of places to eat. But the rest of it is rubbish. All these people. So much congestion. The government does nothing at all. In fact, it only encourages them. We pay taxes but who listens to us? Our RWA has had to file a court case to get that slum next door removed. It's filthy; their children shit out on the street; you can't even walk in the colony park because of the stink from across the wall. Who knows what disease we might get? Last year, we paid extra to raise the boundary wall; you hear of theft and murder in the news all the time. Gurgaon is a mess, I tell you.

Across the wall from Anuj Gupta's apartment, Sarita Devi, 32, sits outside her *jhuggi* (shack) chopping onions and potatoes, every now and then swatting at the mosquitoes that swarm up from the open drain that runs alongside. Even at dusk, the tin-walled shed in which she lives with her husband and three children is stifling; they have a watercooler but when the power cuts out—which it does frequently—there's no respite from the heat. The cooler and a television set are Sarita's prized possessions, purchased from her earnings as a domestic worker cleaning homes and washing dishes in the Malibu Towne apartments across the razor wire-topped wall. She earns a steady wage, all the more essential because her husband Manoj does not have a steady job, as she describes:

He used to work in a factory but they fired him when he missed a few days because he had to go back to the village to help his brother. Then he thought he'd start his own business and sell vegetables, but that didn't work out either. He then became a helper to a mason on a construction site and got good money but, for the last two years, there's been no demand. Now he drives a cycle rickshaw but that doesn't bring in much. If I didn't work, how would we feed ourselves? Everything keeps getting more and more expensive.

Tilting her head toward the tiny room behind her, a tidy space in which pots and pans gleam on a shelf beside wall calendars showing Hindu deities and a bed—the sole item of furniture—is pushed up against the wall, Sarita continues:

> And this *jhuggi*: we bought it and built it up with our own hands but they keep saying that our *basti* (settlement) is illegal and we'll be evicted. We've been here for twenty years, how can they remove us? The Councilor says he'll look after us, but you can't trust anyone these days; they're only looking after their own interest. If we had to leave this place, where would we go? Where would I work? As it is, life is hard. I queue up at 5:30 every morning to fill up pots of water; every day there's a fight at the tap. The toilets are so filthy it turns my stomach. In the monsoon last year, the drains overflowed and sewage entered my *jhuggi*. My youngest daughter had diarrhea for two months. We spent three thousand rupees on getting her treated. But I say, all right, at least we're not starving. At least we have a roof over our heads. This is the fate of poor people. What can we do? But if they take away even this, what's left for us? I can't sleep at night I'm so worried.

Though only a wall separates them, Anuj Gupta and Sarita Devi seem to inhabit different worlds. The contrast is most vivid in the physical spaces in which they live. Gupta's home is at least twenty times larger than Sarita's shack. It has running hot and cold water, three bathrooms, and six air conditioners. The apartment block sits amidst lush lawns and frangipani trees, a swimming pool, and children's play area. Sarita's home is squashed between other shacks, each a tangle of tin sheets and rough masonry, along a potholed lane bisected by a drain where young children squat to relieve themselves and pigs snuffle around in the wet muck. Dogs root through heaps of waste, dirty plastic bags, and decaying organic matter. There's nowhere to play so kids crowd the street, dodging between passing rickshaws and motorbikes. At each end of the lane there is a public tap that supplies water for two hours in the morning and evening; it's usually at low pressure and the number of waiting people high. There is one mobile toilet with ten cubicles for the entire lane of more than two hundred households: by common consent only women use it; men defecate in a scrubby wasteland nearby.

Despite these obvious differences, Gupta and Sarita's observations about the places in which they live also contain some telling parallels. Take water and electricity, for instance. Both complain about shortages and unreliability. However, Gupta is able to buy his way out of the problem, whereas Sarita must make do with the little that comes her way. Both are concerned about their children's well-being and the burden of disease to which they are exposed. However, while Gupta junior has access to the best health care, Sarita's child almost died because of a preventable gastrointestinal infection caused by drinking contaminated water. Both worry about safety. Gupta frets about burglaries and violent crime, anarchy on the streets, and ensures that his college-going daughter is chauffeured everywhere. Sarita spends the second half of her working day wondering if her eight-year-old daughter came home from school all right and whether her neighbor is keeping an eye on her as she promised. Along with this, however, is an ever-looming anxiety that Gupta will never have to face: Sarita's fear about losing her home. More than the precariousness of her husband's earnings, it is the threat of eviction that constitutes the core of Sarita's worries: that this modest yet precious home will be razed by bulldozers, its contents scattered, her family's life shattered.

Gupta's and Sarita's lives crisscross in other ways. For one, Sarita cleans his apartment, wiping down its marble floors with rose-scented detergent, dusting the knickknacks on its shelves, washing dishes. For another, when Gupta's chauffeur impatiently honks at a rickshaw to move out of his way on the street outside, he could be honking at Sarita's husband, Manoj. And of course, the *jhuggi basti* that so disgusts Gupta, and that his RWA has mobilized to evict, happens to be Sarita's home.

What Is the Environment?

How do we interpret these sometimes conflicting, sometimes converging narratives in terms of environmentalism? For most people in cities in the Global South, the popular understanding of "environment"—one that cuts across social classes—centers on its meaning as *habitat*. That is, the surrounding landscape within which one lives—its physical characteristics, social relations, as well as the ideas and sentiments associated with it—represents the sum total of one's environment. The concerns that emerge from this environment may vary: for Gupta, spatial order and physical safety matter most, whereas Sarita's priorities are security of shelter, job opportunities for her husband, and her children's health. The "structure of feelings" that the environment evokes for them may be different: pride, disgust, fear, comfort, hope. Yet both share the notion of "environment" as related to amenities and infrastructure—water, electricity, housing, sanitation, roads—that constitute the essentials of a decent life in the city. Several of these are understood to be public goods and a shortfall in their provisioning is felt keenly as a breach of the contract between state and citizens. The language of claiming these "environmental" amenities therefore uses the vocabulary of civic rights.

This all-encompassing notion of "environment" as habitat—a tangible place imbued with intangible yet powerful relationships governed by the state—is narrowed down within the field of *urban studies*. Here, the "environment" is defined in terms of the characteristics of physical space, especially land use—density of built-up areas, quality of housing, extent of green cover, but also infrastructure in the form of transport and sanitation. Within this literature, then, one comes across an explicit discussion of "environmental problems": air pollution from factories and motor vehicles, water pollution from untreated sewage and industrial effluents, the shrinking of green areas and the congestion of the built environment. Historically, urban studies took a wide-angle view aimed at understanding (and engineering) an ideal relationship between environment and society, including within its sweep the moral as well as material well-being of city dwellers as shaped by their physical setting. The work of scholars and planners such as Patrick Geddes, Lewis Mumford, and Jane Jacobs exemplifies this perspective (Tyrwhitt 1947; Jacobs 1961; Mumford 1961). However, "environmental issues" in contemporary urban studies are usually studied in isolation and the socioeconomic processes in which they are embedded are treated as given. And when changes in urban land use and political economy are the focus of analysis, ecological aspects are mentioned only in passing, if at all (e.g., Desai 2012).

Public perception largely regards polluted air and water as externalities, unfortunate by-products of the wealth-generating processes that make cities engines of economic growth to which municipal authorities turn a blind eye because of the powerful players involved or because they are pressed with providing more "basic needs" such as water, roads, and waste disposal. This is the case with almost all towns in India where industrial manufacturing is a major part of the economy (Varghese et al. 1998). Only when urban elites distance themselves from dirty production processes—for instance, when the economic base of a city shifts from industrial manufacturing to services—do they mobilize against particular environmental ills, especially those that affect their "quality of life."[1] Besides air and water pollution, green areas and, increasingly, wetlands are also the focus of analysis and action but, notably, ecological arguments are mobilized in a manner that brackets them off from a wider consideration of environmental flows (Baviskar 2017). Gupta, for example, is active in a campaign to protect a patch of wilderness on the edge of Gurgaon that is threatened by real estate developers. The campaign highlights the ecological importance of the area as a refuge for biodiversity and as catchment for groundwater. However, the campaigners do not recognize or address the fact that they themselves are largely responsible for what they define as Gurgaon's environmental problems: they were the first to buy and occupy the luxury apartments that were built on farmers' fields and village commons; their water-intensive lifestyle plunged aquifers into the dark zone; buildings and roads that cater to them created heat islands that cry out for relief. Protecting the last remnant of Gurgaon's greenery then is a token, a talisman to denote the desire for a landscape that includes urban forests as well as shopping malls, never mind the contradictions.

Thus, it is a select social group that explicitly invokes ecological arguments and mobilizes them to pursue interests that they deem to be "environmental." In practice, environmentalism as an ideology centered on ecological protection

---

[1] In the late 1990s and early 2000s, public interest litigation against pollution led the Delhi High Court to order the closure of older manufacturing industries in the metropolis. This was not, however, an instance of environmental concerns overriding economic ones. The extensive lands vacated by mills were profitably repurposed as higher-value real estate for building offices, shopping and entertainment malls, and luxury residences (Baviskar et al. 2006).

for the benefit of present and future human generations, nonhuman species, and planetary biophysical processes is a resource available primarily to those who have the cultural capital to leverage it. This cultural capital is not immobile or impregnable: in some cases, it has been eroded by sections of the rural underclass through their struggles to secure rights to land, forests, water, fisheries, and other resources. Their claims have succeeded when they have highlighted how ecologically sustainable their practices are and how their cause aligns with social justice. Such social movements, classified in the literature as the "environmentalism of the poor," have curbed the power of dominant institutions that frame and prosecute environmental agendas by deploying the language of scientific rationality and economic efficiency to dispossess vulnerable populations (Guha and Martinez-Alier 1997). Yet, in the urban context, it has been far harder to challenge the cultural capital of these institutions and the elite social groups associated with them. While the dominant framing of urban "environmental issues" pertaining to noxious externalities has sometimes been successfully used by poor people to oppose specific projects—the location of a solid waste incinerator near African American and Hispanic neighborhoods in South-Central Los Angeles, for example—it has rarely been questioned or replaced by another that represents poor people's priorities (Bullard 1990; Di Chiro 1995). Those who share Sarita's social location are scarcely able to articulate their rights to water and sanitation—or, for that matter, secure shelter and jobs—in the vocabulary of environmentalism.[2] Furthermore, they are rarely able to use an ecological frame to mount a critique of the resource-intensive lifestyles of the residents of Malibu Towne.

Why is this? There are two aspects to this puzzle. First, ecology is often hard to see in the city. The concentration of concrete and tarmac, brick and glass, seems to squeeze it out of existence. The built environment overwhelmingly appears to be an artifact of human manufacture, of materials transformed by technology. For minds socialized to separate "nature" from "culture"—a long-standing intellectual distinction made across the world—it seems evident that there is not a lot of ecology in the city, except in the attenuated form of gardens, birds, and insects (Williams 1980).[3] Unlike rural landscapes where nature is palpably visible in the form of soil and water, plants and animals, and is valued as a productive resource, the urban environment fails to yield such recognizable indices to which productive value may be ascribed. In rural areas, social movements have been able to fuse the vocabulary of citizenship and environmentalism to fight for productive resources

---

[2] The only exception to this rule is the recent attempt by waste-pickers who collect and recycle paper, plastic, and other scrap, to demand a place in the city because of the environmental services they provide. I shall discuss their case later in this essay.

[3] Thus, a recent book on Bengaluru, *Nature in the City* (Nagendra 2016) focuses entirely on trees, public parks, and private gardens without considering the larger set of biophysical processes and material transformations that urban "nature" encompasses. In this book, as in commonsense understandings, "nature" is uncritically equated with green spaces.

and livelihoods as embodied in and deriving from nature. In urban areas, productive resources—factories, firms, and financial capital—seem unrelated to nature. Visible only in isolated units (and not as the underlying foundation of economic and social well-being) and valued primarily in terms of consumption, urban ecology becomes a concern mainly for elites who can afford to pursue such "minor," "nonessential" causes.

The second reason why the urban poor have not been able to wield environmentalism as a discursive resource to secure their interests has to do with legitimacy. Environmental debates are almost always framed in terms of *public interest*. That is, an environmental good is held to be universally beneficial, transcending the interests of particular sections of society. Poor people's quest to secure the environmental resources that matter most to them—shelter and sanitation, for instance—are viewed by the state and its reference publics as particularistic interests, of concern primarily to the affected group. This is especially so when these interests come up against those of more powerful groups who claim to represent the wider social good. In the last fifteen years in Delhi, public interest litigation by environmentalists and RWAs like that of Gupta's colony has brought about the eviction of settlements, such as the one that Sarita inhabits, on the grounds that they were an "environmental nuisance" or were polluting the river Yamuna. The courts came down on the side of "clean and green Delhi," dismissing pleas that their orders would deprive vulnerable groups of basic shelter and subsistence. Instead, they castigated slum dwellers for occupying land illegally, a crime born out of compulsion in a city that provides little affordable, legal housing for its underclass near places of work. At the same time, portraying the poor who lived along the riverbanks as environmental villains in the Yamuna case was a spectacular miscarriage of justice since the untreated waste released into the river came from better-off neighborhoods that were connected to the sewage system. Evicting squatter settlements along the embankments didn't solve the pollution problem; it only allowed land in Central Delhi to be made available for redevelopment (Baviskar 2011b). A similar pattern of public interest environmental activism adversely and unfairly affecting the poor prevailed in the case of Delhi's drive to deal with air pollution caused by vehicular traffic (Véron 2006). On the other hand, the same courts have condoned environmental violations by powerful corporate organizations and by the government, on the grounds that these projects involved a lot of money or were "prestigious." Thus, a clutch of shopping malls and luxury hotels on a tract of forest in South Delhi, and a grand temple complex and luxury apartments on the Yamuna floodplain in East Delhi, were retrospectively legalized. Ghertner (2015) argues that, along with the financial capital at stake, what moved the judges was the notion of environment-as-aesthetics: glittering shopping malls and opulent luxury hotels *look* good, they enhance the appearance of a "world-class city," so they are to be preferred over forests and floodplains. Court decisions reflect the hierarchy in how public/private interests around the environment come to be organized in the public mind: first come "economic growth" and "national prestige" as represented by corporate capital (state and private), while clean air and water or green areas languish far below. And at the bottom, stigmatized by their lack of cultural capital, flounder the urban poor.

The notion that "what is good for General Motors is good for America" continues to dominate perceptions of public interest across the world. Economic growth is the hegemonic ideology of national development; ecological issues are relegated to second-order concerns. The state promotes capital-intensive projects that generate short-term revenues, licit and illicit, for state actors as well as "infrastructure" for further economic growth: this is believed to be synonymous with the "public good." When the state does take note of environmental damage by powerful actors—as was the case in March 2016 when the politically well-connected Art of Living Foundation organized a giant event on the floodplain of the river Yamuna in Delhi—it first makes minatory noises and then soft pedals on punitive action.[4] These modes of thinking also regard the urban poor as illegitimate or, at best, irrelevant actors in environmental matters. By the rules of this game, capitalism usually trumps ecology, and equity is a particularly low-value card. So, in an unequal city, elite notions of environmental good prevail as the public interest, and elite projects override ecological concerns. The pursuit of apparently universally desired goals by the state and public-minded citizens grievously hurts the most vulnerable residents. How could the rural poor occasionally overcome a similar hegemonic stranglehold? As noted above, it requires considerable cultural capital to be able to insert oneself into an elite conversation bristling with class bias. Even when activist organizations working with the poor muster legal arguments and the facts to support them, they find themselves handicapped. However, in all cases where the rural poor have been successful in asserting their rights to resources, they have done so by mobilizing a counter-narrative about the superiority of their conservationist ethics and practices, often performing the role of the "virtuous peasant" or "ecologically noble savage" (Baviskar 1995).[5] Organizations of the urban poor find it very difficult to marshal similar moral claims that marry ecology with justice. For there they are: slum dwellers, living in squalor and condemned for it. When their claims to their habitat are unsupported by law or long usage (compared to their rural counterparts) and they cannot demonstrate a conservationist ethic, how do they assert their ecological virtue? Among their numbers, only waste pickers and recyclers are now trying to repackage their

---

[4] Analyses of environmental administration and litigation show a clear and consistent record of corporate firms being let off the hook for environmental crimes. Compounding the state–industry nexus is the fact that those crimes could not have been committed without the complicity of the state. This willingness to turn a blind eye to corporate crimes and, when forced to take cognizance, to condone rather than condemn, pervades the political economy of environmental law (Narain et al. 2014; Sethi 2016).

[5] This is also the case with the 2006 Scheduled Tribes and Other Traditional Forest Dwellers (Recognition of Forest Rights) Act, achieved after sustained political mobilization.

public image as upholders of urban ecology and it remains to be seen whether they will be able to use environmentalism as a lever to prise open the door to the status and security so long denied to them (CERAG 2009). For the rest of the urban poor, the notion that the public interest might include their rights to space, shelter, jobs, and civic amenities remains not only out of reach but out of the realm of possibility.

Order in the City

The responsibility for creating and managing cities so that all citizens have access to a healthy environment rests mainly with governments. Political survival requires that the state secure legitimacy for its rule by supervising stable conditions for capital accumulation. Historically, the state's push to regulate urban environments has been prompted by the desire to shape model citizens who would be willing subjects in this project of rule (Joyce 2003). The threat of a restive urban underclass seizing power has been around for centuries, but was realized most dramatically during the French Revolution in 1789, leading to a drastic rethinking of how city spaces were used and organized. Baron Haussmann's demolition of dense neighborhoods and the insertion of wide boulevards throughout Paris in 1853–1869 was aimed at making the city "more governable, prosperous, healthy, and architecturally imposing" but, above all, "safe against popular insurrections" (Scott 1998:59, 61). The logic of urban planning was to simultaneously create spatial and social order. In colonial India, this imperative was at work in the redesign of Delhi (Gupta 1981) and Lucknow (Oldenburg 1984) after widespread revolt against British rule in 1857. When the British built their imperial capital of New Delhi, spatial segregation between white and native populations, as well as a strict code assigning housing on the basis of rank and status, was the norm (Legg 2007). The strategy of regulating physical spaces for social control, whether in the form of explicit policies of apartheid or the unspoken yet all-pervasive rules of class and ethnicity that govern gated communities and the like, continues to prevail into contemporary times (Caldeira 1992; Fischer et al. 2014).

Along with the imperative of social control, there were other, more liberal ideas of social welfare at work in the imposition of spatial order. The notion that urban spaces should be designed for the physical, social, and moral improvement of all citizens, especially the most deprived, lay behind the first experiments in town planning initiated by Victorian philanthropists in Britain (Macqueen 2011). Social engineering via spatial fixes was attempted on a larger scale when Ebenezer Howard's Garden City model was adopted in Britain and the United States, and was subsequently imported to the colonies, influencing the layout of Lahore, New Delhi, Quezon City, Canberra, and parts of Sao Paulo (Howard 1902/1946; Glover 2013). Concerns about public health were central to these plans: since the early nineteenth century, the threat of contagious disease epidemics had led to increasing attention to the quality of urban water, air, and waste disposal, as well as to monitoring hygiene in places of public dining and homes. Soliciting the cooperation of citizens was crucial for this enterprise. It was continuously reiterated that cities could be clean and healthy only if their citizens were; civic compliance was sought through laws penalizing the "nuisance" of littering, urinating, and spitting in public. Thus the two organizing principles of environmental regulation were hazard and nuisance.

From its inception, then, urban environmental management has been governed by anxieties around health, hazard, and social order (see Kaviraj 1997; Chakrabarty 2002). It has privileged an aesthetic that values capital-intensive buildings and manicured green areas (Baviskar 2018). These concerns and sensibilities have precedence over issues of life and livelihood that are central to "the environmentalism of the poor." Since angry and resentful poorer sections may constitute a threat to political order, an array of disciplining techniques in the work economy, social welfare system, and public spaces are deployed to keep them on the defensive and defuse collective mobilization (Chatterjee 2004). Thus, a 1988 cholera and gastroenteritis epidemic that killed more than 150 people in East Delhi slums led to improved municipal supply of water and, eventually, sewerage (Hazarika 1988), but larger questions about citywide distributive justice in access to water and sanitation went unaddressed, as did the issue of vulnerability to disease aggravated by the poverty of people who had been evicted from their homes in Central Delhi and forcibly settled on flood-prone land on the edge of the city. This continues to be the case even today; a brief phase of populist welfare policies in the early 1990s gave way to two decades of economic liberalization policies that have worsened inequalities in Indian cities and created a harsher, more intolerant climate for poor people's livelihood security.[6] For the government, resource politics in the city is about regulating spaces and managing social order such that economic growth and accumulation can continue without disruption from below. In this context, it is bourgeois environmentalists with their quality of life concerns that decide what constitutes an environmental issue (see also Mawdsley 2004).

Conclusion

As I have argued above, environmentalist action in Delhi—and in several other cities in the Global South—is guided neither by the principle of ecological sustainability nor social justice. The dominant form of action has been driven by the ideology of bourgeois environmentalism which has had perverse effects on

---

[6] Saajha Manch, "Manual for People's Planning: A View from Below of Problems of Urban Existence in Delhi" (unpublished manuscript, 2001). It remains to be seen whether the Aam Aadmi Party government in Delhi, elected in 2015 with overwhelming support from poorer sections of society, will succeed in improving living and working conditions for its supporters.

air and water quality, and has penalized the poor while ignoring the culpability of other classes and their "luxury emissions" and discharges (Baviskar 2011a, b). In some instances, ecological values *have* been pursued, as in campaigns to protect wetlands and areas of wilderness, but in a manner hostile to the poor communities who rely on these areas for shelter and subsistence. Only in exceptional cases do we find urban environmental action aimed at securing ecological sustainability as well as social justice. Remarkably, this is achieved by one of the most deprived and discriminated against social groups in urban India; namely, those who collect and sort solid waste for recycling (Gill 2010; Gidwani 2013).

Those who gather and process urban waste bring into the environmental frame a notion that has been missing so far: *urban metabolism* or ecology as the sum of stocks and flows of materials and energy, which includes those embodied in the built environment as well (Demaria and Schindler 2015). If urban environmentalism were to be based on such a metabolic matrix, the fact of the city as "nature's metropolis" would overcome assumptions about the nature/culture divide (Cronon 1991). By showing the presence of "natural resources" in productive practices in the city, and by revealing the glaring inequalities in the ownership and distribution of wealth derived from nature, such an analysis would allow ecology to be made accountable to equity. (I refer here to the notion of urban metabolism as a *metaphor* for understanding environmental politics, and not as an actual model for computing quantified/monetized flows.) If Sarita and her fellow *basti*-dwellers could show how light their ecological footprint is compared to Gupta's resource-guzzling lifestyle, and how the latter's privileges have been facilitated by preferential treatment from the state, they too could claim environmentalism as an ideological resource, just as it has been used by some of their rural counterparts to challenge forestry, mining, and dam projects. At the same time, the systematic misrecognition of public interest would stand revealed for what it is: the pursuit of interests and ideas that serve powerful private players.

To be truly disruptive, however, such an ecological framing would have to be supplemented by a wider definition of "environmental resources." For instance, *space* is an environmental good that may also generate livelihoods— the space of the street enables not only walking and sociality but also vending goods and selling services (de Certeau 1984; Baviskar 2011a). The social value of streets cannot be approximated by a computation of ground rent, but neither can it be subsumed within a mapping of material and energy flows. Such a calculus would have to be incorporated within a cultural matrix of what counts, where the concerns of the poorest citizens—those who lack access to private spaces and resources and rely all the more on urban commons—would have priority. This is the challenge posed by dominant notions of the city: what exactly qualifies a resource/problem to be classified as "environmental"? If it is its ecological component, where does ecology begin and end and what does it encompass? If it is about "natural resources" like land and water, how do we

discern them in the highly mediated forms in which they appear? Is a munici-
pality's crackdown on street vendors an environmental issue? Is a *basti's* bid to
get piped drinking water an environmental campaign? These questions about
definition are important because they allow and disallow not only *what* can be
talked about but *who* can do the talking.

References

Baviskar, A. 1995. In the Belly of the River: Tribal Conflicts over Development in the
Narmada Valley. Delhi: Oxford Univ. Press.
———. 2011a. Cows, Cars and Cycle-Rickshaws: Bourgeois Environmentalism and
the Battle for Delhi's Streets. In: Elite and Everyman: The Cultural Politics of
the Indian Middle Classes, ed. A. Baviskar and R. Ray, pp. 391–418. New Delhi:
Routledge.
———. 2011b. What the Eye Does Not See: River Yamuna in the Imagination of Delhi.
Econ. Polit. Wkly. **46**:45–53.
———. 2018. Shades of Green: Remaking Urban Nature and Its Publics in Delhi, India.
In: Grounding Urban Natures: Histories and Futures of Urban Political Ecologies,
ed. H. Ernstson and S. Sorlin. Cambridge, MA: MIT Press, in press.
Baviskar, A., S. Sinha, and K. Philip. 2006. Rethinking Indian Environmentalism:
Industrial Pollution in Delhi and Fisheries in Kerala. In: Forging Environmentalism:
Justice, Livelihood and Contested Environments, ed. J. Bauer, pp. 189–256. New
York: ME Sharpe.
Bullard, R. D. 1990. Dumping in Dixie: Race, Class, and Environmental Quality.
Boulder: Westview Press.
Caldeira, T. P. R. 1992. City of Walls: Crime, Segregation, and Citizenship in São
Paulo. Berkeley: Univ. of California Press.
CERAG. 2009. Cooling Agents: An Examination of the Role of the Informal Recycling
Sector in Mitigating Climate Change. New Delhi: Chintan Environmental Research
and Action Group, Safai Sena and The Advocacy Project.
Chakrabarty, D. 2002. Of Garbage, Modernity, and the Citizen's Gaze. In: Habitations
of Modernity: Essays in the Wake of Subaltern Studies, pp. 65–79. Chicago: Univ.
of Chicago Press.
Chatterjee, P. 2004. The Politics of the Governed: Reflections on Popular Politics in
Most of the World. New York: Columbia Univ. Press.
Cronon, W. 1991. Nature's Metropolis: Chicago and the Great West. New York: W.
W. Norton.
de Certeau, M. 1984. The Practice of Everyday Life. Berkeley: Univ. of California Press.
Demaria, F., and S. Schindler. 2015. Contesting Urban Metabolism: Struggles over
Waste-to-Energy in Delhi, India. *Antipode* **48**:293–313.
Desai, R. 2012. Governing the Urban Poor: Riverfront Development, Slum Resettlement
and the Politics of Inclusion in Ahmedabad. *Econ. Polit. Wkly.* **47**:49–56.
Di Chiro, G. 1995. Nature as Community: The Convergence of Environment and Social
Justice. In: Uncommon Ground: Rethinking the Human Place in Nature, ed. C. W.,
pp. 298–320. New York: W. W. Norton.
Fischer, B., B. McCann, and J. Auyero. 2014. Cities from Scratch: Poverty and
Informality in Urban Latin America. Durham, NC: Duke Univ. Press.
Ghertner, D. A. 2015. Rule by Aesthetics? World-Class City Making in Delhi. New
York: Oxford Univ. Press.

Gidwani, V. 2013. Value Struggles: Waste Work and Urban Ecology in Delhi. In: Ecologies of Urbanism in India: Metropolitan Civility and Sustainability, ed. A. Rademacher and K. Sivaramakrishnan. Hong Kong: Hong Kong Univ. Press.

Gill, K. 2010. Of Poverty and Plastic: Scavenging and Scrap Trading Entrepreneurs in India's Urban Informal Economy. New Delhi: Oxford Univ. Press.

Glover, W. 2013. The Troubled Passage from 'Village Communities' to Planned New Town Developments in Mid-Twentieth-Century South Asia. In: Ecologies of Urbanism in India: Metropolitan Civility and Sustainability, ed. A. Rademacher and K. Sivaramakrishnan, pp. 93–118. Hong Kong: Hong Kong Univ. Press.

Guha, R., and J. Martinez-Alier. 1997. Varieties of Environmentalism: Essays North and South. London: Earthscan.

Gupta, N. 1981. Delhi between Two Empires, 1803-1931. Delhi: Oxford Univ. Press.

Hazarika, S. 1988. 157 Dead in New Delhi Epidemic (from July 27). *New York Times.* http://www.nytimes.com/1988/07/28/world/157-dead-in-new-delhi-epidemic.html. (accessed November 21, 2016).

Howard, E. 1902/1946. Garden Cities of Tomorrow. London: Faber and Faber.

Jacobs, J. 1961. The Death and Life of Great American Cities. New York: Random House.

Joyce, P. 2003. The Rule of Freedom: Liberalism and the Modern City. London: Verso.

Kaviraj, S. 1997. Filth and the Public Sphere: Concepts and Practices About Space in Calcutta. *Public Culture* **10**:83–113.

Legg, S. 2007. Spaces of Colonialism: Delhi's Urban Governmentalities. Oxford: Wiley-Blackwell.

Macqueen, A. 2011. The King of Sunlight: How William Lever Cleaned up the World. London: Corgi.

Mawdsley, E. 2004. India's Middle Classes and the Environment. *Dev. Change* **35**:79–103.

Mumford, L. 1961. The City in History: Its Origins, Its Transformations, and Its Prospects. New York: Harcourt, Brace and World.

Nagendra, H. 2016. Nature in the City: Bengaluru in the Past, Present, and Future. New Delhi: Oxford Univ. Press.

Narain, S., C. Bhushan, and R. Mahapatra. 2014. Bhopal Gas Tragedy: After 30 Years. New Delhi: Centre for Science and Environment.

Oldenburg, V. T. 1984. The Making of Colonial Lucknow, 1856-1877. Princeton: Princeton Univ. Press.

Scott, J. C. 1998. Seeing Like a State: How Certain Schemes to Improve the Human Condition Have Failed. New Haven: Yale Univ. Press.

Sethi, N. 2016. Government Cancels Rs 200-Crore Green Fine on Adani. *Business Standard.* http://www.business-standard.com/article/current-affairs/govt-cancels-rs-200-crore-green-fine-on-adani-116070101477_1.htm. (accessed Sept. 12, 2017).

Tyrwhitt, J. 1947. Patrick Geddes in India. London: L. Humphries.

Varghese, G. K., P. R. J. Pradeep, and V. Katariya. 1998. Small Towns Big Mess. Down To Earth. http://www.downtoearth.org.in/coverage/small-towns-big-mess-22723#3. (accessed Aug. 31, 2017)

Véron, R. 2006. Remaking Urban Environments: The Political Ecology of Air Pollution in Delhi. *Environ. Plann. A* **38**:2093–2109.

Williams, R. 1980. Ideas of Nature. In: Problems in Materialism and Culture, pp. 67–85. London: Verso.

6

Nature of Cities and Nature in Cities

Prospects for Conservation and Design of Urban Nature in Human Habitat

Nancy B. Grimm and Seth Schindler

Abstract

Cities have scarcely been considered in the environmentalist agenda, except when they are invoked as examples of environmental degradation. This should change because human beings and their creations, of which cities might be the most prominent or obvious, are an inextricable part of the natural world. The way our species has distributed itself on the planet and how it will do so in the future has great implications for its impacts on the global environment, not all of which are gloom and doom. This chapter introduces a broader conception of nature in cities and nature of cities, inclusive of issues relevant to both the Global North and South. Cities certainly confront challenges owing to old and decrepit infrastructure, or needed infrastructure that remains unbuilt as cities expand with massive in-migration. Infrastructure of all types, including its social, ecological, and technological dimensions, provides services to urban residents but can also yield disservices, hazards, and risk. These features demand new thinking about services in cities, the role of nature, the meaning of conservation and possibility of restoration, as well as urban design that is appropriate for urban systems in the Global North and South. A broader concept of nature in cities, as well as a vision of the nature of cities, holds promise for justice, diversity, and sustainability of this future habitat for most of humanity.

Introduction: Cities and Their Changing Environments

The human imprint on the natural environment is massive, extending to nearly every piece of the Earth's land surface, oceans, and atmosphere (Vitousek et al. 1997; Kareiva et al. 2007). This impact is such that a new geologic epoch is under consideration—the Anthropocene—to reflect the pervasiveness of human influence on patterns and processes of the Earth system (Zalasiewicz et al. 2011; Ellis and Trachtenberg 2014). Humans have rapidly come to dominate all ecosystems (Boyden 2004; Ellis 2015) and the continued pressure of the extractive, manipulative, and productive activities of this keystone species portends severe consequences for Earth's species and nonhuman nature unless the trajectory is deliberately changed.

Of all ecosystems on Earth, cities arguably display the imprint of human design and intention to the greatest extent. Cities are places where human activities are concentrated, where the majority of the world's human population makes its home, and where the creative work of people is most evident (Grimm et al. 2008; Seto et al. 2012). Yet many cities in the Global South exhibit startling inequality, and many southern urban ecosystems are under tremendous stress (Baviskar, this volume; Swilling and Annecke 2012; Schindler 2017). Cities feature the infrastructures constructed for provision of food, water, shelter, energy supply, and transportation, and these constitute the form of an urban area. There is an unseen portion, such as the built infrastructure that is deliberately buried (e.g., sewers, electrical cables, or gas lines) as well as the social, political, and institutional components of society. Cities also feature urban nature. Encouraged or even designed, urban nature provides food or shade, or simply recreation and enjoyment. Preserved, produced, or "restored" (as in a remnant forest patch or a coastal wetland), as well as unrecognized or "accidental," urban nature persists or is reestablished by ecological and evolutionary processes in hidden places.

Cities are home to the majority of the world's population, and that proportion is projected to increase to 70% by 2050 and perhaps to near 90% by the end of this century (UN 2014). The history of urbanization, beginning 7,500 years ago with urban settlements in Mesopotamia (Redman 1999), has been a story of increasing perceived independence from nature, harnessing the benefits of technology and exploiting emerging hierarchies of wealth and power to replace the services once delivered by surrounding ecosystems—or the need to access them directly (Elmqvist et al. 2013b).

Cities have historically been dynamic centers of innovation and culture. At the same time, the emergence of impersonal and instrumental social relations in cities and the concomitant weakening of "traditional" social relations and institutions characteristic of agricultural communities has been met with trepidation in many societies. These changes in social order or social norms coupled with the obvious environmental problems accompanying a sudden rise in population density in rapidly urbanizing areas mean that cities are commonly perceived as dirty, dangerous, and centers of vice. Examples fitting both extreme images can be found among the world's cities. In fact, large cities have been shown to be centers of innovation out of proportion to their population (Bettencourt et al. 2007). Environmental conditions in many cities of the developed world have improved dramatically as industrial activities have relocated and governmental regulations on air and water quality have been implemented. A trend toward greening of these cities is reflected in the expansion of parklands and open space, "million tree" programs, and installation of "green infrastructure" projects (Tzoulas et al. 2007; Pataki et al. 2011; Hansen and Pauleit 2014). Yet in other parts of the world, particularly cities in decolonized societies, the extreme rapidity of urban expansion driven by massive migrations to cities by the rural poor and high intrinsic rates of increase in urban populations (Fox 2012) have outpaced the capacity of local governments to provide basic infrastructure (UN 2014). Such cities are suffering from air pollution, poor sanitation, and lack of clean water, among other stresses. The growth of urban populations without a concomitant expansion of infrastructure and services has indeed contributed to the production of dangerous and unhealthy urban environments.

The urban century—one in which we will see the movement of the vast majority of the Earth's human population to cities—coincides with other accelerating changes in the environment. Perhaps most urgent among these are climate change and increases in the frequency and severity of extreme events. The resulting collision course is one that presents opportunities and in which an ecologist's perspective—along with the perspectives of social scientists, planners, designers, engineers, and builders—has potential to move cities along a trajectory toward greater livability, resilience to extreme events, and sustainability (Childers et al. 2015; McPhearson et al. 2016). The ecologist who clings to an environmental protectionist view, however, presents a perspective in conflict with the needs of the twenty-first (urban) century. At the very least, protection of unaltered nature (if there is such a thing) involves trade-offs that may be untenable within urban centers and even suburban areas. Further, most urban ecosystems and urbanization processes are resource intensive; hence cities displace environmental problems to commodity frontiers and waste sinks, extending trade-offs beyond city boundaries to regional and global hinterlands.

An urban ecosystem includes all of the swatches of green that may be recognizable from an airplane window, here interchangeably termed green infrastructure or urban nature, but also the designed and built parts—the gray infrastructure—and the designers and builders themselves. This entire urban ecosystem is worthy of intentional care by urban residents and managers. It also is deserving of the attention of ecologists, environmentalists, designers, engineers, and others who are engaged in the process of envisioning, building, reforming, retrofitting, and managing the human habitat of the twenty-first century—the city.

We will explore the current and projected trends in urbanization and the issues those trends raise concerning equity, access, and services provided by cities in the Global North and South. The distinction between industrialized states in the North and developing countries in the South was made by the Independent Commission on International Development Issues (Brandt 1980). While we recognize the shortcomings of this distinction given the increasingly blurry boundaries between North and South, and the existence of many cities and neighborhoods that disrupt this neat classification, we retain it as a heuristic device reflecting an objective trend. In short, cities in the Global South persistently exhibit higher levels of inequality with regard to urban nature and services, as well as incomplete infrastructure systems that serve only a fraction of the population.

We review the causes of environmental degradation in cities, which are well understood. We then ask, what is the role of urban infrastructure and urban nature in providing services to people or protecting them from hazards? What are the prospects for managing unplanned-for risks of the future? Finally, we consider how the traditional views of conservation and restoration must be modified to apply to cities, by explicit acknowledgment of the importance of human design and human intention in the concept of urban nature. The concepts of diversity, justice, and sustainability are woven throughout the chapter.

Trends in Urban Development

The urban population is growing as the rural population stabilizes or even declines. The number of very large (>10 million) and large (5–10 million) cities is increasing globally, although cities of 1–5 million inhabitants have the fastest growth rate. An excellent special feature in *Science* illustrates some of these trends (Wigginton et al. 2016). As urban populations grow, the extent of urban areas grows even faster (Seto et al. 2012) so that more land is being transformed to urban uses than ever before. In 1950, 24% of the world's 233 countries were urbanized (i.e., the urban population was greater than the rural population) and only 8% had urban populations that were >75%. In 2014, these proportions increased to 63% and 33%, respectively; by 2050, over 80% of countries are projected to have more than half of their population living in cities, with about half of these countries being >75% urbanized.

Perhaps obvious, but still sobering, this grand transition to urban living will essentially be complete by the end of the century. When the vast majority of the world's population is concentrated in cities, urban land use will most likely occupy less than 5–10% of Earth's land surface, leaving vast areas sparsely populated (Brondizio and Le Tourneau 2016). Does this represent the most sustainable configuration for both Earth's ecosystems and its human population? Currently, the mean population density on Earth[1] is 97 individuals/km^2 and if all of these people were concentrated into cities occupying 5% of the land surface, mean urban population density would be 970 people/km^2—orders of magnitude lower than the world's densest cities and comparable to most U.S. cities today (Figure 6.1). Even with global population rising to 11 billion by the century's end as some of the highest projections suggest (Gerland

---

[1] Assuming population of 7.3 billion and land surface area of 150 million km^2, of which half is habitable.

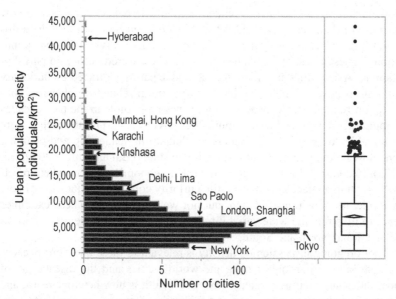

Figure 6.1 Frequency histogram of population density for the world's cities, showing several well-known cities. Data from Demographic 2016; city definitions were for large metropolitan regions including, in some cases, multiple cities. Standard box plot with mean, median, confidence intervals, and outliers is shown at right.

et al. 2014), if concentrated across the world's cities (we use the range 5–10% of the Earth's habitable land surface), population density would be ~725–1550 people/km². The world's cities today vary tremendously in their population size and density. Median urban population density is 5800 people/km², equivalent to the population density of Shanghai, China (Fig. 6.1), with a minimum of 500 people/km² in Knoxville, TN, U.S.A., and maximum of >44,000 people/km² in Dhaka, Bangladesh. Nevertheless, this arithmetic exercise illustrates that projections for city size and land area could feasibly accommodate the urban population surge, all else equal. Of course, all is not equal, a point we shall explore in more detail later.

A second important question is raised that is perhaps more central to the goals of this volume. What will the environment of cities, and the external ecosystems on which they depend, be like when the vast majority of the world's population lives in cities? Must expanding cities in previously rural countries experience the stages of environmental degradation and hazard seen in the early twentieth century in today's urbanized countries, or will they implement measures to improve local urban conditions? Will recent trends toward urban greening, urban agriculture, and open space preservation take hold in these developing cities, or will the simple provision of basic clean air, water, and sanitation overwhelm them? Will all urban populations have access to the external resources required to keep them functioning?

How will cities vary in their dependence upon external ecosystems, and how distant will their reach extend? What measures can be taken to reduce this dependence while ensuring the adequate provision of services to urban populations? These questions have been asked and answered before, in part. For example, Asian cities in China, Korea, and Japan have traced evolutionary paths from poverty-related problems to pollution-related problems to consumption-related problems at different paces and points in the past decades of rapid urbanization (Bai and Imura 2000). As cities aspire to reach a more sustainable "ecocity" phase, Bai argues that strong leadership and planning can allow them to circumvent less desirable states (Bai and Imura 2000; Bai 2003). In contrast, for many cities in weaker economic regions, where meeting basic human needs in cities is still a priority concern, the question of how to accelerate transitions is urgent but not well understood. These examples serve to illustrate the wide variability in developmental challenges for cities of the Global North and South.

The foregoing discussion on density is relevant to these questions because the distribution of people among the world's cities and the distribution of those cities and their associated power and wealth is now heterogeneous, and the resulting inequality between and within large regions of the Earth (i.e., Global North and South) may persist or worsen. Today's transition to urban living is not experienced evenly across the globe, with trends favoring rapid urbanization in the Global South and fastest growth in African, Asian, and Latin American cities with less than one million inhabitants (UN 2014). North America, the Caribbean, and Europe are already >75% urban; most increases in the urban population are expected to occur in low-income and lower- to middle-income countries. For instance, low-income countries are now 30% urban and expected to become 40% urban, while lower- to middle-income countries that are now 39% urban will rise to 57% urban by 2050 (UN 2014). Most of these lower-income countries are in the Global South and many are in warmer climates, where problems of excess heat, inadequate sanitation, and vector-borne diseases already, and will increasingly, challenge public health.

Other trends associated with urbanization exacerbate problems of resource scarcity and environmental deterioration. One in particular is the proliferation of informal settlements in the world's fast-growing cities. The expansion of these settlements can be attributed to migration from rural areas by people seeking refuge from conflict and/or environmental degradation, and natural increase (i.e., births minus deaths) (Fox 2012; UN 2014). Informal settlements are commonly situated in peri-urban areas or hazard-prone areas, such as on steep slopes, or in stream or river floodplains. As a result, informal settlements are particularly vulnerable to global environmental change (Romero Lankao and Qin 2011). In addition to their disproportionate exposure to environmental hazards, residents of informal settlements typically lack access to resources (e.g., drinking water and electricity), services (e.g., waste collection and public transportation), and economic opportunity (Figure 6.2).

Figure 6.2 Informal settlements, sometimes called slums, favelas, or shantytowns are a common feature of rapidly urbanizing regions in the Global South: (a) A town in Northern Mexico where residents have appropriated power by hooking in their own lines (photo by Nancy B. Grimm). (b) Favela da Rocinha, a shantytown built on a steep slope in Rio de Janeiro, Brazil (photo by Donatas Dabravolskas). (c) This stormwater "drain" in San Juan, Puerto Rico does not connect to a storm sewer system (representing inadequate infrastructure) and rapidly overflows during rainstorms (photo by Nancy B. Grimm). (d) An informal settlement in Marrakech, Morocco, with a discontinued sewer infrastructure project (photo by Seth Schindler).

Today, there are thousands to tens of thousands of cities distributed widely across the habitable world that vary in population size, demographics, city age, urban growth rate, history, spatial extent, percentage open space, urban form, as well as in the sociocultural, geographical, ecological, and political environment in which they are evolving. As yet, there is no consistent or agreed-upon typology of cities, no "urban biome" (Pincetl 2015) that can be used to predict how cities might drive or respond to global environmental change. But we know that human activity drives environmental change and, under the current sociotechnical regime, the consumption and fossil-fuel burning that is concentrated in cities contributes to climate change. Collectively, cities are responsible for over two-thirds of greenhouse gas emissions, for example, although those in the Global North are vastly more prominent contributors. We can expect that as urbanizing countries continue to develop, they will have improved access to fuel and transportation and will consume greater amounts of meat, all of which are associated with increased intensity of fossil fuel burning. Cities also are important contributors to land degradation associated with food production and extraction of natural resources, given that they largely depend upon external ecosystems for these services.

Urban areas increasingly experience the impacts of global environmental change. Many of the most populous and rapidly growing cities worldwide are located in low-lying coastal areas and along river floodplains (Mansur et al. 2016). While affording the benefits of water delivery and access to transportation and trade, these regions are at increased risk from sea-level rise and coastal, riverine, and urban flooding, which are exacerbated by extreme weather-related events like typhoons or hurricanes, tsunamis, river flooding, and storm surges (De Sherbinin et al. 2007). Others are located in water-scarce regions, such as the North American West, where water must be appropriated—often from afar—and where extreme drought and heat events are substantial risks (IPCC 2012; Pachauri et al. 2014). Indeed, urbanization and climate change are on a collision course. Extreme events are the most immediate way that people experience climate change and urban areas are particularly vulnerable to such events, given their location, concentration of people, and increasingly complex and interdependent infrastructures (IPCC 2012). Infrastructure offers the possibility for protection from extreme events, but evidence suggests the vulnerability of outdated infrastructure (in the developed world) and the inadequacy or even absence of infrastructure (in the rapidly urbanizing world) to provide resilience in the face of an uncertain future climate. We assert that a resilient infrastructure is one that incorporates ecological, social, *and* technological elements. This concept of social-ecological-technological systems (SETS) (Grimm et al. 2016; Redman and Miller 2016) is foundational to a new sustainability research network focused on urban resilience to extreme events (McPhearson et al. 2016), recognizing the capacity of people and their institutions to not only harness and invent new technological solutions but to coproduce services with the natural environment.

The Urban Social-Ecological-Technological System

The creation of an urban ecosystem is first and foremost about transforming land from what was there previously to a new system state (Grimm et al. 2017). The reasons for this transformation may be obvious, but should be stated: people need places to live and work (shelter), a means to get around (transportation), a steady supply of food and water, and energy sources to carry out their work and enjoy the comforts of modern life. Waste products need to be removed rapidly and efficiently and protection afforded from natural hazards. Such needs are met by infrastructure: the basic fabric of the urban built system (Ramaswami et al. 2016). These infrastructure systems have specific purposes, usually the provision of basic needs (e.g., water provisioning) or the solution to environmental challenges associated with concentrations of people (e.g., waste removal).

Initially, urbanization might involve pushing back nature, disrupting and degrading existing ecosystems, but cities continue to be ecosystems nevertheless (Golubiewski 2012), albeit dramatically transformed ecosystems. Sometimes nature is left in place or invited back in: some rivers, lakes, forest patches, or coastal wetlands may remain relatively unaltered; gardens and parks are planted with trees, shrubs, grass, and flowers; artificial ponds are created, and feeders are set out to attract birds. These components of the urban ecosystem continue to perform their functions (meaning: the ecological processes associated with their existence, such as primary production or pollination), perhaps less robustly than before, but blithely unconscious of human intent.

Together, the elements we think of as nature, the blue, green, and turquoise infrastructure (Childers et al. 2015), as well as the built environment and the beings who built and arranged it all, make up an urban SETS. Over time, urban SETS evolve and change, experiencing increases or decreases in human population, introductions of new technology, changes to urban form, continued loss or degradation of urban nature, or sometimes reversals of those trends through restoration of wild nature. Yet, there is a basic urban identity: the prevalence of built infrastructure; the concentration of people, their activity, and the products of their enterprise; the dependence upon external systems; and the production of wastes that are discharged to air, soil, and water and often transported to external ecosystems.

The Urban Environment: Benefits and Services

Cities provide for the basic needs of their inhabitants via built infrastructure (an expanded definition of infrastructure includes the knowledge systems and institutions responsible for decisions about infrastructure, as well as the biophysical underpinnings of the system). Ecosystem services, usually defined as the benefits derived by society from natural ecosystems, also contribute to human needs and well-being (Elmqvist et al. 2013a; Haase et al. 2014; Grimm et al. 2016). Does the alteration of the preexisting system during urbanization always mean that nature's services are lost or degraded? Are infrastructure services necessarily more or less valuable than nature's services? Can services from green infrastructure (nature-based or ecosystem-based services) replace the services of gray infrastructure, and if so, under what conditions? Because most literature on ecosystem services in cities primarily addresses the services derived from ecological functions of urban nature, such as parks and remnant patches (e.g., Gómez-Baggethun and Barton 2013; Haase et al. 2014; McPhearson et al. 2015; but see Grimm et al. 2016)), there are scant data available to compare the multiple values of services delivered between the built and nonbuilt, although such a comparison is a common means of valuation of nature's benefits (i.e., replacement value) (NRC 2004). A simple overview of services (Table 6.1) suggests that services from infrastructure and services from nature are interdependent and extend to scales much larger than the city itself, thus revealing its ecological footprint (Rees and Wackernagel 1996). Scholars of urban ecosystems are converging on ideas of ecosystem services in cities that recognize the following:

- Ecosystem services of urban nature are different from those of nonurban nature; they often are considered to be degraded.
- People are actively involved in the production, management, and extraction of ecosystem services from urban nature (Reyers et al. 2013; Andersson et al. 2014).
- Ecosystem services are often inequitably delivered.
- Built infrastructure provides services that in some cases substitute for local ecosystem services (i.e., shelter; see Table 6.1).
- Most ecosystem services in cities are the product of ecosystem processes that are modified by infrastructure and the built environment.
- Services in urban SETS are often dependent upon processes that take place far from the urban center where they are consumed.
- Urban residents are usually unaware of the prominent role external ecosystems play in providing the benefits that they enjoy in their cities.
- Protection of urban nature is challenged by the high cost of land in cities (a conformational and historical inertia), competing needs (multiple other urban problems), and a lack of connection with and understanding of the potential benefits of urban nature.

Making the conservationist case for nature preservation in cities based on their value in providing benefits to people is thus complicated by at least two issues. First is the broad perception that urban nature is degraded and that it is therefore unable to provide services. Although demonstrably oversimplified, until recently this has been a common view. The second challenge is the complexity of what constitutes "services" from urban nature and urban infrastructure, whether gray, green, or in between (Grimm et al. 2016), and whether they can substitute for one another. The service-providing elements listed in Table 6.1

Table 6.1 A list of purposefully contrasting elements of built infrastructure and (non-urban) nature that provide some basic services, defined as benefits derived by people from either built infrastructure or nature. Service delivery to urban populations nearly always involves some combination of the two extremes. Note that all services ultimately are derived from nonurban ecosystems as the original source (e.g., water supply) or for raw materials to construct infrastructure. Most services from built infrastructure require large inputs of energy to construct, maintain, or operate.

Service	Built infrastructure	Nonurban nature
Water supply	Dams, wells, interbasin transfers (pipes, canals)	Streams, springs, rivers, lakes
Water delivery	Canals, pipes, plumbing	Streams, springs, rivers; gravity
Water quality assurance	Water treatment plants	Protected lakes and reservoirs, wetlands, rivers
Shelter	Housing, other buildings	Caves, trees*
Food provision	Food processing and storage plants, delivery systems	Farms, orchards, animal populations
Transportation	Roads, canals, public transit lines	Rivers, lakes, oceans,* land routes,* and human-powered or passive transport systems
Energy supply	Power grid, power plants, delivery systems	Fire and biofuel, sun,* wind*
Protection from flooding	Sea walls, river levees, drainage canals	Coastal wetlands, dunes, floodplains, natural terraces
Sanitation, waste removal and processing	Sewers, wastewater treatment plants, solid waste incinerators	Rivers,* soils*
Recreation and experience of nature	Parks, zoos, gyms, gardens, swimming pools, cinema, television, virtual reality	Forests, deserts, grasslands, rivers, lakes, streams, beaches, etc.

*minimally useable without built infrastructure

are purposefully contrasting; clearly, no one would propose that the use of rivers and soils should replace sewerage and wastewater treatment for sanitation. Yet, in some cities constructed wetlands (which contain elements of both built environment and nature and thus are hybrid green-gray infrastructure) are used to further purify water below wastewater treatment plants (Sanchez et al. 2016). In contrast, cities like New York are choosing (or at least debating) the restoration of coastal wetlands, dunes, and oyster beds (green infrastructure) for protection over sea walls (gray infrastructure) (Rosenzweig et al. 2011).

Nature-as-infrastructure should not be embraced as a second-best option for residents of informal settlements, and hence a justification for unequal exposure to environmental hazards. Nevertheless, there is a diverse range of ways in which informal settlements and their residents are related to infrastructure and integrated into the urban fabric (Bishop and Phillips 2014). In some instances, residents of informal settlements develop complex infrastructure systems incrementally on an ad hoc basis (Silver 2014) and informal yet visible and institutionalized systems mediate the relationship between formal and informal infrastructure systems (Demaria and Schindler 2016). Thus, as urban residents experiment with the materiality of the city and incrementally expand infrastructure systems in innovative ways (driven by need, availability of resources, and capacity), a complex urban landscape emerges. This landscape is characterized by the interconnection of hybrid infrastructure systems and uneven distribution of access, which combine core gray infrastructure systems with informal subsystems that in turn incorporate nonhuman nature in a range of ways (Björkman 2015). The planting of trees to provide shade, the development of community gardens and compost to support urban food production— even the installation of multiple, distributed, constructed wetlands to reduce discharge of pollutants—all may provide moderate solutions to problems of urban heat, food security, and waste disposal in absence of infrastructure development by local governments. In sum, a wide range of potential services is being discovered and increasingly promoted, albeit not necessarily by municipal authorities—as alternatives to the unifunctional, gray infrastructure approach.

The Urban Environment: Disservices, Hazards, and Risks

Urban nature provides quantifiable benefits to residents of the world's cities (Haase et al. 2014), as does built infrastructure. Both urban nature and built environment, however, can be associated with harmful outcomes to people. Design of infrastructure and urban nature must consider both benefits and costs in terms of human health and well-being as well as trade-offs and synergies between distinct classes of services (Bennett et al. 2009). In most cities of the Global North, air quality has dramatically improved over the past decades. Yet, despite careful planning, many growing Asian cities are struggling with air pollution resulting from expanding vehicle ownership and increased consumption of fossil fuels. Worsening air pollution continues to present a serious public health risk as industrialization fuels urban growth in other parts of the world.

Changes to the land surface also may have negative outcomes. For example, the coverage of large areas of land with heat-absorbing materials like asphalt and concrete, along with concentrated energy use in urban areas, can result in local climate change called the urban heat island that is far in excess of any increases in temperature yet experienced globally. The same impervious surfaces rapidly convey rainwater that falls on them, resulting in damaging floods. Even urban nature can be detrimental to human health and well-being: wetlands may attract insect pests, unmanaged streams and lakes can become polluted or eutrophic, unpoliced parks and open space may become dumping grounds or provide cover for criminals.

Often, it is the absence of human intervention that degrades urban nature, given that pollutants and waste produced in cities can become concentrated in areas that are not actively managed. Streams and rivers are integrators of the landscapes they drain, and lakes and wetlands as low places in the landscape receive the materials transported by streams and rivers. For example, Hale et al. (2014) report a dominant signal of residential landscape fertilization in stormwater of Phoenix, Arizona (U.S.A.) whereas Kaushal et al. (2008) show that leakage from sanitary sewers dominates the chemistry of stormwater in Baltimore, Maryland (U.S.A.). While entire cities can be subject to environmental hazards (Mansur et al. 2016), they are commonly borne disproportionately by the poor (Baviskar, this volume; Bullard 2000; Zeiderman 2016).

Sanitary solid waste disposal in the world's slums is haphazard at best, putting populations at risk of disease associated with inadequate sanitation (UN 2014). Complex informal systems of solid waste management have evolved in most cities in the Global South, as waste management has been somewhat of an afterthought for municipal officials who prioritize economic growth. Informal-sector waste workers tend to mediate the flow of recyclable material from small and inefficient formal-sector waste streams to formal and informal-sector recyclers. Informal-sector waste workers were historically not recognized by authorities in many cities, but their contribution to waste management was tacitly encouraged. This changed as the proportion of recyclable material in municipal solid waste increased over the course of a decade of increasing commodity prices. The result has been conflict over access to waste between large-scale, capital-intensive, waste-management firms and informal-sector waste workers (Demaria and Schindler 2016). The challenge is to integrate these systems in ways that ensure livelihoods for waste workers as well as sanitary and environmentally sustainable waste-management outcomes. Of course, sustainable outcomes with regard to solid waste are not assured in the Global North, given that some waste management consists of shipping the hazards elsewhere, often to poor regions.

Alteration of the natural environment and installation of infrastructure are sometimes intended to provide a measure of protection for urban residents against hazards. The service "protection from harm" is encompassed in a massive literature on disaster risk reduction. Hazards to which urban populations are exposed include natural hazards, such as storm surge or tidal flooding in coastal cities, riverine flooding, earthquakes, or fire; and anthropogenic hazards (either caused or exacerbated by human activity), such as urban flooding, water pollution, or contamination from wastes. In most cases, gray infrastructure is built to withstand all but the rarest of such events. Yet when these protections fail, the consequences can be severe. There is increasing interest in using urban nature to bolster or even replace the protective function of gray infrastructure through the use of what are called ecosystem-based approaches, nature-based solutions, or ecosystem-based adaptation (Royal Society 2014), and in the United States are generally encompassed by the preferred term green infrastructure. Green infrastructure includes stormwater retention features such as rain gardens, green roofs, bioswales, wetlands, and retention basins, but also street trees and parks maintained for aesthetic reasons.

Risks to urbanites from hazards are increasing. Urbanization itself sometimes removes natural protective ecosystems, such as coastal wetlands. Exposure to coastal flooding under future climate scenarios without protective ecosystems is estimated to be double the exposure with such ecosystems left intact (Arkema et al. 2013). As the frequency and magnitude of extreme weather-related events increases (Munich RE 2017), our sense of security from infrastructure designed for a 1% probability event is likely misplaced. Under climate change, the future is uncertain (Milly et al. 2008; Miller et al. 2011) and a 1% event may actually happen as frequently as once in five or ten years. A recent study using proxy records for historical sea levels (to AD 800) shows that coastal floods that occurred once every 500 years in the past are now occurring once every 24 years in New York City (Reed et al. 2015). Thus, the capacity of extant urban infrastructure to provide protection against hazards that are increasing in frequency and/or magnitude is to some extent itself uncertain. Finally, hazard risk is increasing in some areas of the world because more people are settling in exposed places, and the rate of population increase in these areas is outstripping the capacity of local governments to provide services. Thus, there are both detriments and potential benefits to urban nature, the former usually arising from lack of management and the latter, which will be taken up in the final section of this chapter, representing a suite of potential solutions to the challenges of global environmental change.

Conservation and Restoration in Cities

Nature preservation, conservation, and restoration are the challenges that an environmentalist takes on. But what is it that we want to preserve or restore in cities? What is nature in cities? Do we include green infrastructure such as a green roof or a curb cut or a rain garden in this conception? Does our backyard qualify? Is the neighborhood park or a community garden worthy of preservation? Or are just the forest patches, remnant grasslands, unburied streams, or surviving wetlands eligible for the environmentalist's concern? In fact, the traditional nature protectionist doesn't even ask these questions; these pieces of urban nature are "fake nature" (Ross et al. 2015) and unworthy of preservation or restoration.

Cities are thought to have "novel nature" (Pincetl 2015) or "middle nature" (Tanner et al. 2014), concepts that recognize the idea that culture and nature are coevolving (Boyden 2004; Barthel et al. 2010). Much of urban nature fits the description of "designer ecosystems" proposed by Ross et al. (2015), reflecting the control that people as the designers exert and the extent to which they intentionally alter pattern and process. But some urban nature is "accidental"; that is, it arises independently of human intention (Palta et al. 2017). Examples include the grass and weed expansion in abandoned home lots (Ripplinger et al. 2016) and wetlands that have arisen in the once-dry channel of the Salt River in Phoenix as a result of increased outdoor water use and altered hydrodynamics

in the river channel (Palta et al. 2017). Promoters of the intrinsic value of nature as a sole rationale for its preservation, "nature protectionists" (Miller et al.
2011), may have trouble with these concepts and may refuse to acknowledge
urban nature as nature. A value inculcated in many conservation groups, and
possibly in much of ecology, is that nature is virtuous and humanity profane
(Ross et al. 2015), and that humans are a threat to nature. If this is so, how can
urban nature, with only the occasional "accidental" escape from control by
humanity, be thought of as nature?

When we ask these questions, it is instructive to think about how our species
came to be so dominant on the planet. In a brilliant and fascinating conceptual
treatise, Ellis (2015) argues that it is "sociocultural niche construction" and the
long-term intertwined processes of human cultural evolution and engineering
of the environment that underlie our species' transformation of the biosphere.
He states: "It is no longer possible to understand, predict, or successfully manage ecological pattern, process, or change without understanding why and how
humans reshape these over the long term" (Ellis 2015:287.) Whether one agrees
with the mechanism or not, Ellis's admonition to incorporate an understanding
of social and cultural drivers is most certainly relevant to the coevolution of
society and urban nature (see also Boyden 2004), and is probably good advice
for nonurban nature as well. We should target urban nature in our conservation
efforts, both for its intrinsic value as nature and for its utility as a provider of
services to urban residents. The intrinsic value of urban nature may be called
into question, but as Hartig and Kahn (2016) note, we must both "experience
nature in cities and experience cities as natural."

In an urban world, daily exposure to nature is reduced compared to that
of early hunter-gatherers or even early agrarian societies. This reduction is
particularly severe in the United States, where awareness of the perils of the
so-called "nature-deficit disorder" (Louv 2008) has spurred the "No Child Left
Inside" movement. Research shows a clear benefit to people's health and well-
being from nature exposure. Opportunities for people to interact with nature
can benefit their health through improvements in air quality, encouragement
of physical activity, increased social interactions leading to a greater sense of
community, and stress reduction with direct impacts on health and performance
(Hartig et al. 2014). Even views of green space from classrooms have been
shown to reduce stress and improve scholastic performance in high-school stu-
dents (Li and Sullivan 2016). Recognizing this, many cities are working to
both reduce urban sprawl and increase the amount of green space, two objectives that may at times conflict with one another. The challenges of "renatur-
ing" cities (Hartig and Kahn 2016)—affording people greater opportunity to
interact with nature—include not just enhancing the amount and quality of
urban nature but ensuring access to it, a way for people to connect (Shanahan
et al. 2015). Citizen science or "civic ecology" (Krasny and Tidball 2012) engages people in gardens, tree planting, or as friends of parks, resulting in positive outcomes from contact with nature as well as helping to change people's

thinking about their relationships with nature. A study of community gardens in Stockholm has shown how people's interaction with ecological systems has evolved over many years, resulting in a "socioecological memory" that builds resilience (Barthel et al. 2010).

Management of urban nature has mostly focused on remnant ecosystems (which may be refuges for species that cannot survive in urban habitats), parks, and open space. This approach targets a small portion of the total urban land area and as a result, does not address the larger issue of the absence of nature from the fabric of cities and city residents' daily lives. To some extent individual landowners address this through their own decisions about local nature (i.e., in their gardens or yards), but a broader, landscape-scale management paradigm is seldom applied. Our understanding of the spatial and temporal scales of how ecosystem services are provided is underdeveloped (Andersson et al. 2014) and failure to understand the history of ecosystem service change may lead to inaccurate assessments of trade-offs (Tomscha and Gergel 2016). Recent work suggests that large tracts of undeveloped land are necessary to preserve ecosystem services while small, distributed bits of nature may be needed to ensure nature contact for urban residents (Stott et al. 2015). Does this necessarily apply, however, to all situations and all cities? For example, how urban nature can be woven into the too-rapidly developing fabric of explosive urban growth in Asia and Africa is an open question, but we have little to go on (McHale et al. 2013) with respect to cultural attitudes that shape ecosystem services, the kinds of designs that might be effective in these settings, the interactions between immigrants and the places in which they settle, and whether the governance structures exist to ensure the place of urban nature in an ever-expanding city.

The conservation of urban nature raises multiple issues of environmental equity and justice (see also Baviskar this volume). For example, the creation of extensive natural amenities can drive poor people from their neighborhoods as land prices increase or homes are demolished to make way for parks. The largest housing demolition in Delhi's history resulted in the displacement of approximately 150,000 people, whose community on the banks of the Yamuna River ostensibly posed an environmental hazard and was ultimately transformed into a park (Bhan 2009). In San Juan, Puerto Rico, proposed dredging of a long-abandoned canal and the creation of a wide riparian area will be put in place at the expense of several homes. This large riparian park could be a new, attractive amenity in a city at risk of gentrification. However, a local community organization has been working to ensure that the displaced homeowners will be relocated within their neighborhoods and that access to new housing and amenities will be restricted to current residents, mostly poor homeowners in eight neighborhoods that border the canal.

Across many cities, vegetative cover, which provides shade and respite from heat, habitat for birds, and other benefits, varies tremendously among neighborhoods. Hope et al. (2003) found that vegetation abundance was correlated with wealth in Phoenix, Arizona (U.S.A.); this relationship was recently supported for multiple neighborhoods in Southern California, where tree species' richness was significantly higher in wealthy than in less affluent neighborhoods (Avolio et al. 2014). Schwarz et al. (2015) went so far as to say the "trees grow on money" when they found support for the relationship between tree cover and income across seven U.S. cities.

Who benefits from the conservation of urban nature? The above studies suggest that in many cities it has been the wealthy—presumably people living in already clean, healthy neighborhoods with few other problems. The trend toward urban greening, it appears, is often unevenly applied. Further, the question must be asked, both at scales of individual cities as well as across the broader spectrum of world urbanization, whether conservation of urban nature can be given a high priority when confronted with the plethora of other issues with which the urban poor and the world's slums have to cope (Baviskar, this volume). We believe the answer lies in fully embracing the utilitarian view of urban nature, but even more importantly the view of social-ecological-technological coproduction of services as a means of building resilience to shocks and stresses. We address this in the next section on design.

Restoration has been conceptually challenging for ecologists confronted with altered ecosystems with no clear "natural" or "reference" analog. This is certainly the case for urban systems, and yet there continue to be efforts at restoration in cities. Efforts to restore streams, especially the excellent work of Australian stream ecologists and hydrologists (Walsh et al. 2005, 2012; Burns et al. 2012), involve the reduction of damaging peak flows that are a consequence of impervious surfaces in cities that are directly connected (through water flow) to streams. Yet, much of the literature on stream restoration concentrates on scarcely urbanized watersheds, since impervious areas over ~30% are considered too "far gone" to be restored. Note that most cities have >50% impervious cover, and many European cities are in the range of 70–80% impervious cover. While these restoration efforts should certainly continue, there also is reason to rethink the concept of restoration in highly altered urban environments. Instead, designing waterways with combinations of technological and ecological features specifically intended to be multifunctional (i.e., deliver multiple benefits) —that is, intentionally creating urban SETS infrastructure—may ultimately represent a more sustainable pathway (Ahern 2011; Larson et al. 2013).

Design of Urban Nature for Services and Resilience

Imagine a city that is a part of nature, not apart from nature. —P. Mittenmeier[2]

The challenges facing cities of both Global North and South are massive, and against that backdrop, focusing on environmentalism—in the traditional

---

[2]   https://twitter.com/Conserve_WA/status/647100689756262402 (accessed Sept. 27, 2017).

sense—can seem the height of hubris. Yet the strengthening of ecological elements in cities may be key to enhancing their chances of weathering these challenges. First, however, we have a long way to go in changing the attitudes and deeply embedded biases of ecologists, managers, and urban residents about what is nature and what is natural, in cities. Second, urban residents must be sensitized to the impact that their consumption patterns have on their regional and global hinterlands, that is, their role as drivers of global change. We need concerted efforts at environmental education that take on ecology in and beyond the city (Elser et al. 2003; Banks et al. 2005; Bestelmeyer et al. 2015) and civic ecology opportunities that engage citizens (Krasny and Tidball 2012), improving their environmental literacy, and perhaps moving them closer to stewardship that reconnects urbanites with the biosphere (Andersson et al. 2014). In both cities of the North and South, the interaction of residents, decision makers, and planners with regard to the design and promotion of urban nature's services holds potential to foster resilience and sustainability in the face of an uncertain future.

Designing Urban Nature in Northern Cities

It is probably no accident that the rise of interest in green infrastructure and nature-based solutions has occurred in cities of the United States and Europe, where transition from the industrial city to the sanitary city occurred during the last century (Grove 2009). These cities have enjoyed improved local environmental conditions including cleaner air, expanded and protected parks and open space, the best available wastewater treatment, and so on. Several cities are at the forefront of conceiving and implementing changes to institutions like zoning and park management, restoring urban streams, improving public transportation, offering bike lanes and pedestrian zones, and other positive trends for enhancing livability in cities. Initiatives like the World Wildlife Federation's Earth Hour City Challenge asks cities to reduce their emissions, and the Rockefeller Foundation's 100 Resilient Cities provides funds for cities to hire a Resilience Officer to implement changes in city structure to enhance resilience. There is much to be learned from Scandinavian cities, which consistently are at the top of the world's greenest and most eco-friendly and sustainable cities. Indeed, developing concepts of using coproduced ecosystem services from urban nature and its stewards to build urban resilience through the strengthening of people's connection to their local environment and promotion of ecological processes that underpin services emanate from urban ecology programs in Sweden (e.g., Andersson et al. 2014).

In many cities of the Global North, however, particularly where the transition to urban living occurred in the last century, urban infrastructure is aging and in need of replacement. A recent report in the United States gives infrastructure a failing grade for providing adequate protections for city populations (ASCE 2013). Furthermore, zoning and other protections are proving inadequate in the United States. In many flood-prone areas, new development has sprawled into flood plains, where federally funded levees and national flood insurance afford a false sense of security (for an analysis of recent flooding in Louisiana, see Colten 2016). Traditional risk-based engineering approaches to infrastructure design that focus on minimizing the risk of failure by investing in hard, structural, resistant elements—fail-safe designs—are inappropriate in the fast-changing environment of the Anthropocene. Instead, more ecologically based designs, which may be viewed as safe-to-fail (Ahern 2011), allow for some failure but minimize its consequences (Park et al. 2013). These designs should be appropriate to place, equitable (neither disproportionately benefiting nor putting at risk any particular segment of the population), and incorporate ecological as well as technological elements. For example, the combination of sea walls in highly built-up segments of a coastal city with restoration of marshes along less built-up coastlines may help to reduce the impacts of storm surge on neighborhoods.

Building resilience to climate extremes, that is, the capacity of SETS to experience and weather shocks from extreme climatic events without losing fundamental structure and function (Walker et al. 2004; Folke 2006), is urgently needed if we are to avoid, or even minimize, the debilitating impacts of such events in uncertain futures. This concept of resilience emphasizes multifunctional, diverse, participatory, and flexible solutions, which are best achieved by incorporating nature into design (Ahern 2011; Childers et al. 2014; Grimm et al. 2016).

Designing Urban Nature in Southern Cities

Rapidly growing cities in the Global South present serious social and ecological challenges, particularly as they attempt to meet basic needs for the urban poor and in doing so, exacerbate air and water pollution. The rapidity of urban growth has resulted in increasing flows of energy and material to sustain cities and this has strained local and regional ecosystems. Most cities in the Global South exhibit significant informal expansion, and in some cases municipal authorities have sought to connect informal settlements with formal infrastructure systems, while elsewhere they have sought to inhibit residents of informal settlements from accessing formal infrastructure systems. Regardless of whether municipal authorities decide to integrate or isolate informal settlements, the design of infrastructure and the extent to which it is resilient to future shocks and stresses is likely to differ from northern cities. Furthermore, it is important to note that there is no compelling reason to assume that solutions that have worked in the North can be effectively transplanted to the South (McHale et al. 2015).

There is an unequivocal necessity to enhance well-being and reduce environmental degradation in cities of the South. In developing regions, nearly one-third of urban residents live in informal settlements or slums, and in most of these cases the basic infrastructure for water delivery and waste removal is inadequate or absent (UN 2014). Small-scale, distributed, green infrastructure that fits the particular setting may be a cost-effective solution that improves resilience (Schäffler and Swilling 2013), certainly as opposed to the alternative, which is often the sluggish and uneven expansion of gray infrastructure. One problem, however, is that green infrastructure is typically not valued as infrastructure and its potential benefits in alleviating poverty, creating jobs, and ameliorating pollution are scarcely known (Schäffler and Swilling 2013). Instead, ecological elements are seen as a nuisance (if they are not managed) or a luxury (if they are) rather than as infrastructure. Thus, a reconceptualization of what constitutes viable infrastructure is needed.

The trajectory of urbanization in the Global South is unlikely to mirror the urban experience of North America and Europe (Roy 2009; McHale et al. 2013). This is in part because there is neither time nor resources to construct massive water treatment works, stormwater drainage systems, or wastewater treatment facilities. And, as we have seen in the case of Louisiana in the United States, a false sense of security can be associated with protective infrastructure constructed today based on traditional formulas of minimizing failure probability, given future increases in frequency and magnitude of extreme events. The urgency of the infrastructure needs in these situations argues strongly for the flexible, multifunctional, low-cost, replaceable, safe-to-fail type of options that ecosystem-based or hybrid solutions can potentially provide. In other words, planners are not faced with an either/or choice between gray and green infrastructure, but they should try to incorporate elements of each to develop unique hybrid and flexible systems that are city-specific and address localized contexts and challenges. Flexibility can be achieved through experimentation and continual monitoring and adjustment, as advocated by Ahern et al. (2014). In their transdisciplinary design and planning model, ecosystem services goals are stated and prioritized (decided by all stakeholders), an experiment is designed (which means a project is actually constructed), indicators and metrics are chosen to measure goals, monitoring is used to assess efficacy, and the process is iterated so that adjustments can be made when goals are not met. This kind of process is very similar to adaptive management used in natural resource management, but the design and construction of infrastructure has seldom involved all or even one or two of these steps. The pervasive view that infrastructure must conform to a standard design, usually gray or highly engineered and seemingly impervious to failure, must be reconsidered in favor of the more flexible alternatives that green infrastructure or urban nature can provide.

Personal Reflections

Schindler: Delhi's ecology is under tremendous strain and this generates hazards to which the poor are disproportionately exposed. The city and its ecology have changed tremendously since my first visit in 2006. Wasteland has become elite retail space, forests have become illicit landfills, landfills are slated to become parks. One colony I have visited regularly over the years is home to scavengers that collect recyclable material from an adjacent mountainous landfill. The houses are made of tarpaulins and wooden beams and, until recently, the lanes were dirt which meant that they became impassable during the monsoon season. The colony was long and thin, sandwiched between rows of two-storey brick buildings on one side, and a major road on the other. However, between the colony and the road there was a meager strip of greenery that provided residents with some space to momentarily escape from their cramped quarters and relax or play cricket. This space has been razed to make way for the construction of a metro station, and a massive incinerator now looms behind the brick houses. The community is being squeezed and its residents are losing a competition with the metro for space and with the incinerator for waste. For residents there is no escape from a localized ecology characterized by the absence of green space, leachate that flows out from the landfill toward the Yamuna River, and poor ambient air quality. Their exposure to environmental hazards is compounded by insecure livelihoods and lack of tenure security. Sadly, this sort of multifaceted vulnerability is all too common, and the livability of the future city depends on the emergence of novel human–environment interactions. Urban nature is diverse and can take many forms, but when conceived as more than a resource whose exploitation can facilitate economic growth, it has the potential to augment livelihoods and enhance well-being.

Grimm: I have lived in the Phoenix, Arizona metropolitan area for nearly forty years, although I grew up in the verdant eastern portion of the United States. As I drafted this paper, I was enjoying a mini-sabbatical in Stockholm, the home of the only urban national park in the world. Walking through those woods with their majestic, several hundred year-old oak trees, I marveled at the social structures and care in this socioecological system that have, over centuries, allowed these creatures to persist (Barthel et al. 2010). Phoenix seems to be a wholly designed and engineered city; even its flora is largely imported (Hope et al. 2003). Yet exposure, enjoyment, and recreation for its citizens and education about the desert environment are made possible by the recent formation of a community organization, the Central Arizona Conservation Alliance (CAZCA), dedicated to protection, education, research, and restoration of the Phoenix area desert mountain parks—a cluster of volcanic outcrops that have escaped the voracious sprawl of this desert metropolis primarily because of unsuitability for rapid housing development. CAZCA may function as the middle governance entity envisioned in Andersson et al. (2007), because it can coordinate the individual efforts of park managers, community groups, and conservation organizations and act as a representative in working to achieve protection goals through municipal and county governments. Indeed, a vision of county parks that ring the Phoenix metropolitan area with a hiking trail, the Maricopa Trail, passing through them uninterrupted, could emulate the urban national park of Stockholm. Instead of centuries-old oaks, the area features centuries-old giant cacti—saguaros—and other unique vegetation of the Sonoran Desert. This is urban nature on the wild end of the spectrum, but the necessity for social engagement among the four million inhabitants of the Phoenix metro area makes it urban nature all the same.

Conclusions

Our reflections capture threads woven through this chapter, contrasting the challenges associated with ensuring fundamental environmental rights in cities of the South and the continuing trend of reintroducing nature (and ensuring its conservation) in cities of the North. Both challenges benefit from a systems perspective that recognizes the interwoven SETS dimensions of cities, their infrastructure, and inhabitants. As the world's population has become increasingly urban, the distribution of wealth, urban density, and pathways of urban development have become extremely heterogeneous, leading to inevitable inequalities in the distribution of environmental problems both within and among the world's cities. Increasing risk from global environmental change adds further stress. Amplifying the benefits and services that can be provided by a SETS infrastructure, while reducing the hazards and disservices, is a key imperative for cities in the face of such changes. Design of a SETS infrastructure that incorporates urban nature appropriate to the setting is likely to be more flexible, multifunctional, replaceable, safe-to-fail, and cost-effective than the monumental gray infrastructure that characterizes urban development of past centuries. Both in terms of replacing the aging infrastructures of older cities and providing new infrastructures for rapidly urbanizing regions, multiple services and greater resilience may be expected from a SETS-sensitive design that incorporates urban nature.

Acknowledgments

Ideas for this paper were nurtured in discussions with colleagues in the Urban Resilience to Extremes Sustainability Research Network (UREx SRN, supported by the U.S. National Science Foundation grant number 1444755) and with colleagues at the Stockholm Resilience Center. Support was provided by CAP LTER (NSF 1026865) during Grimm's sabbatical. The paper was improved by a review from Xuemei Bai and by rich and stimulating discussions with members of the urban group at the Ernst Strüngmann Forum.

References

Ahern, J. 2011. From Fail-Safe to Safe-to-Fail: Sustainability and Resilience in the New Urban World. *Landscape Urban Plann.* **100**:341–343.

Ahern, J., S. Cilliers, and J. Niemelä. 2014. The Concept of Ecosystem Services in Adaptive Urban Planning and Design: A Framework for Supporting Innovation. *Landscape Urban Plann.* **125**:254–259.

Andersson, E., S. Barthel, and K. Ahrné. 2007. Measuring Social-Ecological Dynamics Behind the Generation of Ecosystem Services. *Ecol. Appl.* **17**:1267–1278.

Andersson, E., S. Barthel, S. Borgström, et al. 2014. Reconnecting Cities to the Biosphere: Stewardship of Green Infrastructure and Urban Ecosystem Services. *Ambio* **43**:445–453.

Arkema, K. K., G. Guannel, G. Verutes, et al. 2013. Coastal Habitats Shield People and Property from Sea-Level Rise and Storms. *Nat. Clim. Chang.* **3**:913–918.

ASCE. 2013. Report Card for America's Infrastructure. American Society of Civil Engineers. https://www.infrastructurereportcard.org/. (accessed Sept. 29, 2017).

Avolio, M. L., D. E. Pataki, S. Pincetl, et al. 2014. Understanding Preferences for Tree Attributes: The Relative Effects of Socio-Economic and Local Environmental Factors. *Urban Ecosyst.* **18**:73–86.

Bai, X. M. 2003. The Process and Mechanism of Urban Environmental Change: An Evolutionary View. *Int. J. Environ. Pollut.* **19**:528–541.

Bai, X. M., and H. Imura. 2000. A Comparative Study of Urban Environment in East Asia: Stage Model of Urban Environmental Evolution. *Int. Rev. Environ. Strat.* **1**:135–158.

Banks, D. L., M. M. Elser, and C. Saltz. 2005. Analysis of the K–12 Component of the Central Arizona–Phoenix Long–Term Ecological Research (CAP LTER) Project 1998 to 2002. *Environ. Educ. Res.* **11**:649–663.

Barthel, S., C. Folke, and J. Colding. 2010. Social–Ecological Memory in Urban Gardens: Retaining the Capacity for Management of Ecosystem Services. *Global Environ. Change* **20**:255–265.

Bennett, E. M., G. D. Peterson, and L. J. Gordon. 2009. Understanding Relationships among Multiple Ecosystem Services. *Ecol. Lett.* **12**:1394–1404.

Bestelmeyer, S. V., M. M. Elser, K. V. Spellman, et al. 2015. Collaboration, Interdisciplinary Thinking, and Communication: New Approaches to K–12 Ecology Education. *Front. Ecol. Environ.* **13**:37–43.

Bettencourt, L. M. A., J. Lobo, D. Helbing, C. Kuhnert, and G. B. West. 2007. Growth, Innovation, Scaling, and the Pace of Life in Cities. *PNAS* **104**:7301–7306.

Bhan, G. 2009. This Is No Longer the City I Once Knew: Evictions, the Urban Poor and the Right to the City in Millennial Delhi. *Environ. Urban.* **21**:127–142.

Bishop, R., and J. W. Phillips. 2014. The Urban Problematic II. *Theory Cult. Soc.* **31**:121–136.

Björkman, L. 2015. Pipe Politics, Contested Waters : Embedded Infrastructures of Millennial Mumbai. Durham, NC: Duke Univ. Press.

Boyden, S. V. 2004. The Biology of Civilisation: Understanding Human Culture as a Force in Nature. Sydney: Univ. of New South Wales Press.

Brandt, W. 1980. North–South: A Programme for Survival: Report of the Independent Commission. Cambridge, MA: MIT Press.

Brondizio, E. S., and F.-M. Le Tourneau. 2016. Environmental Governance for All. *Science* **352**:1272–1273.

Bullard, R. 2000. Dumping in Dixie : Race, Class and Environmental Quality. Boulder, CO: Westview Press.

Burns, M. J., T. D. Fletcher, C. J. Walsh, A. R. Ladson, and B. E. Hatt. 2012. Hydrologic Shortcomings of Conventional Urban Stormwater Management and Opportunities for Reform. *Landscape Urban Plann.* **105**:230–240.

Childers, D. L., M. L. Cadenasso, J. M. Grove, et al. 2015. An Ecology for Cities: A Transformational Nexus of Design and Ecology to Advance Climate Change Resilience and Urban Sustainability. *Sustainability* 7:3774–3791.

Childers, D. L., S. T. A. Pickett, J. M. Grove, L. Ogden, and A. Whitmer. 2014. Advancing Urban Sustainability Theory and Action: Challenges and Opportunities. *Landscape Urban Plann.* **125**:320–328.

Colten, C. E. 2016. Suburban Sprawl and Poor Preparation Worsened Flood Damage in Louisiana. http://theconversation.com/suburban-sprawl-and-poor-preparation-worsened-flood-damage-in-louisiana-64087. (accessed Sept. 29, 2017).

Demaria, F., and S. Schindler. 2016. Contesting Urban Metabolism: Struggles over Waste-to-Energy in Delhi, India. *Antipode* **48**:293–313.

De Sherbinin, A., A. Schiller, and A. Pulsipher. 2007. The Vulnerability of Global Cities to Climate Hazards. *Environ. Urban.* **19**:39–64.

Ellis, E. C. 2015. Ecology in an Anthropogenic Biosphere. *Ecol. Monogr.* **85**:287–331.

Ellis, M. A., and Z. Trachtenberg. 2014. Which Anthropocene Is It to Be? Beyond Geology to a Moral and Public Discourse. *Earth's Future* **2**:122–125.

Elmqvist, T., M. Fragkias, J. Goodness, et al., eds. 2013a. Urbanization, Biodiversity and Ecosystem Services: Challenges and Opportunities. Dordrecht: Springer Netherlands.

Elmqvist, T., C. L. Redman, S. Barthel, and R. Costanza. 2013b. History or Urbanization and the Missing Ecology. In: Urbanization, Biodiversity and Ecosystem Services: Challenges and Opportunities, ed. T. Elmqvist et al., pp. 13–30. Dordrecht: Springer Netherlands.

Elser, M. M., B. Musheno, and C. Saltz. 2003. Backyard Ecology. *Sci. Teach.* **70**:44–45.

Folke, C. 2006. Resilience: The Emergence of a Perspective for Social–Ecological Systems Analyses. *Global Environ. Change* **16**:253–267.

Fox, S. 2012. Urbanization as a Global Historical Process: Theory and Evidence from Sub-Saharan Africa. *Popul. Dev. Rev.* **38**:285–310.

Gerland, P., A. E. Raftery, H. Sevčíková, et al. 2014. World Population Stabilization Unlikely This Century. *Science* **346**:234–237.

Golubiewski, N. E. 2012. Is There a Metabolism of an Urban Ecosystem? An Ecological Critique. *Ambio* **41**:751–764.

Gómez-Baggethun, E., and D. N. Barton. 2013. Classifying and Valuing Ecosystem Services for Urban Planning. *Ecol. Econ.* **86**:235–245.

Grimm, N. B., E. M. Cook, R. L. Hale, and D. M. Iwaniec. 2016. A Broader Framing of Ecosystem Services in Cities: Benefits and Challenges of Built, Natural, or Hybrid System Function. In: Handbook on Urbanization and Global Environmental Change, ed. K. C. Seto et al. London and New York: Routledge. https://www.routledgehandbooks.com/doi/10.4324/9781315849256.ch14. (accessed Sept. 29, 2017).

Grimm, N. B., S. H. Faeth, N. E. Golubiewski, et al. 2008. Global Change and the Ecology of Cities. *Science* **319**:756–760.

Grimm, N. B., S. T. A. Pickett, R. L. Hale, and M. L. Cadenasso. 2017. Does the Ecological Concept of Disturbance Have Utility in Urban Social-Ecological-Technological Systems? *Ecosyst. Health Sustain.* **3**:e01255.

Grove, J. M. 2009. Cities: Managing Densely Settled Social-Ecological Systems. In: Principles of Ecosystem Stewardship, ed. C. Folke et al., pp. 281–294. New York: Springer.

Haase, D., N. Larondelle, E. Andersson, et al. 2014. A Quantitative Review of Urban Ecosystem Service Assessments: Concepts, Models, and Implementation. *Ambio* **43**:413–433.

Hale, R. L., L. Turnbull, S. Earl, et al. 2014. Sources and Transport of Nitrogen in Arid Urban Watersheds. *Environ. Sci. Technol.* **48**:6211–6219.

Hansen, R., and S. Pauleit. 2014. From Multifunctionality to Multiple Ecosystem Services? A Conceptual Framework for Multifunctionality in Green Infrastructure Planning for Urban Areas. *Ambio* **43**:516–529.

Hartig, T., and P. H. Kahn. 2016. Living in Cities, Naturally. *Science* **352**:938–940.

Hartig, T., R. Mitchell, S. de Vries, and H. Frumkin. 2014. Nature and Health. *Annu. Rev. Public Health* **35**:207–228.

Hope, D., C. Gries, W. X. Zhu, et al. 2003. Socioeconomics Drive Urban Plant Diversity. *PNAS* **100**:8788–8792.

IPCC. 2012. Managing the Risks of Extreme Events and Disasters to Advance Climate Change Adaptation. A Special Report of Working Groups I and II of the Intergovernmental Panel on Climate Change, C. B. Field et al., series ed. Cambridge: Cambridge Univ. Press.

Kareiva, P., S. Watts, R. I. McDonald, and T. Boucher. 2007. Domesticated Nature: Shaping Landscapes and Ecosystems for Human Welfare. *Science* **316**:1866–1869.

Kaushal, S. S., P. M. Groffman, L. E. Band, et al. 2008. Interaction between Urbanization and Climate Variability Amplifies Watershed Nitrate Export in Maryland. *Environ. Sci. Technol.* **42**:5872–5878.

Krasny, M. E., and K. G. Tidball. 2012. Civic Ecology: A Pathway for Earth Stewardship in Cities. *Front. Ecol. Environ.* **10**:267–273.

Larson, E. K., S. Earl, E. M. Hagen, et al. 2013. Beyond Restoration and into Design: Hydrologic Alterations in Aridland Cities. In: Resilience in Ecology and Urban Design, ed. S. T. A. Pickett et al., pp. 183–210. Dordrecht: Springer Science+Business Media.

Li, D., and W. C. Sullivan. 2016. Impact of Views to School Landscapes on Recovery from Stress and Mental Fatigue. *Landscape Urban Plann.* **148**:149–158.

Louv, R. 2008. Last Child in the Woods: Saving Our Children from Nature Deficit Disorder. Chapel Hill, NC: Algonquin Books.

Mansur, A. V., E. S. Brondizio, S. Roy, et al. 2016. An Assessment of Urban Vulnerability in the Amazon Delta and Estuary: A Multi-Criterion Index of Flood Exposure, Socio-Economic Conditions and Infrastructure. *Sustain. Sci.* **11**:625–643.

McHale, M. R., D. N. Bunn, S. T. A. Pickett, and W. Twine. 2013. Urban Ecology in a Developing World: Why Advanced Socioecological Theory Needs Africa. *Front. Ecol. Environ.* **11**:556–564.

McHale, M. R., S. T. A. Pickett, O. Barbosa, et al. 2015. The New Global Urban Realm: Complex, Connected, Diffuse, and Diverse Social-Ecological Systems. *Sustainability* **7**:5211–5240.

McPhearson, T., E. Andersson, T. Elmqvist, and N. Frantzeskaki. 2015. Resilience of and through Urban Ecosystem Services. *Ecosyst. Serv.* **12**:152–156.

McPhearson, T., S. T. A. Pickett, N. B. Grimm, et al. 2016. Advancing Urban Ecology toward a Science of Cities. *Bioscience* **66**:198–212.

Miller, T. R., B. A. Minteer, and L.-C. Malan. 2011. The New Conservation Debate: The View from Practical Ethics. *Biol. Conserv.* **144**:948–957.

Milly, P. C. D., J. L. Betancourt, M. Falkenmark, et al. 2008. Stationarity Is Dead: Whither Water Management? *Science* **319**:573–574.

Munich RE. 2017. Natural Catastrophes: Analyses, Assessments, Positions. https://www.munichre.com/topics-online/en/2017/topics-geo/overview-natural-catastrophe-2016.

NRC. 2004. Valuing Ecosystem Services: Toward Better Environmental Decision-Making. Washington, DC: National Research Council, National Academies Press.

Pachauri, R. K., M. R. Allen, V. R. Barros, et al. 2014. Climate Change 2014: Synthesis Report. Contribution of Working Groups I, II and III to the Fifth Assessment Report of the Intergovernmental Panel on Climate Change. Cambridge: Cambridge Univ. Press.

Palta, M. M., N. B. Grimm, and P. M. Groffman. 2017. "Accidental" Urban Wetlands: Ecosystem Functions in Unexpected Places. *Front. Ecol. Environ.* **15**:248–256.

Park, J., T. P. Seager, P. S. C. Rao, M. Convertino, and I. Linkov. 2013. Integrating Risk and Resilience Approaches to Catastrophe Management in Engineering Systems. *Risk Anal.* **33**:356–367.

Pataki, D. E., M. M. Carreiro, J. Cherrier, et al. 2011. Coupling Biogeochemical Cycles in Urban Environments: Ecosystem Services, Green Solutions, and Misconceptions. *Front. Ecol. Environ.* **9**:27–36.

Pincetl, S. 2015. Cities as Novel Biomes: Recognizing Urban Ecosystem Services as Anthropogenic. *Front. Ecol. Evol.* **3**:1–5.

Ramaswami, A., A. G. Russell, P. J. Culligan, K. R. Sharma, and E. Kumar. 2016. Meta-Principles for Developing Smart, Sustainable, and Healthy Cities. *Science* **352**:940–943.

Redman, C. L. 1999. Human Impact on Ancient Environments. Tucson: Univ. of Arizona Press.

Redman, C. L., and T. R. Miller. 2016. The Technosphere and Earth Stewardship. In: Earth Stewardship, vol. 2, ed. R. Rozzi et al., pp. 269–279. Cham: Springer International Switzerland.

Reed, A. J., M. E. Mann, K. A. Emanuel, et al. 2015. Increased Threat of Tropical Cyclones and Coastal Flooding to New York City During the Anthropogenic Era. *PNAS* **112**:12610–12615.

Rees, W., and M. Wackernagel. 1996. Urban Ecological Footprints: Why Cities Cannot Be Sustainable and Why They Are a Key to Sustainability. *Environ. Impact Assess. Rev.* **16**:223–248.

Reyers, B., R. Biggs, G. S. Cumming, et al. 2013. Getting the Measure of Ecosystem Services: A Social–Ecological Approach. *Front. Ecol. Environ.* **11**:268–273.

Ripplinger, J., J. Franklin, and S. L. Collins. 2016. When the Economic Engine Stalls: A Multi-Scale Comparison of Vegetation Dynamics in Pre- and Post-Recession Phoenix, Arizona, USA. *Landscape Urban Plann.* **153**:140–148.

Romero Lankao, P., and H. Qin. 2011. Conceptualizing Urban Vulnerability to Global Climate and Environmental Change. *Curr. Opin. Environ. Sustain.* **3**:142–149.

Rosenzweig, C., W. D. Solecki, R. Blake, et al. 2011. Developing Coastal Adaptation to Climate Change in the New York City Infrastructure-Shed: Process, Approach, Tools, and Strategies. *Clim. Change* **106**:93–127.

Ross, M. R. V., E. S. Bernhardt, M. W. Doyle, and J. B. Heffernan. 2015. Designer Ecosystems: Incorporating Design Approaches into Applied Ecology. *Annu. Rev. Environ. Resour.* **40**:419–443.

Roy, A. 2009. The 21st-Century Metropolis: New Geographies of Theory. *Reg. Stud.* **43**:819–830.

Royal Society. 2014. Resilience to Extreme Weather. https://royalsociety.org/topics-policy/projects/resilience-extreme-weather/. (accessed Oct. 6, 2017).

Sanchez, C. A., D. L. Childers, L. Turnbull, R. F. Upham, and N. Weller. 2016. Aridland Constructed Treatment Wetlands II: Plant Mediation of Surface Hydrology Enhances Nitrogen Removal. *Ecol. Eng.* **97**:658–665.

Schäffler, A., and M. Swilling. 2013. Valuing Green Infrastructure in an Urban Environment under Pressure: The Johannesburg Case. *Ecol. Econ.* **86**:246–257.

Schindler, S. 2017. Towards a Paradigm of Southern Urbanism. *City* 21:47–64.

Schwarz, K., M. Fragkias, C. G. Boone, et al. 2015. Trees Grow on Money: Urban Tree Canopy Cover and Environmental Justice. *PLoS One* 10:e0122051.

Seto, K. C., B. Guneralp, and L. R. Hutyra. 2012. Global Forecasts of Urban Expansion to 2030 and Direct Impacts on Biodiversity and Carbon Pools. *PNAS* 109:16083–16088.

Shanahan, D. F., R. A. Fuller, R. Bush, B. B. Lin, and K. J. Gaston. 2015. The Health Benefits of Urban Nature: How Much Do We Need? *Bioscience* 65:476–485.

Silver, J. 2014. Incremental Infrastructures: Material Improvisation and Social Collaboration across Post-Colonial Accra. *Urban Geogr.* 35:788–804.

Stott, I., M. Soga, R. Inger, and K. J. Gaston. 2015. Land Sparing Is Crucial for Urban Ecosystem Services. *Front. Ecol. Environ.* 13:387–393.

Swilling, M., and E. Annecke. 2012. Just Transitions: Explorations of Sustainability in an Unfair World. New York: United Nations Univ. Press.

Tanner, C. J., F. R. Adler, N. B. Grimm, et al. 2014. Urban Ecology: Advancing Science and Society. *Front. Ecol. Environ.* 12:574–581.

Tomscha, S. A., and S. E. Gergel. 2016. Ecosystem Service Trade-Offs and Synergies Misunderstood without Landscape History. *Ecol. Soc.* 21:art43.

Tzoulas, K., K. Korpela, S. Venn, et al. 2007. Promoting Ecosystem and Human Health in Urban Areas Using Green Infrastructure: A Literature Review. *Landscape Urban Plann.* 81:167–178.

UN. 2014. World Urbanization Prospects 2014: Highlights. United Nations Publications. https://esa.un.org/unpd/wup/publications/files/wup2014-highlights.Pdf. (accessed Sept. 29, 2017).

Vitousek, P. M., H. A. Mooney, J. Lubchenco, and J. M. Melillo. 1997. Human Domination of Earth's Ecosystems. *Science* 277:494–499.

Walker, B., C. S. Holling, S. R. Carpenter, and A. Kinzig. 2004. Resilience, Adaptability and Transformability in Social–Ecological Systems. *Ecol. Soc.* 9:5.

Walsh, C. J., T. D. Fletcher, and M. J. Burns. 2012. Urban Stormwater Runoff: A New Class of Environmental Flow Problem. *PLoS One* 7:10.1371/journal.pone.0045814.

Walsh, C. J., T. D. Fletcher, and A. R. Ladson. 2005. Stream Restoration in Urban Catchments through Redesigning Stormwater Systems: Looking to the Catchment to Save the Stream. *J. N. Am. Benthol. Soc.* 24:690–705.

Wigginton, N. S., J. Fahrenkamp-Uppenbrink, B. Wible, and D. Malakoff. 2016. Cities Are the Future. *Science* 352:904–905.

Zalasiewicz, J., M. Williams, A. Haywood, and M. Ellis. 2011. The Anthropocene: A New Epoch of Geological Time? *Phil. Trans. R. Soc. A* 369:835–841.

Zeiderman, A. 2016. Endangered City: The Politics of Security and Risk in Bogotá. Durham, NC: Duke Univ. Press.

7

Urban Environments and Environmentalisms

Xuemei Bai, Eduardo S. Brondizio, Robert D. Bullard,
Gareth A. S. Edwards, Nancy B. Grimm,
Anna Lora-Wainwright, Begüm Özkaynak, and Seth Schindler

Abstract

Within academia, professional practice, and stakeholder groups concerned with environmental issues, urban environment carries many meanings. This chapter demonstrates the framing of environmental problems, especially as concerns cities, is driven by different normative and theoretical positions. The various social, economic, and political contexts in place play a strong role in shaping the perception of how the problems are conceived, how they gain support, and who will be involved. Often, common dichotomized perspectives underpin the conceptual and analytical framing used to examine urban socioenvironmental problems. To advance both future research and practice, this chapter argues that a more inclusive definition of the urban environment is needed and proposes a broad and inclusive framing that recognizes that these different, seemingly contradicting views actually reflect the various aspects of its multifaceted nature.

What Are Urban Environments and What Is Urban Environmentalism?

The urban environment means different things to different academic disciplines, professional practices, and stakeholders. Conservation biologists and some urban ecologists are predominantly concerned with the urban environment that provides habitat for flora and fauna. For urban environmental managers, air and water pollution, flooding, and other hazards often constitute the

Group photos (top left to bottom right) Xuemei Bai, Eduardo Brondizio and Xuemei Bai, Victoria Reyes-Garcia and Nancy Grimm, Robert Bullard, Gareth Edwards, Anna Lora-Wainwright, Nancy Grimm, Eduardo Brondizio, Begüm Özkaynak, Seth Schindler, Robert Bullard, Xuemei Bai, Eduardo Brondizio and Xuemei Bai, Begüm Özkaynak, Nancy Grimm, Gareth Edwards, Robert Bullard, Anna Lora-Wainwright, Eduardo Brondizio, Nancy Grimm and Seth Schindler, Xuemei Bai primary concern. For slum dwellers in developing cities, the urban environment often entails the provision of fundamental needs in their immediate living environment: access to shelter, clean water, food sources, and sanitation. For those concerned about resource consumption or climate change mitigation, focus is on the extent of resources consumed or the amount of greenhouse gases emitted by the city.

We adopt a broad and inclusive framing of the urban environment, recognizing that these very different, seemingly contradicting views reflect different aspects of its multifaceted nature. Our conceptualization of the urban environment includes four dimensions:

1. The nonhuman nature of cities, such as parks, green areas, and urban biodiversity
2. The level of provisioning of and access to basic services, such as clean water or sanitation
3. Protection from hazards or adverse ambient conditions, such as flooding and air and water pollution
4. Impacts beyond cities such as extraction of natural resources for construction and emissions of pollutants and greenhouse gas

These dimensions differ within and across cities, often reflecting political, economic, and demographic hierarchies as well as differences in levels of "development" or income distribution (Bai and Imura 2000; O'Connor et al. 2001; UNDP 2013). Because of this, research that focuses on a single aspect is unlikely to generate sufficient guidelines for planners, decision makers, and the citizenry. Even within a city, the composition and relative importance of these issues change over time, reflecting dynamic internal and external social, economic, and environmental factors (Bai 2003). The mode and intensity of interactions with external regions also vary across cities (Guedes et al. 2009; Kennedy et al. 2007; Metson et al. 2015). Many fluxes of resources, pollutants, and other materials across city boundaries are much larger than that within their own boundaries (Kaye et al. 2006; Metson et al. 2012). Thus, the environmental impacts as well as the "responsibility" of cities (e.g., concern for their supply chains) extend far beyond their physical or administrative boundaries. Hence, when assessing the overall impacts of urban environments and policies, we need to think regionally and globally as well as locally.

Urban Environmentalism

Under an inclusive framing of the urban environment, almost all cities face one or more environmental "problems." Once an issue is identified as an environmental problem, the issue becomes one of "environmental*ism*," implying some active engagement with the environment for the purpose of solving said problem. Urban environmentalism can be triggered by certain environmental disasters and associated health impacts, as demonstrated by the Minamata disease in Japan, smog pollution in China, and municipal waste disposal in areas inhabited by people of color in Houston (Bullard 2000). Urban environmentalism is also found in "urban greening" trends in many cities of the Global North, where communities and local governments work to increase the number of trees or parks (Grimm and Schindler, this volume). To some extent, dimensions of urban environmentalism mirror those of the urban environment: concern for the conservation of nonhuman nature; activism to ensure that the basic needs and rights of people to clean water, clean air, and sanitation are met; interventions designed to solve pollution problems and expose unequal exposure to hazards or pollution; and cross-boundary concerns that reflect impacts on nature or communities in external regions due to a city's activities.

Different underlying perspectives or frames determine whether environmental issues are identified as problems, and the ways in which they are ultimately addressed. These sometimes can seem to be contradictory. For instance, the nature conservationist's agenda to save large tracts for parks that can support species diversity may conflict with the environmental justice activist's concern for access to affordable housing in hazard-free areas. Even if these concerns are not in direct tension, there is often competition for funds to support implementation of different agendas. In terms of cross-boundary concerns, a range of scales should be considered. The export of air pollutants from fast-growing urban areas, for example, impacts regional forest productivity (Innes and Haron 2000) which. in turn, may affect livelihoods. In *Environmentalism of the Poor*, Martinez-Alier (2002) highlights the differences between environmental concern as a postmaterialist luxury in the Global North and as fundamental to livelihoods in the Global South. Indeed, consumption patterns in cities of the Global North have driven resource extraction in the Global South, which has effectively transferred a range of environmental problems from North to South (Burger et al. 2012; Deutsch et al. 2013).

Given the linkage between different perspectives and framings of urban environmental issues and the various approaches to environmentalism, we seek two objectives in this chapter. First, we present and review five dominant framings of the urban environment that exist in the literature and examine commonly used dichotomies that influence these framings: urban–rural, Global North–Global South, the brown–green agendas, environmental commons, and private common property rights. Second, based on the assumption that a transformative change is needed for urban futures that are sustainable, diverse, and just, we explore whether it is possible to address multiple concerns through plural framings and environmentalisms, and to understand the role that urban constituents and global–local interactions might play in bringing about such desirable change.

Our discussion is organized according to four main sections: In the first, we examine five conceptual and analytical framings of the urban environment. Next, we discuss the way issues of sustainability, justice, and diversity in urban environments are addressed in these alternative framings, and expose the commonalities, intersections, and divergences of different "environmental-isms" as mobilizers. In the third and fourth sections, our focus shifts to tensions, trade-offs, and synergies among different agendas or framings of urban environment. Here we consider the challenges posed by social, economic, and political power structures and legacies that underpin opportunities for transformative change in cities. In conclusion, we highlight the need for collaboration around integrated conceptual framings and analytical tools for reimagining urban futures.

The Urban Environment: Conceptual and Analytical Framings

Before going into the conceptual and analytical framings of the urban environment, it is important to frame the framing; that is, to understand the factors that influence the ways in which issues are framed as environmental problems. Conceptual and analytical framings are socially constructed: exactly who is able to identify and frame a problem or ignore issues considered problematic by others is determined by sociopolitical power structures (Baviskar, this volume). In addition to the rather obvious influence of different theoretical or disciplinary perspectives, influences that stem from underlying value systems, the socioeconomic context, and lived experiences (with the caveat that even in the same urban environment, people can have drastically different lived experiences) play a role in determining what is perceived as the dominant problem and the type of conceptual and analytical framing used to devise a solution. In this discussion, we refer to four common dichotomies that influence conceptual and analytical framings.

First, the *urban–rural framing* is one of the most commonly used dichotomies to analyze the urban environment. From an environmental justice perspective, rural residents criticize city constituencies for overexploiting rural resources while ignoring the needs of rural areas for basic access to services and investments provided by the city (Brondizio 2016). Depending on the region of the world, rural and urban economies often compete, each claiming to be the economic engines of entire regions. The difference between rural and urban areas is, however, not as clear-cut as is often asserted and is better conceived as a spectrum rather than as a dichotomy. It is also important to consider how the concept of "rural" and "urban" has changed over time as well as differs across regions. For instance, rural areas were once defined mainly by agriculture and subsistence. However, an increasing number of so-called "rural" areas (defined in terms of administrative jurisdictions) are now industrialized or dedicated to resource extraction, or are sustained by service industries that are embedded in "urban" networks. Conversely, some urban areas currently encompass practices such as allotment farming, which previously may have been regarded as the mainstay of rural areas. Finally, trends of suburbanization and ex-urbanization prevalent in many areas extend "urban" processes (e.g., land conversion, hydrologic modification, and resource pressures) to much larger areas than the original city.

Second, is the *Global North versus Global South*. In its final report, the Independent Commission on International Development Issues (Brandt 1980) highlighted inequality as one of the primary challenges for humanity. It noted that dividing the world into two groups was a simplification, but maintained that "in general terms, and although neither is a permanent grouping, North and South are broadly synonymous with 'rich' and 'poor,' 'developed' and 'developing' " (Brandt 1980:31). Although dichotomizing North and South is a gross simplification, this framing does draw attention to the persistence of fundamentally unequal power relations and living standards, as well as inequalities with regard to knowledge production. Most urban theory has been developed in cities from the Global North (Roy 2009), and thus it is imperative to think critically before models from the North are applied to the South. Counter-trends and diversity within each of these categories must not also be ignored. For instance, many cities and regions in rapidly developing countries like China are difficult to classify according to this binary system, showing well that poverty is not only a feature of the so-called Global South, nor is wealth only a feature of the Global North. As discussed below and elsewhere in this volume (see Baviskar as well as Grimm and Schindler, this volume), the environmental justice literature shows that inequality within cities is as much a feature of the Global South as in many regions of the Global North. Thus, it is analytically more useful to think about a spectrum that stretches between Global North and Global South than a clear-cut dichotomy. Such a spectrum takes into account diversity both between and within cities.

Third, we must consider the *brown versus the green agenda* in cities. In the urban environment, the so-called brown agenda addresses issues such as access to housing, safe water, sanitation, and remediation of air and water pollution, whereas the green agenda focuses on promoting urban green spaces, reducing greenhouse gas emissions, planting trees, and so forth. The conflicts or trade-offs between these two agendas can be very real, especially when competing for political attention or limited budgets, as it is often the *power dynamic* that identifies which agendas/concerns get addressed and supported. It is also important to recognize that synergy is possible between these two agendas: air pollution control and reduced greenhouse gas emission may, for example, coexist in cities, contributing to both the brown and green agendas. Thus, exploring such co-benefits is sometimes more important than prioritization. Different power bases of these agendas, governance, and institutional structures, however, often hamper the real synergies from being realized.

Fourth is the *private rights versus environmental commons*. City landscapes are marked by the intersection of private and public property rights, yet function as a result of the flow of common-pool resources and public goods. Instead of a clear-cut distinction in property systems, urban landscapes represent complex matrices and bundles of rights, such as those which define spaces or who has the right to access, manage, withdraw, exclude, and/or alienate property and benefit from different types of resources. The spatial allocation of rights and access to common and public goods is mediated by social and political structures and is uneven across class, color, origin, and ethnicity lines. Yet while the distribution of harms may be associated with social inequalities, it also transcends property boundaries with consequences for urban collectives as a whole. For instance, lack of access to sanitation in one part of the city affects waterways that are used by diverse social groups. While still common in urban analysis and planning, simple typologies of property systems and rights are insufficient to interpret the collective action dilemmas posed by urban environmental problems.

Five Conceptual and Analytical Framings of the Urban Environment

Conceptual and analytical frameworks provide a common structure and language to support the analysis of a given phenomenon and/or problem. Conceptual frameworks can be considered as metatheoretical tools: they explicitly identify relationships and directionality between components of a phenomenon without necessarily posing a predefined causality between them (Figure 7.1). Conceptual and analytical frameworks can be organized at

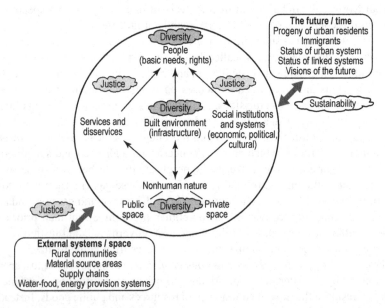

Figure 7.1 Depiction of frameworks and how they intersect with justice, diversity, and sustainability.

different levels of generality, from showing broad components and relationships underlying a phenomenon (e.g., land use and cover change) to describing more specific processes (e.g., land-use intensification). Conceptual and analytical frameworks can be schematic (e.g., systems-based flow charts such as the Millennium Assessment framework) or narrative (e.g., a proposition defining a political ecological approach). Over the past two decades, they have been instrumental as tools for interdisciplinary collaboration around complex and cross-scale problems. Here, we propose and review five groups of conceptual and analytical frameworks used to examine different, but interrelated types of urban processes and problems.

Structure of the Urban System: The SETS Framework

To which urban environment does urban environmentalism apply? One framing of the urban environment derives from the ecosystem concept (Pickett and Cadenasso 2002) but is extensively modified to encompass the urban system's components of people, their varied systems, and the built environment: the SETS (social-ecological-technological systems) framework (Grimm et al. 2016; McPhearson et al. 2016; Grimm and Schindler, this volume). In SETS, cities are systems with social, ecological, and technological components that interact within a boundary as well as across this boundary with external systems (for an earlier manifestation of this framing, see Wang and Ma 1984). In SETS, the boundary must be defined, although it is less important where it is located; ecosystem scientists often locate boundaries for convenience of measurement. For example, in the United States, a "metropolitan statistical area" that includes the inner city, suburbs, and exurban areas may delineate an appropriate boundary for quantifying flows of materials or movements of people because of the availability of data for this unit, but it may not necessarily match the boundaries of biophysical systems upon which it depends, such as a watershed.

The ecological components of cities (i.e., urban nature) provide both benefits and detriments (services and disservices) to people in cities, often reflecting social differences in relative economic and political power. Built infrastructural components of cities are designed and managed to provide specific services. Both urban nature and these built elements are enabled, designed, and managed by people via their political institutions, economic systems, and so forth. Some urban nature and built elements exist in public spaces (e.g., parks, rivers, lakes, water treatment facilities), while others are in private spaces. Thus, the SETS framing of urban environment encompasses green-brown, private-public dichotomies. It also allows the analysis of inequitable distributions of services and disservices as well as differential access to nature and the built environment, which are primary concerns in urban environmental justice. Interactions with external systems are a defining feature of urban SETS. Some of these interactions produce negative impacts on those linked, external systems (e.g., resource extraction from or pollution to distant areas). Thus environmental justice for cities, understood through the framing of SETS, might extend to intersystem justice because it incorporates these external interactions.

The SETS framing is flexible and allows us to develop a new conception of the urban environment, since by definition the environment incorporates the service-delivering components of cities, whether natural or built. Potentially, this frame permits us to diagnose environmental inequity, alter thinking about the environment, and explore innovations that represent potential solutions to environmental problems and thus contribute to a transition to sustainability.

Understanding External Dependency and Impact: Urban Metabolism

There is a long history of conceptualizing cities in terms of their "metabolism," but current scholarship owes a significant debt to Abel Wolman's (1965) framing. Although Wolman inspired a proliferation of scholarship in a range of fields that conceived of cities as having metabolism analogous to an organism, considerable divergence and scant communication among disciplines was the result (Bai et al. 2016; Castán Broto et al. 2012). Here, we identify two main approaches to the concept of metabolism that emerged independently and without any notable dialogue.

Industrial ecologists have used the term "urban metabolism" (or "social metabolism") to describe input, distribution, and output in terms of energy and materials that sustain a human settlement. This approach focuses on the circulation and absorption of some resources into the built environment as "stocks" and the transformation of others into waste. Terradas (2001) notes, for instance, that in Europe, a city with one million inhabitants requires a daily input of 11,500 tons of fuel, 320,000 tons of water, 31,000 tons of oxygen, and 2000 tons of food. The same city will also produce 300,000 tons of wastewater, 25,000 tons of CO_2, and 1600 tons of solid waste as output. These figures tend to increase along with increasing levels of per capita consumption. In addition to quantifying the flows, research has focused on understanding the socio-economic determinants of the flows and their distribution (Bai 2016). This framing, therefore, is useful for those seeking to situate cities within larger ecosystems, as well as city-level policy makers trying to reduce resource use within cities, understand structural determinants, and promote a sustainable mode of living.

Urban political ecology conceives of urban metabolism in a similar way but goes beyond an accounting of stocks and flows to ask why a city draws in resources and expels wastes as it does. Grounded in critical theory, it seeks to transgress the rigid dualism that separates society and nature, and show how unequal power relations shape urban resource distribution. Urban political ecology frames cities as metabolisms through which flows of resources circulate to create cities as "socionatural" entities that are characterized by multiple and often contradictory relationships; these shape their social, economic, political, and infrastructural form (Swyngedouw 1996). Although these metabolisms are contingent on these relationships, they are often configured with the objective of accumulating capital, and this results in poor social and ecological outcomes (Gandy 2004). There is a clear concern with issues of justice, which has also been influenced by the pioneering work of environmental justice scholars (e.g., Bullard 1990; Schlosberg 2007). In contrast to the urban metabolism analysis by industrial ecologists, whose scale of analysis is typically the city, the suburbs, and more recently the household, much of this work is predominantly situated at the microscale and focuses on particular neighborhoods or political conflicts (Demaria and Schindler 2016; McFarlane 2013). Recently, some studies have shown that intercity distribution of the flows and their determining factors (Bai 2016; Lin et al. 2013) can be linked to diversity and justice concerns.

Scholars have sought to bridge these fields and expand the scope of urban metabolism research in an effort to incorporate insights from ecology. Newell and Cousins (2014) refer to the "metabolism of the urban ecosystem," as a means of identifying the points of intersection between the existing approaches to the concept of metabolism in cities and the relationship with the broader world. Similarly, Bai (2016) argues for the reconciliation of industrial ecology and urban ecology: the former benefits from being closely linked to conceptual advances within urban ecosystem studies, whereas the latter needs explicitly to include anthropogenic materials and energy flows. The large amount of empirical evidence that results can reveal key ecosystem characteristics of cities. Perhaps an underlying motivation for whole-system metabolism studies of industrial ecology and related mass-balance studies in urban ecosystem ecology is to understand how dependent the city is on external ecosystems; a consequent hypothesis holds that reducing this dependence will promote sustainability (e.g., Metson et al. 2015).

Industrial ecology and urban political ecology are in a state of flux with regard to the conceptualization of urban metabolism. Nonetheless, there is increasing willingness among scholars to expand the scope of their analysis in an effort to represent the complexity of cities as human-dominated, complex systems. To be successful, these hybrid approaches should include a focus on ecosystems as well as the contingent nature of urban metabolisms. For instance, solid waste can be interred in a landfill or incinerated, and the ways in which material throughput is managed will have more or less sustainable and socially just outcomes. Rather than a "natural" phenomenon, the configuration of an urban metabolism is unpredictable and commonly contested. Thus, metabolism can signal both an objective relationship between a city and its surroundings and be used as a heuristic device to draw attention to conflicts over access to resources and services. Both perspectives should inform analyses of the urban environment, which should be multiscalar and attend to regional, citywide (eco)systems as well as microscale events.

136 X. Bai et al.

Dynamic, Evolutionary, and Complex Urban Environments

The phenomenon of urbanization represents one of the biggest social transformations in human history (Bai et al. 2014) and is a dominant driver of global environmental change. Grimm et al. (2008) characterize current and future cities as protagonists in the Anthropocene, a human-dominated geological epoch where dynamism, nonlinear change, uncertainty, and complexity prevail (Bai et al. 2016; Brondizio et al. 2016). Thus, as many have argued, the theoretical principles of complexity science are most appropriate for understanding systems in this context.

As human-dominated, complex, dynamic, evolving systems (Alberti et al. 2008; Bai 2003, 2016; Batty 2007, 2016; Grimm et al. 2013), cities exhibit alternative stable states and emergent properties; they are not predictable in a deterministic sense and carry the potential for multiple possible futures (Bai et al. 2016). Each trajectory is somehow shaped by a unique combination of endogenous and exogenous forces, reflecting both pressures from outside the system as well as responses from within the city (Bai 2003). Thus, although the study of patterns and trends of urbanization of the past may generate a general set of expectations, history cannot be used in a predictive sense given the likelihood of nonlinear and abrupt changes. For instance, Bai and Imura (2000) have shown how the diverse environmental profiles and trajectories of Asian cities, as a case in point, are linked to their development stages, and how an evolutionary perspective can help explain the commonalities and differences.

Important aspects of urban dynamism tell the story of urbanization today: as mass movements of the human population, often the result of uneven distribution of opportunities and beneficial or detrimental environmental conditions; as the phenomenon of shrinking cities in the Global North and expanding slums in the Global South, and diverse and novel urban forms, with expanding suburban and exurban development. Deterioration of rural environments, often a driver of migration to cities, is a concern that falls under the diversity dimension of environmentalism. Shrinking cities embody the reverse trend: loss of population from former centers of industry and/or commerce, often owing to economic globalization, can present opportunities for environmental improvement (Haase 2008). The consistency of short- and long-term goals in urban contexts could also be assessed from an urban dynamism point of view (Zetter and Hassan 2002). As LeGates and Stout (2003:228) note:

> What looks like a good solution to urban environment to one generation, such as a massive highway construction program to solve transportation problems, building identical suburban tract housing to deal with the housing crisis, or creating a nuclear power grid to meet urban energy needs, may not look so to the next generation.

Viewing cities as complex, adaptive, evolving systems acknowledges a range of possible futures and, in this sense, offers a framework that strongly incorporates sustainability. From this perspective, it is undoubtedly possible for a city to be more or less sustainable, and managing the resilience of urban systems is one way to guide transitions toward more desirable futures. Resilience is a system property that reflects the capacity to undergo shocks without substantially changing structure, function, feedback relationships, or its fundamental identity (i.e., a resilient city would still be recognized as a city after a major disturbance) (Walker et al. 2004). Awareness of this complex, dynamic nature of cities supports the envisioning of plausible (based on current trajectories) and desirable (managed transitions) futures as well as transformative futures (Iwaniec and Wiek 2014), which may result from game-changing drivers, unanticipated shocks, or even deliberate interventions (Bai et al. 2016). At the same time, however, resilience cannot be simply accepted *prima facie* as a "good thing," since resilience could act to sustain undesirable socioenvironmental states (Anderson 2015; Turner 2014).

Environmental Justice

The environmental justice framing starts from the perspective that environmental issues are fundamentally questions related to justice, and thus focuses on how environmental harm (externalities or "bad stuff") and environmental benefits ("good stuff") are distributed through policies and actions. This framing (a) incorporates the principle of the "right" of all individuals to be equally protected from environmental harm, (b) assesses cumulative impact as opposed to assessing the impact of one chemical at a time, (c) shifts burden of proof to polluters who do harm, (d) redresses disproportionate impact through "targeted" action and resources, and (e) has historically adopted a public health model of prevention over a cure.

In the Global North, particularly the United States, where the environmental justice movement emerged through the work of Robert Bullard and collaborators (e.g., Bullard 2000), the environmental justice framing points to issues common to other parts of the world, but more frequently employs a race and class-based lens to point out the injustices of how environmental harms and benefits are distributed. In the Global South, this form of environmentalism tends to focus on social justice, claims to recognition and participation, and efforts to defend indigenous land rights and preserve their livelihoods against mining, dams, land grabs, oil and gas exploitation. The manner by which the environmental justice framing emphasizes social justice concerns (which may be broader than specific urban environmental challenges) means that it aligns well with the livelihood focus of the "environmentalism of the poor" (Guha 2000; Guha and Martinez-Alier 1997; Martinez-Alier 2002). The plight of slum dwellers for clean water (see Baviskar, this volume) illustrates a form of "environmentalism of the poor" as it applies to urban situations. In the end, neither the northern nor southern perspective subscribes to the "traditional" idea that environmentalism is rooted in a romanticized view of nature as something to be enjoyed and protected and separated from humans (Grimm and Schindler, this volume).

It is thus important to avoid simplistic oppositions between northern and southern environmentalisms and to be critical of deploying this dichotomy to imply that these present coherent agendas or social movements. As much as "northern environmentalism" cannot be labeled as driven by affluence or self-indulgence, one should not slip into the trap of suggesting that the poor in the Global South are necessarily more likely to act in pro-environmental ways. In fact, local communities in the North or South have mixed views of development, and they may be victims of pollution as much as they are complicit in it or even perpetrate it themselves. Finally, we should be cognizant of the complex dialogue created as environmental concepts move between academia and activist circles as well as across regions, through multiple networks and learning processes (Martinez-Alier et al. 2014, 2016).

Cities as Solutions

The environmental challenges faced by and created by cities are well known. Still, cities can also be framed as agents of positive change. They are locations of extraordinary social, political, and economic power and, as such, can play a pivotal role in transitioning to more sustainable modes of living. As Rees (1995:42) stated:

> Paradoxically...while there is no hope for the city per se to achieve sustainability independent of its vast and scattered global hinterland, it is in cities that the greatest opportunities exist to make the changes necessary for general sustainability.

In this sense, cities are well-suited focal points for experimentation and operationalization of sustainability (Selman 1996). The notion of cities as loci for sustainability experiments is increasingly being adopted by urban scholars as well as design and engineering professionals (Bai et al. 2010; Bulkeley et al. 2015). New concepts and approaches, for instance, to climate change are well underway (e.g., smart cities, green cities, ecocities, low or zero-carbon cities). The role that cities play in climate change mitigation and adaptation is widely recognized (Bulkeley et al. 2015; Revi et al. 2014; Rosenzweig et al. 2010). Actions at the urban scale by a range of actors might be central to the ability of the world to circumvent the worst effects of climate change and forge productive solutions to the challenges it poses.

This framing needs to take into account the diversity of urban functions as well as social, economic, and political contexts. For instance, social justice and sustainability are affected by the size, population density, and extant infrastructure of cities as well as the type of political organization and level of inequality of its citizens. Although research shows that larger cities drive more innovation (Bettencourt et al. 2007), there is no single optimal urban morphology or population density with regard to outcomes that are both ecologically sustainable and socially just. Some cities show a positive relationship between density and inequality, whereas others with dispersed populations exhibit economic inequality and unevenly distributed environmental hazards. Social and ecological outcomes are determined by many structural factors that are mediated or transformed by urban morphology. Thus, it is difficult, if not impossible, to establish universal prescriptions or normative assumptions with regard to city size and density in pursuit of ecologically sustainable and socially just outcomes.

Intersection or Diversity of Environmentalisms as a Mobilizer

In urban contexts, an important way to generate justice and sustainability is through the interaction and productive engagement of a diversity of *environmentalisms*. To illustrate why this is productive, consider the example raised by Amita Baviskar in Chapter 5 (this volume). Baviskar frames her argument around two central narratives, both of which represent a form of environmentalism located in a particular place: Gurgaon, India. Anuj Gupta stands as a paradigmatic case of the "traditional affluent environmentalism" while Sarita Devi represents the paradigmatic case of Joan Martinez-Alier's "environmentalism of the poor." Baviskar's analysis shows that these two environmentalisms revolve around issues of access to adequate, clean environmental resources; distribution of environmental resources, social and economic power, and political agency; and avoidance of harm.

The environmentalisms presented by Baviskar can be articulated with the environmentalism espoused by the environmental justice movement—from its roots as a mobilizing force against the racist and discriminatory positioning of toxic waste facilities in the southern United States (Bullard 2000:29; 2008:89), to more recent mobilizations that seek to frame climate change as inherently an issue of justice (Schlosberg and Collins 2014). Indeed, one of the key aspects of the environmental justice movement has been its commitment to a plural notion of justice (Schlosberg 2007), one that has recently tended to seek larger normative frameworks in which to situate its analysis, most notably Amartya Sen's "capabilities approach" to justice. The implication of this shift has been an increasing tendency to theorize the "justice" of environmental justice in terms of well-being, an arguable departure from the earlier environmental justice movement, which sought to frame its justice demands in terms of established, liberal notions of justice (Edwards et al. 2015).

In other words, mobilizing the actors' portrait in Baviskar's account requires framings that take into account diversity and justice as intrinsically linked to sustainability in cities. The challenge in this approach is for localized interest groups to cohere into a movement that mobilizes the environmentalisms of both Gupta and Devi in unison, rather than playing them off against each other, as has historically been the case (and, indeed, how Baviskar's analysis seems to suggest the protagonists view their positions). How, for instance, can Gupta's concern for water access or energy security be connected to the concerns of Devi for green spaces, clean air, and landscape aesthetics in a fast-developing urban area?

The Promise and Peril of Data in the Urban Environment

One of the key barriers in achieving a sustainable and just urban environment is the lack of reliable data, which is absent in some instances, and inaccurate or incomplete in others. Although there is a proliferation of schemes and projects that generate data on cities, these efforts typically prioritize standardization over accuracy.

The standardization of inaccurate, incomplete, and incommensurable city-level data has the potential to disadvantage already marginalized populations and neighborhoods. In his now classic research on environmental racism in the United States, Robert Bullard (1990) showed that African-Americans were disproportionately exposed to environmental hazards. Their disproportionate exposure persisted unabated because, prior to the emergence of the environmental justice movement, data that exposed the unequal distribution of environmental hazards were lacking. The inexorable move toward representing cities as "big data" risks cementing similarly unequal urban systems by rendering informal flows and stocks illegal, not to mention people. One of the pillars of the environmental justice movement that emerged in the United States is that the people most exposed to environmental hazards must be able to speak for themselves. Self-representation becomes more challenging as the standardization of city-level data becomes an increasingly technocratic affair.

Global Capital Circulation and Environmental Agendas

Cities are embedded in a global context, connected in various ways to webs of regional and local resources as well as economic flows, power relations, conventions and regulations, population movement, and various types of displacements of pollution and waste. Cities influence and are influenced by these proximate and distant forces nested within a multitude of feedbacks and interactions that ultimately affect specific places and groups of people (Kok et al. 2006).

The actions of global institutions, such as multinationals, intergovernmental agencies, and regional and international bodies, are key drivers at global and regional levels. Multinational corporations relocated production from industrial centers in the Global North to emerging economies, thus giving rise to the so-called "new international division of labor" (Fröbel et al. 1980) as well as a redistribution of environmental impacts (Steffen et al. 2015). This precipitated a crisis in many industrial cities, to which municipal authorities have struggled to respond (Bluestone and Harrison 1982; Schindler 2016), as well as in emerging economies, where cities suddenly faced new and complex challenges with regard to the emergent urban environment. The result has been paradoxical: cities are the key scale at which national and global events unfold (see Brenner 2000), yet global policy frameworks (e.g., Habitat III) and intercity competition make it increasingly difficult to define objectives and set priorities at the local level (Cohen 2004). Overall, local strategies have to be formed and reformed, based on the logic of macro-level factors as to what is, or is not, feasible, and actors' responses and political judgments about which values and interests they most wish to promote.

Mollenkopf (2003), while studying the urban political realm, argues that one should not underestimate the importance of urban politics and community action, and the role of agency should never fade out of the analytical picture. There is a broad literature that stresses the fact that grassroots mobilization has been a crucial factor in the shaping of cities (Esteva and Prakash 2004; Mollenkopf 2003). Of course, the plausibility or success of such reactions relies on the stakes at hand as well as on the social actors' ability to counterbalance pressures from the larger context. In general, stakeholders differ in their agency capability, given local conditions. Whereas some stakeholders may have considerable lobbying power, be well versed in interactive action, and significantly oppose the system, others may be more easily affected by external influences.

Overall, it is possible to argue that national and international forces provide the framework conditions (both to benefit from and to react to) and then localities do something about these systemic impediments: they resist, cooperate, form alliances, adapt. What actually happens is a result of the dialectic of structural change and actors' responses. It is thus important to understand the local dynamics as well. Overall, the nature of the linkage between different constituency interests and the roles that actors play at different scales must be understood for multilevel governance.

In linking the role of agents for change (in particular, the potentials and limits of different constituents in cities) to the questions of sustainability, justice, and diversity, three related areas must be considered:

1. Who defines what change is desirable, and who has the power and knowledge to effect change? This question speaks to the themes of justice and diversity. A related question is: How should the sustainability of these changes be measured? Change happens at different scales, and different constituents may aim to effect change at different scales; this has to do, in part, with the limits (both real and perceived) to their power to effect change. Interventions that may be deemed sustainable or just at one scale may change completely when zooming in or out. Providing clean water to one region may seem just and sustainable,

but it is a question that must be evaluated in the context of the larger socioecological system.

2. Regional variation and interdependence must be taken seriously in urban analysis and planning. The political, legal, social, and environmental opportunities available in one place may be starkly different from those in another. Cultural and moral norms affect the potential of cities and their inhabitants as agents of change, both positively and negatively. The stigma of "rocking the boat," for instance, may work to silence mobilization unless a major critical event pushes people to act. Conversely, a strongly felt sense of equality and rights might push people to ask for better cities to live in, even where laws and regulations themselves have not protected these rights effectively to date.

3. Intraregional and intra-urban variation is equally important in urban analysis and planning. Regional urban systems are often organized by economic, political, and demographic hierarchies, marked by differences in services, resources, and quality of life. Likewise, different social groups are unevenly positioned to benefit from opportunities for gain and may experience different levels of environmental harm (see, e.g., Baviskar, this volume; Bullard 1990).

For constituencies that are typically endowed with little or no power, how might they be able to speak up and gain attention, or make their approach and demands heard and realized? The answers may sound idealistic and unrealistic, but they are worth considering as ultimate aims. The most important aim resonates in the 17 principles of environmental justice, established at the First National People of Color Environmental Leadership Summit (1991): *the people most affected must be able to speak for themselves* [paraphrased]. To do so, they must be endowed with better tools, better networks, and more power to be heard. If formal institutions do little to assist them, alternatives must be considered. For instance, the Environmental Justice Atlas project, directed by Joan Martinez-Alier, documents and maps the global distribution of environmental conflicts. Among other things, it shows how the terms that are used by local communities and campaigners to highlight the conflicts or injustices they face may travel to other localities, be adopted by academics, or vice versa. It also demonstrates how academic concepts may be embraced as tools by campaigners to mobilize knowledge and gain recognition (Martinez-Alier et al. 2014, 2016).

Another way of mobilizing knowledge is to enable people to collect their own data and, where necessary, to challenge scientific parameters. This citizen science represents a growing trend: affected communities gather evidence to challenge data collected by industry or local governments (San Sebastián and Hurtig 2005). Louisiana's bucket brigades provide an excellent example: they adopted a simple tool to measure air pollution among fenceline communities exposed to contamination and challenged the scant and inaccurate data collected by industries and the environmental protection agency (see Allen 2003; Lerner and Bullard 2006). The strategy has inspired similar efforts in other parts of the world. For instance in China, a Beijing-based NGO affixed air pollution monitors on kites. Similar crowd-sourcing efforts have resulted in the creation and constant updating of maps of pollution to name and shame offenders. More often than not, their claims to legitimacy are based more on moral grounds than on scientific or legal arguments.

Of course, there are difficulties faced in the model of popular epidemiology, which deserves particular attention. In their study of the childhood leukemia cluster in Woburn, Massachusetts, Phil Brown and Edwin Mikkelsen (1997) defied classic epidemiology methods and proposed mapping illness that appeared in very small clusters—considered statistically irrelevant by classic epidemiology—onto local pollution. This approach seems very similar to Robert Bullard's mapping of the overlaps between exposure to environmental harm and communities of color (Bullard 1990). The difference is that whereas the latter is difficult to deny, the former is still subject to scientific scrutiny. After all, just because cancer and pollution coexist in a small cluster does not provide incontrovertible proof, by epidemiological standards, that the cause of illness is pollution. The same challenge affects China's cancer villages. In some cases, even if the entire village population died of cancer, it would still not be statistically significant because data need to be aggregated at a larger scale, and at that scale, these small samples become invisible (Lora-Wainwright and Chen 2013).

Another important obstacle to communities acting as drivers of change is this: not only may their capacity and tools be limited, they may have come to accept their position since they have experienced the same sort of injustice for decades or more. There may be legal or political grounds for complaint, but unless communities feel empowered and entitled to speak up, they most likely will not do so. For several reasons, they may learn to normalize and accept pollution and injustice, and only challenge them on a very low scale, without demanding deeper change (Lora-Wainwright 2013). The only way for such communities to speak up is if they develop a stronger sense of their rights or have better access to tools to measure their exposure to harm; if they came to see their situation not as an accident of fate but as systemic injustice; and if they became connected to activists and campaigners who have faced similar challenges elsewhere.

How can we bring about change across places and communities? Should we adopt a universalist frame, argue that everyone should have the same rights to live in a clean environment, and expect and demand the same outcomes? The answer should be a strong, resounding, idealistic yes, albeit with a realistic sense that we need to adapt to local conditions and that the timelines for change will vary in different places. For instance, global pressure needs to be exerted on governments as they oversee communities exposed to lead contamination; compliance needs to adhere to a global standard of what is an acceptable level to ensure health. This approach should help campaigners, even though immediate compliance may not always be forthcoming.

To ensure that diverse perspectives are heard, promising tools for urban environmentalists include participatory planning, participatory scenario framing, and other similar approaches. Participatory workshops that bring together different stakeholders to consider options for a neighborhood, a district, a city, or a region offer a means to achieve consensus or, at least, to understand the options, trade-offs, and potential involved. The coproduction of future scenarios by academics, governmental officials, planners, and various groups of people, for example, is a powerful way to convert simple lists of aspirations or (more commonly) problems to be solved into living narratives that illuminate a suite of pathways to desired futures (Iwaniec and Wiek 2014; Özkaynak 2008; Wiek and Iwaniec 2014).

Final Remarks

How are different framings of environmental problems driven by differences in normative and theoretical positions? How might more inclusive framings enable more societally relevant and impactful research and more concerted action/practice?

Our discussion has demonstrated that in cities, the framing of environmental problems is indeed driven by different normative and theoretical positions. In addition, different social, economic, and political contexts play a strong role in shaping the perception of what the problems are and which ones may gain support and for whom. Some common dichotomized perspectives often underlie and influence the conceptual and analytical framing used to examine urban socioenvironmental problems.

However, frequently overlooked are the following: (a) the diversity of urban environmental agendas; (b) the tensions, trade-offs, or synergies among different framings of urban environment and development issues; and (c) the social, economic, and political power structures and legacies which underpin them. At the same time, commonalities or synergies across these agendas are even less commonly acknowledged or applied in practice. We note that there is a lack of fundamentally integrated conceptual framing and analytical tools to understand the complexity of cities and their place within various types of regional and global networks. We do not yet know how to study the complex flows of capital, technology, ideas, and environmental goods and services within and across city and regional boundaries.

More and better data on cities are urgently needed, as is a fundamental reexamination of the nature and method of data collection and analysis. What will be the unit of analysis and data collection in cities? What is left out by conventional urban data collection practices? The data that provide support to legitimize action are often aggregated in a way that masks the vast inequality within urban areas. Is citizen science the solution for more accurate and representative data collection? Or, should the recognition and legitimacy of such an approach, which can be influenced by political power structures and agendas, come first?

In response to the issue of inclusive framings, we propose that the following three approaches may be more relevant in conceptualizing the urban environment, and thus more capable of advancing both research and practice agendas:

1. Utilize a more *inclusive definition* of the urban environment—one that merges justice with the green and brown agendas. We propose a conceptualization that includes four main components: (a) nature in cities, (b) access to basic services, (c) protection from hazard and adverse ambient environmental quality, and (d) cross-boundary influences.
2. Recognize cities as complex, dynamic, and evolving systems with nonlinear trajectories, where various internal and external forces shape multiple possible futures. Since cities are an integrated part of global capital and environmental systems, they contribute to and are subjected to different types of *impact displacements* across and within regions. Such conditions render consideration of inter- and intraregional processes central to urban analysis and planning.
3. Recognize the important role various urban constituencies play in bringing about desirable change, where the potentials of these constituencies are not fully mobilized, and often limited by various factors such as the lack of legitimacy or political voice to act as agents for change.

Many elements of these approaches, including justice and systems-based problems, have been around and discussed for over twenty years, whereas others are relatively new. For those that have long been discussed, we note that discourse was often limited to particular disciplinary or stakeholder groups, and thus a broader engagement is still very much needed.

In our discussion of five different framings, used to examine and act upon urban social and environmental problems, it is important to note that there are tensions between some of these framings. Although the framings typically address specific elements of urban environment, their purposes are often not dissimilar. The diverse nature or urban problems call for plural concepts and approaches with each contributing different but useful insights. It is also important to recognize that scholars and practitioners using different frames often do not talk to each other. More inclusive, open dialogue across disciplines and across concepts and value systems are needed, but may prove difficult to achieve due to persisting epistemological differences.

Looking ahead, few will disagree that a transformative change is needed for urban futures that are sustainable, diverse, and just. Cities are characterized by uncertainties, often subject to shocks that result in unintended and unevenly distributed consequences. Cities are also spaces where multiple possible and desirable futures can be envisioned, which in turn require closer attention to participation and evaluation of benefits and trade-offs. As such, envisioning more diverse, sustainable, and just futures for cities calls for better understanding of the dynamic interactions among different structural components affecting urban systems, including those that structure social inequalities. In closing, several questions emerge as important to research and practice: How do we reimagine the urban future? Can city residents, in their diversity, articulate what futures they want, develop collective visions, and mobilize around common desirable goals? New knowledge and tools, as well as a commitment to engage those who are disproportionately affected by the ever-changing environment, are needed to support and enable such an approach.

References

Alberti, M., J. M. Marzluff, E. Shulenberger, et al. 2008. Integrating Humans into Ecology: Opportunities and Challenges for Studying Urban Ecosystems. In: Urban Ecology: An International Perspective on the Interaction between Humans and Nature, ed. J. M. Marzluff et al., pp. 143–158. Boston: Springer.

Allen, B. L. 2003. Uneasy Alchemy: Citizens and Experts in Louisiana's Chemical Corridor Disputes. Cambridge, MA: MIT Press.

Anderson, B. 2015. What Kind of Thing Is Resilience? *Politics* **35**:60–66.

Bai, X. M. 2003. The Process and Mechanism of Urban Environmental Change: An Evolutionary View. *Int. J. Environ. Pollut.* **19**:528–541.

———. 2016. Eight Energy and Material Flow Characteristics of Urban Ecosystems. *Ambio* **45**:819–830.

Bai, X. M., and H. Imura. 2000. A Comparative Study of Urban Environment in East Asia: Stage Model of Urban Environmental Evolution. *Int. Rev. Environ. Strat.* **1**:135–158.

Bai, X. M., B. Roberts, and J. Chen. 2010. Urban Sustainability Experiments in Asia: Patterns and Pathways. *Environ. Sci. Policy* **13**:312–325.

Bai, X. M., P. Shi, and Y. Liu. 2014. Society: Realizing China's Urban Dream. *Nature* **509**:158–160.

Bai, X. M., S. Van Der Leeuw, K. O'Brien, et al. 2016. Plausible and Desirable Futures in the Anthropocene: A New Research Agenda. *Global Environ. Change* **39**:351–362.

Batty, M. 2007. Cities and Complexity: Understanding Cities with Cellular Automata, Agent-Based Models, and Fractals. Cambridge, MA: MIT Press.

———. 2016. Evolving a Plan: Design and Planning with Complexity. In: Complexity, Cognition, Urban Planning and Design, ed. J. Portugali and E. Stolk, pp. 21–42. Cham: Springer.

Bettencourt, L. M. A., J. Lobo, D. Helbing, C. Kühnert, and G. B. West. 2007. Growth, Innovation, Scaling, and the Pace of Life in Cities. *PNAS* **104**:7301–7306.

Bluestone, B., and B. Harrison. 1982. The Deindustrialization of America: Plant Closing, Community Abandonment, and the Dismantling of Basic Industry. New York: Basic Books.

Brandt, W. 1980. North–South: A Programme for Survival: Report of the Independent Commission on International Development Issues. Cambridge, MA: MIT Press.

Brenner, N. 2000. The Urban Question: Reflections on Henri Lefebvre: Urban Theory and the Politics of Scale. *Int. J. Urban Reg. Res.* **24**:361–378.

Brondizio, E. S. 2016. The Elephant in the Room: Amazonian Cities Deserve More Attention in Climate Change and Sustainability Discussions. Collective Blog: The Nature of Cities. http://www.thenatureofcities.com/2016/02/02/the-elephant-in-the-room-amazonian-cities-deserve-more-attention-in-climate-change-and-sustainability-discussions. (accessed Sept. 29, 2017).

Brondizio, E. S., K. O'Brien, X. M. Bai, et al. 2016. Re-Conceptualizing the Anthropocene: A Call for Collaboration. *Global Environ. Change* **39**:318–327.

Brown, P., and E. J. Mikkelsen. 1997. No Safe Place: Toxic Waste, Leukemia, and Community Action. Berkeley: Univ. of California Press.

Bulkeley, H., V. Castán Broto, and G. A. S. Edwards. 2015. An Urban Politics of Climate Change: Experimentation and the Governing of Socio-Technical Transitions. London: Routledge.

Bullard, R. D. 1990. Dumping in Dixie: Race, Class, and Environmental Quality. Boulder: Westview Press.

———. 2000. Dumping in Dixie: Race, Class, and Environmental Quality, 3rd edition. Boulder: Westview Press.

———. 2008. Dumping in Dixie: Race, Class, and Environmental Quality, 4th edition. Boulder: Westview Press.

Burger, J. R., C. D. Allen, J. H. Brown, et al. 2012. The Macroecology of Sustainability. *PLoS Biol.* **10**:e1001345.

Castán Broto, B. V., A. Allen, and E. Rapoport. 2012. Interdisciplinary Perspectives on Urban Metabolism. *J. Indust. Ecol.* **16**:851–861.

Cohen, B. 2004. Urban Growth in Developing Countries: A Review of Current Trends and a Caution Regarding Existing Forecasts. *World Dev.* **32**:23–51.

Demaria, F., and S. Schindler. 2016. Contesting Urban Metabolism: Struggles over Waste-to-Energy in Delhi, India. *Antipode* **48**:293–313.

Deutsch, L., R. Dyball, and W. Steffen. 2013. Feeding Cities: Food Security and Ecosystem Support in an Urbanizing World. In: Urbanization, Biodiversity and Ecosystem Services: Challenges and Opportunities: A Global Assessment, ed. T. Elmqvist et al., pp. 505–537. Dordrecht: Springer.

Edwards, G. A. S., L. Reid, and C. Hunter. 2015. Environmental Justice, Capabilities, and the Theorization of Well-Being. *Prog. Hum. Geogr.* **40**:754–769.

Esteva, G., and M. S. Prakash. 2004. From Global to Local: Beyond Neoliberalism to the International Hope. In: The Globalisation Reader, ed. F. J. Lechner and J. Boli, pp. 410–416. Oxford: Blackwell.

First National People of Color Environmental Leadership Summit. 1991. The Principles of Environmental Justice. Washington, D.C.: Natural Resources Defense Council. http://www.ejnet.org/ej/principles.html. (accessed Nov. 17, 2017).

Fröbel, F., J. Heinrichs, O. Kreye, and P. Burgess. 1980. The New International Division of Labour: Structural Unemployment in Industrialised Countries and Industrialisation in Developing Countries. Cambridge: Cambridge Univ. Press.

Gandy, M. 2004. Rethinking Urban Metabolism: Water, Space and the Modern City. *City* **8**:363–379.

Grimm, N. B., E. M. Cook, R. L. Hale, and D. M. Iwaniec. 2016. A Broader Framing of Ecosystem Services in Cities: Benefits and Challenges of Built, Natural, or Hybrid System Function. In: The Routledge Handbook of Urbanization and Global Environmental Change. New York: Routledge.

Grimm, N. B., S. H. Faeth, N. E. Golubiewski, et al. 2008. Global Change and the Ecology of Cities. *Science* **319**:756–760.

Grimm, N. B., C. L. Redman, C. G. Boone, et al. 2013. Viewing the Urban Socio-Ecological System through a Sustainability Lens: Lessons and Prospects from the Central Arizona–Phoenix LTER Programme. In: Long Term Socio-Ecological Research: Human-Environment Interactions, vol. 2, ed. S. J. Singh et al., pp. 217–246. Dordrecht: Springer.

Grubler, A., X. Bai, T. Buettner, et al. 2012. Urban Energy Systems. In: Global Energy Assessment: Toward a Sustainable Future, pp. 1307–1400. Cambridge: Cambridge Univ. Press.

Guedes, G., S. Costa, and E. S. Brondizio. 2009. Revisiting the Hierarchy of Urban Areas in the Brazilian Amazon: A Multilevel Approach. *Pop. Environ.* **30**:159–192.

Guha, R. 2000. Environmentalism: A Global History. Oxford: Oxford Univ. Press.

Guha, R., and J. Martinez-Alier. 1997. Varieties of Environmentalism: Essays North and South. London: Earthscan.

Haase, D. 2008. Urban Ecology of Shrinking Cities: An Unrecognised Opportunity? *Nat. Cult.* **3**:1–8.

Innes, J. L., and A. H. Haron. 2000. Air Pollution and the Forests of Developing and Rapidly Industrializing Regions. Report No. 4 of the IUFRO Task Force on Environmental Change. Oxon: CABI Publishing.

Iwaniec, D., and A. Wiek. 2014. Advancing Sustainability Visioning Practice in Planning: The General Plan Update in Phoenix, Arizona. *Plann. Pract. Res.* **29**:543–568.

Kammen, D. M., and D. A. Sunter. 2016. City-Integrated Renewable Energy for Urban Sustainability. *Science* **352**:922–928.

Kaye, J. P., P. M. Groffman, N. B. Grimm, L. A. Baker, and R. V. Pouyat. 2006. A Distinct Urban Biogeochemistry? *Trends Ecol. Evol.* **21**:192–199.

Kennedy, C., J. Cuddihy, and J. Engel-Yan. 2007. The Changing Metabolism of Cities. *J. Indust. Ecol.* **11**:43–59.

Kok, K., D. S. Rothman, and M. Patel. 2006. Multi-Scale Narratives from an IA Perspective: Part I. European and Mediterranean Scenario Development. *Futures* **38**:261–284.

LeGates, R. T., and F. Stout. 2003. The City Reader, 5th edition. London: Routledge.

Lerner, S., and R. D. Bullard. 2006. Diamond: A Struggle for Environmental Justice in Louisiana's Chemical Corridor. Boston: MIT Press.

Lin, T., Y. Yu, X. M. Bai, L. Feng, and J. Wang. 2013. Greenhouse Gas Emissions Accounting of Urban Residential Consumption: A Household Survey Based Approach. *PLoS One* **8**:e55642.

Lora-Wainwright, A. 2013. The Inadequate Life: Rural Industrial Pollution and Lay Epidemiology in China. *China Q.* **214**:302–320.

Lora-Wainwright, A., and A. Chen. 2013. China's Cancer Villages: Contested Evidence and the Politics of Pollution. In: A Companion to the Anthropology of Environmental Health, ed. M. Singer, pp. 396–416. New York: Wiley-Blackwell.

Martinez-Alier, J. 2002. Environmentalism of the Poor: A Study of Ecological Conflicts and Valuation. Cheltenham: Edward Elgar.

Martinez-Alier, J., I. Anguelovski, P. Bond, et al. 2014. Between Activism and Science: Grassroots Concepts for Sustainability Coined by Environmental Justice Organizations. *J. Polit. Ecol.* **21**:19–60.

Martinez-Alier, J., L. Temper, D. Del Bene, and A. Scheidel. 2016. Is There a Global Environmental Justice Movement? *J. Peasant Stud.*1–25.

McFarlane, C. 2013. Metabolic Inequalities in Mumbai: Beyond Telescopic Urbanism. *City* **17**:498–503.

McPhearson, T., S. T. A. Pickett, N. B. Grimm, et al. 2016. Advancing Urban Ecology toward a Science of Cities. *BioScience* **66**:198–212.

Metson, G. S., R. L. Hale, D. M. Iwaniec, et al. 2012. Phosphorus in Phoenix: A Budget and Spatial Representation of Phosphorus in an Urban Ecosystem. *Ecol. App.* **22**:705–721.

Metson, G. S., D. M. Iwaniec, L. A. Baker, et al. 2015. Urban Phosphorus Sustainability: Systemically Incorporating Social, Ecological, and Technological Factors into Phosphorus Flow Analysis. *Environ. Sci. Policy* **47**:1–11.

Mollenkopf, J. 2003. How to Study Urban Political Power. In: The City Reader, ed. R. LeGates and F. Stout, pp. 235–244. London: Routledge.

Newell, J. P., and J. J. Cousins. 2014. The Boundaries of Urban Metabolism: Towards a Political–Industrial Ecology. *Prog. Hum. Geogr.* **39**:702–728.

O'Connor, A., C. Tilly, and L. D. Bobo, eds. 2001. Urban Inequality: Evidence from Four Cities. New York: Russell Sage Foundation.

Özkaynak, B. 2008. Globalisation and Local Resistance: Alternative City Developmental Scenarios on Capital's Global Frontier: The Case of Yalova, Turkey. *Prog. Plann.* **70**:45–97.

Pickett, S. T. A., and M. L. Cadenasso. 2002. The Ecosystem as a Multidimensional Concept: Meaning, Model, and Metaphor. *Ecosystems* **5**:1–10.

Rees, W. 1995. Achieving Sustainability: Reform or Transformation? *J. Plann. Lit.* **9**:343–361.

Revi, A., D. Satterthwaite, F. Aragón-Durand, et al. 2014. Towards Transformative Adaptation in Cities: The IPCC's Fifth Assessment. *Environ. Urban.* **26**:11–28.

Rosenzweig, C., W. Solecki, S. A. Hammer, and S. Mehrotra. 2010. Cities Lead the Way in Climate-Change Action. *Nature* **467**:909–911.

Roy, A. 2009. The 21st Century Metropolis: New Geographies of Theory. *Reg. Stud.* **43**:918–830.

San Sebastián, M., and A. K. Hurtig. 2005. Oil Development and Health in the Amazon Basin of Ecuador: The Popular Epidemiology Process. *Social Sci. Med.* **60**:799–807.

Schindler, S. 2016. Detroit after Bankruptcy: A Case of Degrowth Machine Politics. *Urban Stud.* **53**:818–836.

Schlosberg, D. 2007. Defining Environmental Justice: Theories, Movements, and Nature. Oxford: Oxford Univ. Press.

Schlosberg, D., and L. B. Collins. 2014. From Environmental to Climate Justice: Climate Change and the Discourse of Environmental Justice. *Wiley Interdisc. Rev. Climate Change* **5**:359–374.

Selman, P. 1996. Local Sustainability: Managing and Planning Ecologically Sound Places. New York: St. Martin's Press.

Steffen, W., K. Richardson, J. Rockström, et al. 2015. Planetary Boundaries: Guiding Human Development on a Changing Planet. *Science* **347**:6223.

Swyngedouw, E. 1996. The City as a Hybrid: on Nature, Society and Cyborg Urbanization. *Capital. Nat. Social.* **7**:65–80.

Terradas, J. 2001. Ecologia Urbana. Barcelona: Generalitat de Catalunya, Departament de Medi Ambient.

Turner, M. D. 2014. Political Ecology I: An Alliance with Resilience? *Prog. Hum. Geogr.* **38**:616–623.

UNDP. 2013. Humanity Divided: Confronting Inequality in Developing Countries. New York: United Nations Development Programme.

Walker, B., C. S. Holling, S. R. Carpenter, and A. Kinzig. 2004. Resilience, Adaptability and Transformability in Social-Ecological Systems. *Ecol. Soc.* **9**:5.

Wang, R., and S. M. Ma. 1984. The Social-Economic-Natural Complex Ecosystem. *Acta Ecol. Sin.* **1**:2006.

Wiek, A., and D. Iwaniec. 2014. Quality Criteria for Visions and Visioning in Sustainability Science. *Sust. Sci.* **9**:497–512.

Wolman, A. 1965. The Metabolism of Cities. *Scient. Am.* **213**:179–190.

Zetter, R., and A. M. Hassan. 2002. Urban Economy or Environmental Policy? The Case of Egypt. *J. Environ. Policy Plan.* **4**:169–184.

Energy and Climate Change

8

Social Movements for Climate Justice during the Decline of Global Governance

From International NGOs to Local Communities

Patrick Bond

Abstract

With political foresight, this Ernst Strüngmann Forum considered "how differences in framing environmental problems are driven by differences in normative and theoretical positions, as well as ways in which more inclusive framings might enable more societally relevant and impactful research and more concerted action/practice." When this exercise began, the British electorate's rejection of the European Union and the election of Donald Trump as President of the United States appeared inconceivable. Indeed, both the mid-2015 G7 summit and December 2015 Paris climate conference left the impression that a viable global governance arrangement had been accomplished, and that irrevocable steps toward economic decarbonization were being taken that would potentially save the planet from catastrophic climate change. In opposition to this elite consensus, an international nongovernmental organization (INGO), Friends of the Earth International, along with many "climate justice" movement components, condemned these two crucial instances of global climate governance. The climate justice opposition, however, had no impact whatsoever because the die appeared to have been cast for world climate policy, leaving intact several dangerous features of the Paris strategy: no legally binding responsibilities or accountability mechanisms; inadequate stated aspirations for lowering global temperatures; no liabilities for past greenhouse gas emissions; renewed opportunities to game the emissions-reduction system through state-subsidized carbon trading and offsets, soon moving from the European Union and North America to the emerging markets led by China; and neglect of emissions from military, maritime, and aviation sources. In mid-2017, Trump withdrew the United States from the Paris Agreement. The climate justice answer to both Trump and the top-down policy regime—one overwhelmingly favorable to the United States, from where the strategy emanated—appears to be twofold: an intensification of bottom-up strategies that aim to weaken the state of greenhouse-gas emissions and corporate targets through both direct action (disruptions) and financial divestment. Given that the Paris deal is now in question due to Trump's promise to abrogate U.S. participation in the overarching United Nations Framework Convention on Climate Change (UN-FCCC), the climate justice strategy appears prescient: to undo the damage at local and national scales. As the forces of "extractivism" (especially petroleum and coal mining) are empowered again by Trump, there may be merit to climate justice activists utilizing one of the framing narratives of a "neoliberal nature": natural capital accounting (so as to argue that net losses make fossil fuel extraction economically irrational). For the foreseeable future, the global balance of forces appears extremely adverse—especially with the rise of rightwing populism and the decline of the Latin American center-left regimes—and no system-saving change appears possible at that scale. This could permit a decisive shift of orientation by INGOs toward the climate justice approach, especially because of the potential for unity against Trump at sites like Standing Rock. Evidence of this can be found in how Greenpeace and 350.org have taken up direct action and divestment strategies, respectively, to address climate change and related ecosystem breakdowns effectively and fairly during this rapidly closing window.

Introduction

We're going to rescind all the job-destroying Obama executive actions including the Climate Action Plan....We're going to save the coal industry [and] Keystone Pipeline. We're going to lift moratoriums on energy production in federal areas. We're going to revoke policies that impose unwarranted restrictions on new drilling technologies....We're going to cancel the Paris Climate Agreement and stop all payments of U.S. tax dollars to UN global warming programs.
—Donald Trump (2016)

We are the poor cousins of the global jet set. We exist to challenge the status quo, but we trade in incremental change. Our actions are clearly not sufficient to address the mounting anger and demand for systemic political and economic transformation that we see in cities and communities around the world every day.
— Dhananjayan Sriskandarajah et al. (CIVICUS 2014)

Is the die cast, must at this one throw all thou hast gained be lost?
The Worlds a Lott'ry; He that drawes may win;
Who nothing ventur's, looks for nothing
— Sir Thomas Herbert (1634)

Ālea *iacta est*. On January 10, 49 BC, as he crossed the Rubicon River in Italy, Julius Caesar spoke of casting the die (rolling the dice) in a gamble that could not be reversed. That day he took a crucial step toward conquering Rome, an act that would leave the world changed forever.

By 2015, the importance of addressing climate change was so clear that the same metaphor was invoked in the World Bank's *Turn Down the Heat* series:

"The die is cast. If we do not act now, rising temperatures will endanger crops, freshwater reserves, energy security, and even our health."

The following year, the presidency of Donald Trump and the British electorate's rejection of the European Union appear to have cast the die for the demise of global governance, especially in relation to climate policy. With that comes the likelihood of runaway climate change. The political turn of 2016 sets the stage not only for similar right-wing populist movements gathering pace in other European countries, joining dangerous authoritarian leaders in Turkey and the Philippines, but also an excuse for worsening pollution from the Brazil, Russia, India, China, and South Africa (BRICS) bloc.

What can be done? Is the new political situation appropriate for renewed attention to social-movement resistance, especially in the form of climate justice?

After all, the elite strategy associated with climate policy gambles at the June 2015 G7 summit hosted by Angela Merkel in Elmau and, six months later in Paris, at the United Nations Framework Convention on Climate Change (UNFCCC) would, like Caesar, change the world. This, however, was not due to decisive action, but rather *the opposite*: failure to grapple with climate change as an existential crisis for humanity (Bond 2016). In both sites, the assembled world leaders' economic, geopolitical, technical, ideological, and media powers were dedicated to what they presumed was an irreversible, logical proposition: marginal, market-driven changes augmented by a slight degree of state regulatory assistance will decarbonize the world's energy, land-transport, and production systems as well as protect forests. (No one would deny that nothing of substance was offered at either summit to reduce climate change caused by air transport, shipping, the military, corporate agriculture, overconsumption, and methane-intensive disposal sources.) The self-confidence of those signing the Paris Climate Agreement was a reflection of how far from reality global climate governance had roamed, and how quickly they would be given an unprecedented reality check.

The flaws in the elites' logic would lead to two reactions: (a) an initial leftist critique of the Paris Agreement's reliance upon capitalism's self-correction mechanisms, and hence the downplaying of climate justice; (b) a revival of climate change denial along with the rise of extreme petro-military complex power within the country most guilty of historic greenhouse gas pollution. Ironically, the United States is itself extremely divided as evident from its last presidential election: Trump won the presidency via electoral college, based on ca. 55,000 voters from four "swing states." He lost, however, the popular vote by ca. three million votes, out of the 130 million votes cast. The same month, a poll by Yale and George Mason universities (Leiserowitz et al. 2017) found that 69% of U.S. registered voters endorsed the Paris Agreement (only 13% were opposed) and 78% supported taxes or regulations against greenhouse gas emissions (with 10% opposed). Thus, the grassroots will for "climate action" was in place, even though it was not evident in presidential or congressional leadership.

Regardless of Trump's impact, global climate policy as determined in 2015 had become a very risky toss indeed. Starting at Copenhagen's 15th UNFCCC Conference of the Parties (COP) in 2009, the U.S. State Department's chief climate negotiator, Todd Stern, successfully drove the UN negotiations away from the four essential principles required in a future global governance regime to achieve climate justice (Bond 2012b):

1. Ensure emissions-cut commitments are sufficient to halt runaway climate change.
2. Make the cuts legally binding with accountability mechanisms.
3. Distribute the burden of cuts fairly based on responsibility for causing the crisis.
4. Offer adequate financial compensation to repair weather-related "loss and damage" that occur directly because of that historic liability.

The Elmau goal was for "net zero carbon emissions" by 2100—50 years too late—and instead of full decarbonization, the G7 endorsed "net" strategies; these are based not on direct cuts but instead on offsets, emissions trading, reducing emissions through deforestation and forest degradation (REDD), and carbon sequestration (Reyes 2015). As for the rest of the world, including the high-pollution emerging markets (especially the BRICS), the so-called "bottom-up" pledge-and-review strategy that Stern imposed in Copenhagen was once again endorsed by the major new emitters. Six months after Elmau, at COP21, the "Intended Nationally Determined Contributions" (i.e., *voluntary* pledged cuts) agreed upon by Paris signatories were so low that even if achieved, they would collectively raise the temperature goal set for 2100 more than 3 degrees Celsius, thus catalyzing runaway climate change (Bond 2016).

Given the extreme dangers to civilization and Earth's species inherent in Trump's regressive stance and the Paris and Elmau gambles, the role of a civil society countermovement is vital and must prevail against both climate change denial and the Paris climate policy within the next decade at the latest. But how is this countermovement to emerge?

This chapter assesses the differences between two major civil society forces within climate activism, whose divergences are continually reproduced in global and local settings: (a) international nongovernmental organizations (INGOs), which are part of the global governance regime, and (b) grassroots climate justice activists. Currently, both appear united against Trump's threat, but the more durable divisions between the market-oriented climate politics favored by INGOs and the need for direct, democratic intervention posited by climate justice activists will determine the viability of life on Earth.

The critical question is whether either or both forces will be able to muster the oppositional power necessary to reverse Trump's petro-military politics and the "marketization" of climate policy. Can *global* civil society generate the countermovement required? If not, both the Trump withdrawal from climate governance and the gambles made in Elmau and Paris will likely result in an ecological catastrophe, whether because of climate change denialism or because world elites anticipate that corporate self-survival mechanisms will kick in. As scientists point out, however, the lag times from greenhouse gas emissions mean that market reactions will be too little, too late.

How might INGOs, climate justice activists, or some combination move the world economy and society off the current trajectory? As shown below, civil society forces currently appear bogged down in an interminable conflict over principles, analysis, strategies, tactics, and alliances (the "pasta" problem). The former include the most active Climate Action Network (CAN) members—Worldwide Fund for Nature (WWF) and Greenpeace—but also a notable self-exiled group from CAN, the environmental justice movement Friends of the Earth International (FOEI), which typically allies closely with grassroots movements. The INGOs, even Greenpeace, are much more open to alliances with politicians and, in some cases, corporations and green business federations. To complicate matters, the leaders of CAN's U.S. chapter have embraced climate justice with gusto. The two most savvy INGOs, 350.org and Avaaz, have become known largely through highly creative social media campaigns, and some provide well-recognized, visionary leadership: Bill McKibben from 350.org, Kumi Naidoo from Greenpeace International (2009–2015), and Annie Leonard from Greenpeace U.S.A. Annie Leonard, for example, has probably been the most impactful in combining anti-racist and labor networks with climate justice.

By contrast, climate justice groups are committed to global critique while providing essentially local-level solutions, from militant strategies to "direct action" tactics described as "Blockadia" by their best-known proponent, Klein. To the extent that they tackle corporate power at its financial Achilles' heel, they support the divestment strategy catalyzed by 350.org. Their strength, however, especially in the wake of Paris, is in the use of a disruptive repertoire to defend land, water, and air against polluters. The peak moment of Blockadia was probably the Standing Rock defense of North Dakota "Treaty Land" and water, which the Dakota people had won generations ago and yet was threatened by the Dakota Access Pipe Line (DAPL). By late 2016, opposition to DAPL was formidable and the Obama regime backed down. In February 2017, however, DAPL opponents were routed by the Trump regime and forced to leave the land.

Blockadia activists (depending upon circumstances) point out how the success of their local battles against oil, gas, coal, and major greenhouse gas emitters also benefits humankind and the planet. But the local climate activist movement is so broad—as witnessed in the diversity of signs that appeared at the 400,000-strong New York Peoples March on Climate in September 2014—that all manner of interventions qualify as climate activism. Klein (2014) is correct that "this changes everything."

An authentic nomenclature for climate justice relies in part on the Climate Justice Now! network's 2007 launch at the Bali COP13, in opposition to CAN which was seen as too market oriented. There were five founding principles:

1. Reduce consumption.
2. Enable huge monetary transfers (funded by redirecting military budgets, innovative taxes, and debt cancellation) from North to South, based on historical responsibility and ecological debt, to cover adaptation and mitigation costs.
3. Leave fossil fuels in the ground and invest in energy efficiency using appropriate, safe, clean, and community-led renewable energy sources.
4. Enforce rights-based resource conservation that ensures indigenous land rights and promotes peoples' sovereignty over energy, forests, land, and water.
5. Ensure sustainable family farming, fishing, and peoples' food sovereignty.

By 2010, a conference of 35,000 people in Cochabamba, Bolivia, had developed these into concrete demands (in hundreds of pages of workshop reports), of which the following are of note:

- By 2017, reduce greenhouse gas emissions by 50%.
- Stabilize temperature rises to 1°C and 300 parts per million.
- Acknowledge the climate debt owed by developed countries.
- Achieve full respect for human rights and the inherent rights of indigenous people.
- Universal declaration of rights of Mother Earth to ensure harmony with nature.
- Establish an International Court of Climate Justice.
- Reject carbon markets and commodification of nature and forests through the REDD Programme.
- Promote measures that change consumption patterns in rich countries.
- End intellectual property rights for technologies useful for mitigating climate change.
- Payment of 6% of developed countries' GDP to address climate change.

Some high-profile climate advocates, such as Mary Robinson (a supporter of carbon trading), soon appropriated the concept of climate justice for use in a manner inconsistent with these demands. Other strategies for equity also came into dispute, such as "greenhouse gas development rights" and "contraction and convergence" approaches, which also advocated the sale of surpluses on the markets. Climate justice critics argue that such markets have the tendency to turn a ceiling into a floor. Other concepts such as "common but differentiated responsibilities" between national states and "converging per capita emissions" were much more in the spirit of climate justice, as defined in Cochabamba. Most importantly, as Egardo Lander (2010) explained in his review of Cochabama, the conference brought together the main contemporary struggles in a constructive fusion of interests: justice/equality, war/militarization, free trade, food sovereignty, agribusiness, peasants' rights, struggles against patriarchy, defense of indigenous peoples' rights, migration, the critique of the dominant Eurocentric/colonial patterns of knowledge, and struggles for democracy.

With the contested rise of climate justice narratives in mind, I wish to contribute to the debate about "how differences in framing environmental problems are driven by differences in normative and theoretical positions; and ways in which more inclusive framings might enable more societally relevant and impactful research and more concerted action/practice" (see Lele et al., this volume). The stereotypical premise is that the INGOs are pragmatic and hence correct in their normative approach: deal making within existing UNFCCC constraints. In contrast, climate justice groups are principled, radical, and unbending in their opposition to compromise on a matter as vital as climate change, and are increasingly unwilling to countenance the kinds of compromises that the December 2015 Paris UNFCCC COP 21 summit represented. This is a simple dichotomy, one that begins to break down somewhat upon closer examination (e.g., Greenpeace's direct actions).

In the field of climate politics, however, conditions are becoming so desperate that the more militant, localistic approach may be judged by future generations *as the more pragmatic step required for basic civilizational survival,* especially if the alternative is what can be termed "neoliberal nature," a conceptual framing implicitly adopted by both world elites and many INGOs. The reliance on market solutions is one of the main strategic impulses within what is sometimes termed the theory of ecological modernization, whose other features include technological innovations, efficiencies, and the management of externalities aimed at improving environmental outcomes in a rational manner (for a critical discussion, see Harvey 1996).

The basic thesis is that market imperfections (such as pollution) require market interventions to get the prices right. In what may be its most advanced form of such self-correction within neoliberal capitalism, Deutsche Bank's Pavan Sukhdev initiated "The Economics of Ecosystems and Biodiversity" (TEEB) within the UN Environment Program to "make nature's values visible" and thus "help decision makers recognize the wide range of benefits provided by ecosystems and biodiversity, demonstrate their values in economic terms and, where appropriate, capture those values in decision making." TEEB's search for optimal resource use emphasizes "low-hanging fruit" that can achieve the least costly form of market-facilitated environmental management.

Climate justice networks, by contrast, use contrary framings of environmental justice that are especially hostile to market strategies. To date, they have gathered insufficient strength to counter neoliberal nature advocates, beyond moralizing. One of the most important areas for this debate concerns climate finance, ranging from carbon markets to the Green Climate Fund. As this chapter concludes, climate justice activists may be well advised to take on board *some* of the logic of neoliberal-nature INGOs, even on their own terms, so as to explore limits and confirm the futility of reforming a thoroughly corrupt structure. The more serious INGOs, such as FOEI and Greenpeace, have redoubled efforts to link global and local action, as confirmed by their 2015–16 promotion of local campaigns that incorporate direct action. In other words, the two different environmental narratives have not yet achieved a necessary interconnection to pursue dialectical tensions and perhaps resolutions.

To make this case, the structure of the argument is first aided if we personalize these complex issues by considering climate debates involving several colleagues from Durban, South Africa. This is an ethnography of social struggle vignettes, informed by the fact that personal positionality is vital to the framing narratives chosen by INGOs and climate justice groups. The perspectives of these four individuals, profiled below, to the wider story of climate narrative construction, along with their similar origins, political perspectives, and subsequent placements in an INGO, a national NGO, a community organization, and academia provide an opportunity to assess where the climate justice movement (principles, analysis, strategy, tactics, and alliances) has taken them in relation to the UNFCCC.

Thereafter, we will explore the wider terrain of neoliberal nature. There we find groups that adopt insider positions in relation to global power structures that broadly agree with the conceptual premises behind global incremental change—following market principles—as opposed to climate justice movements that work locally and reject market strategies. A subsequent consideration of vital issues like carbon trading and natural capital accounting clarifies the complexities, and general principles begin to emerge. These principles are not, however, easily reduced to a "sustainable development" rubric. They are more contradictory, as we will see.

Finally, by considering how climate policy analyses, strategies, tactics, and alliances emerge to lend themselves to this dichotomy, we see INGOs and climate justice activists in conflict over markets and technical solutions—or "false solutions" as climate justice activists would argue. This, in turn, allows us to reframe both INGO and grassroots climate justice argumentation. The rise of Trump makes the search for unity all the more urgent, but also more feasible if a world divestment movement picks up momentum.

For the lack of a better phrase, we might term the alternative "ecosocialism," *respectful of the merits of valuing nature* (though not counting it for the sake of marketization), while at same time *confirming the role of decommodifying social movements, including those of indigenous people and ecofeminists, in nature's stewardship.*

Vignettes of Paris Seen from Durban

Four Durban friends of mine (Figure 8.1) are worth introducing: Kumi Naidoo, Bobby Peek, Desmond D'Sa, and Ashwin Desai. The stories they tell about climate politics illustrate the main framing narratives, the structurally delimited locations they occupy at different scales, and the breakthrough potentials.

Naidoo was Greenpeace International's leader from 2009–2015 and now works continentally in Africa on diverse civil society strategies. He also has led South African civil society organizations, the international network CIVICUS, and various initiatives during the mid-2000s global anti-poverty mobilizations. He holds a doctorate in politics from Oxford and is respected by many world leaders.

Peek directs the NGO "groundWork" (working nationally in South Africa). He won the Goldman Prize in 1998 and has established himself as one the world's leading environmental justice experts and practitioners.

Working primarily at the local level with the South Durban Community Environmental Alliance, which he helped found with Peek in 1995, D'Sa has become the city's conscience on matters ranging from climate change to anti-drug, anti-gang, antipollution, and anti-privatization struggles underway in many neighborhoods, especially his toxin-saturated home base of Wentworth. D'Sa was also the recipient of the 2014 Goldman Prize.

Desai is a world-renowned sociologist. A professor at the University of Johannesburg, though mostly resident in Durban, his books on Gandhi, daily life in South African struggles, and sports racism are exceptionally well-read and furiously debated, since he finds every opportunity to slaughter holy cows, including his own traditions on the once-revolutionary left.

These extremely energetic, accomplished activists are about a decade's age apart, ranging from late 40s to late 50s. All of them grew up during apartheid

Figure 8.1   (a) Kumi Naidoo speaking to climate activists in KwaZulu-Natal, (b) Bobby Peek protesting against the BRICS in Durban, (c) Desmond D'Sa in front of a refinery in South Durban, and (d) Ashwin Desai at a book fair.

in Durban, within 20 kilometers of each other (in the suburbs of Chatsworth, Wentworth, Cato Manor, and the downtown Indian Quarter, respectively). They were influenced by highly principled anti-capitalist, anti-racist scholar-activists of the earlier generation, such as the late Fatima Meer and Dennis Brutus. They all fought against the Pretoria regime with exceptional courage. Since freedom was won in 1994, they have regularly come together against injustice, for example, at the 2001 United Nations World Conference Against Racism in Durban and 2002 UN World Summit on Sustainable Development in Johannesburg.

At the Durban climate summit in December 2011, the four friends adopted insider-outsider approaches which included (a) high-profile roles in disruptive events (for which Naidoo and Peek were arrested) inside the lobby of the Durban convention center, (b) leadership of a 10,000-strong march by D'Sa, and (c) a ruthless, scathing critique of the whole process by Desai. With such similar backgrounds, they speak the same language of street-heat politics, they harbor fury at injustices big and small, and they possess enormous charisma that each draws upon regularly, extending from small strategy meetings to academic seminars to mass rallies. They are also regularly frustrated by power, so even when they win minor reforms, they immediately point out the bigger structural enemies they face.

Three of these friends, however, came together—and grew decisively (if temporarily) apart—in Paris in December 2015. Greenpeace International's leader, Naidoo endorsed the deal as "progress," even while viewing the Paris Agreement as "one step on a long road and there are parts of it that frustrate and disappoint me....There's a yawning gap in this deal but it can be bridged by clean technology." Like Greenpeace, the 42-million member clicktivist group Avaaz celebrated: "most importantly, [the Paris deal] sends a clear message to investors everywhere: sinking money into fossil fuels is a dead bet. Renewables are the profit center. Technology will bring us to 100% clean energy is the money-maker of the future."

In contrast, Peek and D'Sa wholeheartedly denounced the Paris Agreement, as had Desai at the same summit four years earlier in Durban (in part because of the admittedly weak counter-summit organized by the other three plus this author; Bond 2012a). After the Durban COP17 concluded, Desai attacked "big name spectacle NGOs" which dominated the main protest march, including Greenpeace:

> The local grassroots organizations were reduced to spectators, and were allowed only the occasional cameo appearance with most often a single line: "Amandla!" [Power!] The march delivered the Minister of International Relations and COP17 president Maita Nkoana-Mashabane to the masses gathered below. She used the opportunity to say how important civil society was and promised to study a memorandum. She was gracious and generous. I could see the NGOs on the truck preening themselves in the glow of this recognition and probably increased funding.

Actually, D'Sa was mightily pleased about the crowd he led to the convention center that day (December 3, 2011) against COP15: in comparative terms it was a very large march for Durban. This was due to the many visiting activists and unfortunately not because Durban residents turned out to participate. When it came to his two-week sojourn in Paris in December 2015, however, D'Sa voiced his appreciation for the civil society mobilizations outside for in the wake of 130 murders by Islamic extremists two weeks before, the inside was hermetically sealed:

> The stark reality is that the people who have the potential to create great changes are being excluded from the process. Instead, an elite minority with access to the COP make decisions for the masses outside. This is quite ironic, as it seems to be the same model which has intensified the crisis. The leaders, who are elected by the people, together with the big corporations are in collaboration, halting the necessary measures needed to stop this runaway climate catastrophe.

For D'Sa, the essential problem was framed as one of participation, self-interest, and power relations. In Peek's recorded comments on the Paris Agreement's failings, the FOEI stance led him to express a North–South critique, namely:

> ...the draft agreement avoids recognition of the climate debt owed to the people of Africa. It sees the need for adaptation, but provides paltry resources. It absolves the imperial powers of any liability for loss and damage resulting from climate change....We call on African governments to negotiate as if our lives mean something. If they cannot put a good deal on the table, we call on them to walk out of the Paris talks.

It was not to be. The African elites joined the world elites. They could have, instead, repeated the precedent of World Trade Organization summits in 1999 and 2003, when Africa's delegates walked out of those events to sabotage the neoliberal agenda. At the 2009 Copenhagen Accord, one African leader—Lumumba di Apeng from Sudan, who coordinated the G77 countries—unsuccessfully attempted a delegitimization strategy so as to gain more concessions. Among the larger INGOs, FOEI (2015) was the only one to condemn the Paris deal, while also criticizing Avaaz for its collaboration (Bond 2016). As FOEI's Asad Rehman explained, in relation to paying for climate damage:

> The political number mentioned for finance has no bearing on the scale of need. It's empty. The iceberg has struck, the ship is going down and the band is still playing to warm applause.

The rural advocacy movement Via Campesina, also possessing global consciousness and anti-imperialist sensibilities, was even more scathing about COP21:

> There is nothing binding for states. National contributions lead us toward a global warming of over 3°C and multinationals are the main beneficiaries. It was essentially a media circus.

The world's best known climate scientist, James Hansen, called Paris, simply, "bullshit." This was not only because of the nonbinding nature of the agreement and its signatories' inability to offer less than the emissions required to keep the temperature rise under 3 degrees—far off the 1.5 degree limit that the delegates claimed to aspire to—but also because Hansen's favorite financial solution—a "cap and dividend" carbon tax—was not contemplated. Even that mild-mannered idea (taxing externalities so as to "make the polluter pay"), raises concerns for climate justice on two levels: (a) whether small, marginal increases in carbon costs are capable of generating the *radical* decarbonization needed (because such taxes lead to marginal, not structural, changes and are passed on to consumers in any case); and (b) whether the commodification of pollution represents the adoption of the neoliberal nature policy strategy often favored by INGOs. What, then, is neoliberal nature?

The Wider Terrain of Struggle: Neoliberal Nature

The very different climate framing narratives and the policy strategies that follow them do not represent a brand-new debate: distinctions in scale politics and the degree of political pragmatism date back decades within environmentalism. Andrew Jamison's 2001 contribution, *The Making of Green Knowledge*, identified a distinct division between the modes of thinking and practice he termed "green business" and "critical ecology movements." The former co-opted environmentalism into the nexus of capital accumulation and flexible regulatory regimes while deploying rhetoric of sustainable development and the "triple bottom line." The green business ontology is grounded in faith in science and technology, instrumental rationality, and market democracy (Jamison 2001).

In contrast, Jamison shows that "critical ecology movements" place emphasis upon the embeddedness of environmental processes with society, state, and market power relations. The various interest groups behind different types of environmental management strategies are highlighted. Their focus is on transformative strategies that also improve human interrelationships, especially tackling racism, sexism, and class power in search of environmental justice. These movements resist the greening of business, demand stronger laws and enforcement, and engage in campaigns against corporations and states which despoil the environment.

Jamison posited four types of environmentalisms: (a) civic work on campaigns and social ecology, (b) professional interventions based upon science and law, (c) militant direct action, and (d) personal environmentalism. Each of these has either reformist or revolutionary currents. Regardless, their politicization of ecology runs counter to green business in virtually all issues and processes, as will be explored further below.

Green business networks have been around for decades, and prominent ones today include the UN Global Compact, World Business Council on Sustainable Development, and World Forum on Natural Capital. Sector after sector, they continue to promote the notion that profit can be reconciled with environmental stewardship. The Marseille-based World Water Council, for example, promotes the commercialization of the most basic element of life, water, as a means to achieve more efficient, sustainable management of the resource. Such networks are dedicated to the strategies of "natural capital accounting" (up to a point, as we will see), payment for ecosystem services, cleaner production, green products, and environmental management systems.

A 2010 list of major environmental INGOs, compiled by a *Greenbiz.com* reporter, that work closely with the more enlightened businesses included the Carbon Trust (with a focus on product carbon footprinting), Ceres (the Global Reporting Initiative), the Clinton Climate Initiative (efficient buildings and waste), Conservation International (biodiversity conservation, product sourcing), EarthShare (workforce charities), Environmental Defense Fund (corporate reforms and efficiencies), GreenBlue (Sustainable Packaging Coalition, CleanGredients and Green2Green), The Nature Conservancy (fresh water, biodiversity, forestry, and land management), Rainforest Alliance (sustainable forestry, agriculture, and tourism), and Rocky Mountain Institute (green business reengineering).

These relationships, however, are sometimes extremely thorny, as when from 2007–2010, the Sierra Club was given USD 25 million by Chesapeake Energy. Apparently, as a result Sierra's leader at the time, Carl Pope, allowed the organization to endorse fracking as a "bridge technology" to lower greenhouse gases, even though methane leakage means that fracking is as bad as, or worse than, coal.

Indeed, to unveil the true character of green business, investigative journalists at "Don't Panic" taped Conservation International (CI) in 2011 blatantly offering Lockheed Martin (or so CI presumed, as the "firm" was represented by Don't Panic undercover reporters) its "greenwashing" public relations support for a partnership rigged to cover up pollution, including the recycling of weaponry for future use.[1] The macro-political context is terribly important, explains Naomi Klein (2013):

> The environmental movement had a series of dazzling victories in the late 60s and in the 70s where the whole legal framework for responding to pollution and to protecting wildlife came into law. It was just victory after victory after victory. And these were what came to be called "command-and-control" pieces of legislation. It was " don't do that." That substance is banned or tightly regulated. It was a top-down regulatory approach. And then it came to screeching halt when Reagan was elected. And he essentially waged war on the environmental

---

[1] The single most uncompromising website to follow critiques of environmental INGOs is http://www.wrongkindofgreen.org/

movement very openly. We started to see some of the language that is common among those deniers—to equate environmentalism with Communism and so on. As the Cold War dwindled, environmentalism became the next target, the next Communism. Now, the movement at that stage could have responded in one of the two ways. It could have fought back and defended the values it stood for at that point, and tried to resist the steamroller that was neoliberalism in its early days. Or it could have adapted itself to this new reality, and changed itself to fit the rise of corporatist government. And it did the latter.

One revealing example of a market-friendly strategy that continues to divide the environmental movement is carbon trading. Misgivings first arose about its pilot in the form of lowering U.S. sulfur dioxide emissions in Southern California, which were slower and less effective than the command-and-control strategies adopted in Germany's Ruhr Valley during the early 1990s. Nevertheless, large environmental INGOs endorsed the idea when presented with it as a deal-breaking demand by U.S. vice president Al Gore at COP3 in Kyoto. Gore promised that Washington would sign the Kyoto Protocol if it included carbon markets as an escape hatch for companies that polluted too much and then desired the right to purchase other companies' pollution permits. The U.S. Senate had already voted 95–0 against endorsing Kyoto.

Even though Gore won this critical concession, there was no change in attitude on Capitol Hill and the United States never ratified the Kyoto Protocol. Yet carbon markets later became one of the most important wedge issues dividing INGOs from the climate justice movement.

The Rocky Terrain of Carbon Markets and Other False Solutions

The overall point of carbon markets is that society can "price pollution" and simultaneously cut costs associated with mitigating greenhouse gases. Moreover, claim proponents, these markets are vital for funding not only innovative carbon-cutting projects in Africa but also for supplying a future guaranteed revenue stream to the Green Climate Fund (GCF), which in turn is supposed to have $100 billion to spend annually on climate-saving projects. GCF's design team cochair, the then South African Planning Minister, Trevor Manuel, argued alongside British economist Nicholas Stern in 2010 that up to half of GCF revenues would logically flow from carbon markets, whose annual trading volume had recently peaked in 2008 at $140 billion (Bond 2012b).

Supporters argue that the use of such "market solutions to market problems" will lower the business costs of transitioning to a post-carbon world. After a cap is placed on total emissions, the idea is that high-polluting corporations and governments can buy ever more costly carbon permits from polluters who do not need so many, or from those willing to part with the permits for a higher price than the profits they make in high-pollution production, energy-generation, agriculture, consumption, disposal or transport.

These markets, however, are in just as much chaos as any financial casino, at a time when faith in bankers—especially confidence they can fairly manage climate-related funding—is badly shaken. In the United States, the national Chicago voluntary carbon market (strongly promoted by Gore) ceased to exist in late 2010 and regional markets crashed. The European Union Emissions Trading Scheme (EU ETS)—the main site of carbon trading—has been moribund since its 2006 and 2008 peaks, when the right to emit extra carbon cost around €30 per ton. Carbon pricing's recent low point was less than €3/ton in the wake of oversupply, various episodes of fraud and hacking, and declining interest in climate change following the 2008–2009 Great Recession.

By 2017, prices remained low and the World Bank (2017) calculated the 2016 global carbon trade at just $32 billion. Of the 15% of world CO_2 equivalent emissions that are covered by either carbon trading or a tax, only a quarter carries a price above $10/ton. The Canadian, Californian, Japanese, and New Zealand carbon trading systems are rare exceptions, with prices ranging from $11–14 per ton. The countries with a carbon price above $25 per ton have achieved this by taxation, not carbon trading: Sweden $126; Switzerland and Liechtenstein $84; Finland $66; Norway $52; France $33; and Denmark $25.

A category of UN-authorized Clean Development Mechanism (CDM) projects was created to allow wealthier countries to engage in emissions reductions initiatives in poor and middle-income countries as a way of eliding direct emissions reductions. Like the global oversupply of carbon credits, however, the price of CDM credits fell to less than €0.50, and to lower supply, the main emerging markets (especially China, India, and Brazil) were no longer allowed to issue them after 2012. China then started eight pilot carbon-trading projects at the local and provincial level, with highly volatile prices which ranged in 2017 from €8/ton in Beijing down to just €0.50/ton in Chongqing. Reflecting the extreme volatility in Chinese financial markets (including stock market crashes in mid-2015 and early 2016), the Shenzhen carbon market fell from a Chinese high of €9.5/ton in early 2013 to just €3.5/ton by mid-2016. These prices are woefully short of making a dent in climate change. According to Joseph Stiglitz and Nicolas Stern's report to the Carbon Pricing Leadership Coalition (Stiglitz and Stern 2017), at least $40–80/$tCO_2$ by 2020 and $50–100/$tCO_2$ by 2030 are needed to lower the rate of emissions to keep below the 2 degree temperature increase targeted at Paris.

Without an ever-lowering cap on emissions, the incentive to increase prices and raise trading volumes does not exist. The overall context remains one of economic stagnation, financial volatility, and shrinking demand for emissions reduction credits. The world faces increasing sources of carbon credit supply in an already glutted market, thanks to the COP negotiators' failure to mandate binding emissions cuts. But another factor remains behind the lax system that the UN, the EU, and other regulatory bodies appear to have adopted. All manner of inappropriate projects appear to be gaining approval, especially in Africa (Bond et al. 2012). As California's carbon market was renewed in 2017, a new round of complaints arose from activists about the scheme's implicit environmental racism (insofar as polluting industries in neighborhoods with predominantly racial minority populations continue emissions because of their purchase of carbon credits).

The carbon market's failures have renewed concern about the "privatization of the air" among climate justice activists. This fear was originally articulated by the Durban Group for Climate Justice (Lohmann 2006) and in the 2009 film by Annie Leonard, *Story of Cap and Trade* (Leonard 2009). Again, aside from FOEI, the INGOs sought reforms, not abolition, of the carbon markets; Greenpeace deprioritized the EU ETS, but the WWF strongly endorsed such markets along with investment in renewables and innovation (Bryant 2016:12–13).

At some point, weaknesses in the carbon trading strategy should be forcefully addressed by INGOs and their justification for ongoing futile reform advocacy reconsidered. This, however, is not the only aspect of neoliberal nature that splits global from local climate justice activists. There are other "false solutions" to the climate and other environmental crises, and many more continue to emerge from the private sector, some in alliance with the business-oriented INGOs, for example:

- controversial forms of so-called "cleaner energy," such as nuclear, "clean coal," fracking shale gas, hydropower, and hydrogen;
- biofuels, biomass, and biochar; and
- geoengineering gimmicks, such as carbon capture and storage; genetically modified trees and other biomass; sulfates in the air to shut out the sun; iron filings in the sea to create algae blooms; artificial microbes to convert plant biomass into fuels, chemicals, and products; and large-scale solar reflection (e.g., industrial-scale plastic-wrap for deserts).

Many of these technical-fix strategies violate the precautionary principle, create land-grab pressure, have excessive capital costs, require increased energy, are unproven in the technological sense, and are years if not decades from implementation. While promoting some obvious technological improvements, such as renewable energy and transport efficiencies, several very small INGOs with a decidedly climate justice orientation (e.g., the ETC Group et al. 2010) confirm their opposition to the more extreme false solutions:

> The shift from petroleum to biomass is, in fact, worsening climate change, increasing deforestation and biodiversity loss, degrading soils and depleting water supplies. Further, the new "bio-based" economy threatens livelihoods, especially in the Global South where it encourages "land grabs."

In McAfee's (2012) view, compensating the poor and other land users for practices that maintain healthy, service-producing ecosystems may be an important part of strategies for sustainable and equitable development. Serious problems arise, however, when such compensation schemes are framed as markets.

If the "net" emissions reduction strategy is not questioned, not only will carbon trading and offsets potentially revive (with all their intrinsic problems unresolved), but a panoply of false solutions will be funded by the GCF. Even when INGOs with a climate justice orientation get involved in global technical advocacy, debilitating problems emerge due to adverse power relations, as the GCF has already demonstrated. Sarah Bracking criticizes *both* the mainstream INGOs and climate justice participants in the GCF who "invested resources and energy into a process that distracts from other types of politics and issue framing" required to address climate finance (Bracking 2015):

> The promise of incremental reform became privileged over strategic withdrawal [from the GCF process], structural change and the insistence on effective government regulation. Representatives of the climate justice movement fought to give substantive weight to the initial radical framings, only for them to be captured in financial logics.

The Uncertain Terrain of Natural Capital Accounting

There is just one case of a neoliberal nature strategy that may have appeal to those with a climate justice orientation: contesting the extraction of fossil fuels (and other raw materials). This can easily be done for sites where it can be demonstrated that drilling for oil or coal does not make sense *economically* and not just in terms of pollution and environmental (including climate) damage, social dislocation, and disrupted spiritual values that are normally the basis for opposition. The main economic argument is that by calculating natural resource depletion associated with extraction, and comparing the outflow of those values with the inflow of retained profits and reinvestment made by the corporations which do the extraction, the overall impact is net negative.

Even though the World Bank has traditionally endorsed extraction (e.g., of fossil fuels), several Bank staff in the group studying Wealth Accounting and the Valuation of Ecosystem Services (WAVES) annually calculate "adjusted net savings" as an augmentation of national economic accounting. This follows the resignation letter of ecological economics founder Herman Daly (1996), in which he scolded the Bank for its failure to comprehend natural capital. WAVES' results are extremely disturbing. For example, the Bank's 2014 *Little Green Data Book* conceded that "88% of Sub-Saharan African countries were found to be depleting their wealth in 2010," with a 12% decline in Africans' per capita wealth that year attributed to the extraction of minerals, energy, and forest products (natural capital) (World Bank 2014:8).

The adjusted net savings measure is the most ambitious attempt to comprehend changes in wealth incorporating nature. Sub-Saharan Africans had the world's second most dramatic loss between gross and adjusted savings (Figure 8.2). For North Africa and the Middle East, gross savings were 27.9%, but adjusted savings were 8.1% thanks mainly to energy depletion being 12.4% of

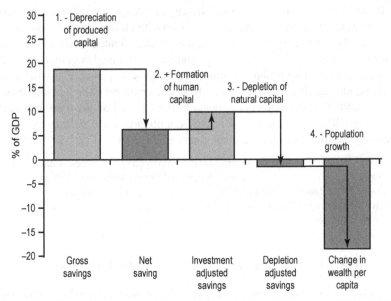

Figure 8.2   Decomposing change in wealth per capita, Sub-Saharan Africa, 2010. Source: World Bank, Wealth database 2014.

gross national income. In contrast, resource-rich wealthy countries (including Canada, the United States, Australia, and Norway) witness sufficient reinvestment by (home-based) corporations, such that their natural capital depletion was outweighed by new physical capital, leaving a net positive outcome (World Bank 2015:12).

Why might climate justice groups dedicated to decommodification of life tolerate such counting exercises, given that they are premised in the monetary valuation of natural resources? After all, in addition to concern about marketization that inexorably follows the monetization of natural values, Sian Sullivan (2013), one powerful critic of natural capital accounting, argues that there are

broader implications of conjuring "nature" in the form of the socioeconomic construct of money. Layer upon layer of abstraction lie between the connected breathing entities comprising aspects of "biodiversity," for example, and their selective calculation as "units" that can constitute "ecosystem service" work and be factored into "natural capital accounts." Once visible as these units, however, "nature" can be put to work as a value-generating asset, just like any other unit of capital. It can become a new source of monetary income (e.g., through Payments for Ecosystem Services and REDD+ carbon credits), and be leveraged as new forms of value-generating capital asset....Indeed, it seems strange, if not delusional, to expect that affirmations of the current economic paradigm will solve these related crises. To invoke Einstein, "we cannot solve our problems with the same thinking we used when we created them."

This is a fruitful and long overdue discussion. In a parallel debate, Robbins and Moore (2015) ask: "Amidst—and despite—its deep-seated rejection of technocratic fixes, can political ecology reconcile itself with ecomodernism?" In the sense, the term includes monetary valuation of nature. They answer cautiously in the affirmative: "We suggest that we join together to render ecomodern political ecology a therapeutic empirical project. Rather than become entrenched in an ongoing battle over the dysfunction of the other group's phobic attachments, then, we would instead explicitly engage them, working together to pose specific questions, open to productive exploration" (one of which might be whether natural capital accounting can be deployed to negate most existing forms of extractivism in Africa).

What climate justice-informed opponents of natural capital accounting have most trouble in criticizing is the need to punish polluters by considering formal monetary liabilities—or some approximation since nature is priceless—so that reparation payments to environment and affected peoples are sufficiently financed, and in the process an incentive is generated not to pollute in future. This is the reason to make at least a rough monetary case for "ecological debt" payments in courts of law.

For example, of Nigeria's $11.5 billion claim against Shell for a 2011 oil spill, more than half is meant to compensate fisherfolk. The liability owed to silicosis-afflicted mineworker victims of Anglo-American and other gold mining houses has begun to reach payment stage. The South African firms Gencor and Cape PLC had to pay $65 million a decade ago to settle South African asbestos lawsuits after they lost their last appeal in the U.K. House of Lords. Similar arguments should be made against the multinational corporations most responsible for what the United Nations terms loss and damage due to climate change. Ideally, over time, this strategy would develop as "fine-and-ban," so that when a corporation makes an egregious error, it is fined punitively for the damage done, and then sent packing.

To be sure, there is a danger that if "fine-and-ban" is not the local state policy, then natural capital accounting will lead, instead, to a "fee" for pollution, with the damage continuing, alongside ongoing payment. That would be the result if a formal market emerged, such as the EU ETS. Naturally, climate justice activists, beginning with the Durban Group for Climate Justice, firmly rejected these in 2004. The distinction should thus be clear between valuing nature for ecological debt payment purposes (a fine-and-ban) and pricing nature for market making (a fee). As Vandana Shiva put it in a 2014 South African talk: "We should use natural capital as a red light to destruction, not as a green light" (Bond 2014).

The "red light" strategy is an example of a potential rapprochement between INGO and climate justice framing strategies, emphasizing technical analysis as well as being useful to anti-extractivist campaigners who want an economic argument against fossil fuel depletion. The "*differences in normative and theoretical positions*" remain, but use of natural capital accounting against extractivism offers one example of *"more inclusive framings [that] might enable more societally relevant and impactful research and more concerted action/practice"* (Lele et al., this volume).

In sum, natural capital accounting is potentially one narrative that might bridge INGOs and climate justice groups, especially in making the economic argument to "leave the oil in the soil, coal in the hole, tar sands in the land, and fracking shale gas under the grass," as Joan Martinez-Alier posited (Martinez-Alier 2014). Instead of extracting such resources when they demonstrably lead to much lower adjusted savings, is there scope for a different narrative that compels a climate debt to be paid to those who suffer climate change and who are also residents of fossil fuel reserve sites? This has been one route taken by Oilwatch members to justify national leaders in places like Ecuador (the Yasuni case) and Nigeria (Ogoniland) to leave fossil fuels untouched (Bond 2012b).

To arrive at that narrative requires one more detour through the philosophies of environmental management: sustainable development.

The Scorched Earth of Sustainable Development Narratives

If there is an alternative worldview to neoliberal nature, most INGO and climate justice narrative shapers and strategists would immediately point to the phrase "sustainable development." The 1987 United Nations Commission, led by Gro Harlem Brundtland, offers a definition still worth returning to (Brundtland Report 1987). Not only does it contain the intergenerational requirement expressed in the first clause of her definition:

Sustainable development is development that meets the needs of the present without compromising the ability of future generations to meet their needs.

the following subclauses observe first "the concept of 'needs,' in particular the essential needs of the world's poor, to which overriding priority should be given," thus generating grounds for social justice advocacy. Second, "the idea of limitations imposed by the state of technology and social organization on the environment's ability to meet present and future needs" repudiates pro-growth assumptions of those who use the words sustainable development in public relations greenwashing.

The idea gained popularity in 1972 with the first Earth Summit in Stockholm and in *The Limits to Growth* (Club of Rome 1972), culminating in the Brundtland Commission and 1992 Rio Earth Summit. Soon, however, sustainability was co-opted by corporations during the 1990s and downgraded in favor of neoliberal ideologues' advocacy of export-led growth and the commodification of nature. Sustainability was raised once again at a 2002 UN Earth Summit in Johannesburg, which unfortunately fused the UN's strategy with the for-profit agendas of privatizers, carbon traders, and mega-corporations which supported the UN Global Compact (which was mostly a fund-raising exercise for a beleaguered institution). Then, in 2012, at the next Rio Earth Summit, sustainability was fused with "green economy" rhetoric, biodiversity offsetting, and market-centric climate change policy. Sustainability had again flowered, but now with a much more direct relationship to neoliberal nature (Büscher et al. 2014). For the 2015–2030 period, sustainable development goals are now the mantra of the UN and many other multilateral agencies, in spite of extensive critique of the realities they elide, such as by the scholar–activist network TheRules.org (2015).

Even if this weak version of the sustainability narrative is contested by climate justice critics—and attacked by the most pollution-intensive fractions of capital—there is no questioning the problem of rampant socioenvironmental unsustainability as the world hits what the Club of Rome (1972) had long warned would be "planetary boundaries." The most serious threat is exhaustion of the carrying capacity for greenhouse gases that cause climate change, and in turn, ocean acidification. There are others: biodiversity loss, stratospheric ozone depletion (abated by the 1987 Montreal Protocol that phased out chlorofluorocarbons by 1996 but leaving atmospheric aerosols as a danger), oceanic degradation and acidification, crises in the biogeochemical nitrogen and phosphorus cycles, other resource input constraints, chemical pollution, freshwater adulteration and evaporation, and shortages of arable land (Mace et al. 2014; Magdoff and Foster 2011; Steffen et al. 2015). So for those in INGOs and climate justice grassroot groups genuinely concerned with global environmental sustainability, the next question is whether the logic of capitalism can generate repairs for the intrinsic damage being done during the "Capitalocene" (Moore 2013)? Seeking sustainablility, many INGOs believe in a "green capitalism" strategy based on arguments by Gore (2009) and Hawken et al. (1999) (for a critique see Tanuro 2014). Yet as Ariel Salleh (2010) argues, a serious consideration of externalized costs should include at least three kinds of surplus extractions, both economic and thermodynamic, never comprehensively incorporated by reformers: (a) the social debt to inadequately paid workers, (b) an embodied debt to women family caregivers, and (c) an ecological debt drawn on nature at large. The more conservative INGOs have simply ignored the logical trajectory of "polluter pays" externalization in the sense pointed out by Salleh.

Concepts of the left dissent from this weak form of sustainability, stressing sustainability as achieved through distributional equity, nonmaterialist values, and a critique (and transcendence) of the capitalist mode of production are:

- The environmental justice vision that African-American activists in North Carolina began to articulate in the 1980s (Bullard 2000).
- "Anti-extractivism" and the "rights of nature" articulated by Ecuadorean and Bolivian activists and constitutions, even if not in public policy as pointed out by Accion Ecologica Colectivo Miradas críticas del Territorio desde el Feminismo (2014).

- Andean indigenous peoples' versions of *buen vivir* (living well) and allied ideas (Biggs 2011).
- "Degrowth" (*décroissance*) (Latouche 2004).
- Post-GDP "well-being" national accounting (Fioramonti 2014), such as Bhutan's Gross National Happiness which emphasises sufficiency
- "The commons" (Linebaugh 2008).
- Ecosocialism (Kovel 2007).
- Strategies for transitioning to genuinely sustainable societies and economies, also hotly debated (see Scoones et al. 2015; Swilling and Annecke 2012).

With such creative options flowering—albeit in a sometimes reformist mode harking back to indigenous conservation, mere accounting reforms, and the slowing (not ending) of capitalism—genuine sustainability ultimately depends on the nature of the critique of unsustainability. Perhaps the most popular systemic analysis comes from Annie Leonard's *Story of Stuff* film and book (Leonard et al. 2007), which link the spectrum of extraction, production, distribution, consumption, and disposal. In her book, *This Changes Everything*, Klein (2014) puts the onus on capitalism for climate change. Martinez-Alier and Spangenberg (2012) express most bluntly what is truly at stake:

> Unsustainable development is not a *market failure* to be fixed but a *market system failure*: expecting results from the market that it cannot deliver, like long-term thinking, environmental consciousness, and social responsibility.

Conclusion: From Dueling Narratives to Practical Fusions

Returning to Durban, here is a revealing question: Can Kumi Naidoo, Bobby Peek, Desmond D'Sa, and Ashwin Desai—ensconced as they are in an INGO, a local NGO, a climate justice community organization, and academia (albeit with Naidoo having moved from Greenpeace to Johannesburg in 2016 to set up the Africans Rising civil society network)—identify common framings for addressing climate change, given the huge wedge between them that opened up during the COP process, especially at COP21 in 2015? The answer remains ambiguous.

At first blush, one factor dating back to the anti-apartheid struggle draws them all together: a deep respect for mass democratic action. That is where one of Naidoo's most important recent statements—made in mid-2014 with dozens of other INGO leaders and strategists at his Rustler's Valley eco-ranch in South Africa—provides hope: On one hand, there is underlying humility in the current generation of INGO leaders. On the other, there is a profound organic intelligence on the part of local climate justice activists who have the potential to take their perspective onto what initially appears to be the extremely hostile terrain of natural capital accounting.

Naidoo and more than thirty others explain why climate justice and similar grassroots forces *are holding the INGOs to account,* in an extraordinarily frank and refreshing confessional (CIVICUS 2014):

A new and increasingly connected generation of women and men activists across the globe question how much of our energy is trapped in the internal bureaucracy and the comfort of our brands and organizations. They move quickly, often without the kinds of structures that slow us down. In doing so, they challenge how much time we—you and I—spend in elite conferences and tracking policy cycles that have little or no outcomes for the poor. They criticize how much we look up to those in power rather than see the world through the eyes of our own people. Many of them, sometimes rightfully, feel we have become just another layer of the system and development industry that perpetuates injustice.

It is that cringeworthy honesty that opens the door to alliances with climate justice groups which want, as Naidoo et al. put it, to "challenge the business-as-usual approach. Prioritize a local community meeting rather than the big glitzy conferences where outcomes are predetermined." To be sure, cynics (like Desai) would point out that the glitzy Elmau and Paris conferences, where such unsatisfactory outcomes were predetermined, gained the endorsement of Greenpeace under Naidoo's leadership. It is also true that Peek and D'Sa continually prepare community activists to intensify multiple Blockadias in South Durban in their attempts to halt neighborhood-destroying truck and ship traffic (partly on grounds of climate change), calling for divestment from the firms involved (with Desai sniping, most often with exceptional insight, from the sidelines).

Yet it is also true that Naidoo's time at Greenpeace was marked by a revival of both militant leadership (e.g., his heroic disruption attempt in the Arctic) and decentralization of resources to the South, and that Greenpeace U.S. under Annie Leonard's lead has fused traditional monkey-wrenching with social and racial justice advocacy for the first time. Linkages of women, Muslims, Latinos, African-Americans, immigrants, indigenous Native Americans, other minorities, the LGBTQ community, poor people, trade unionists, environmentalists, and social justice activists are increasingly common as a result, offering a "social self-defense" which activist Jeremy Brecher (2017) identifies in his survey of anti-Trump struggles at the time of the inauguration.

Trump's decades' worth of extreme real estate corruption, property gambles, debt defaults and full-fledged bankruptcies, refusals to pay suppliers, and tax chiseling have reportedly attracted more than 4,000 lawsuits (Penzenstadler and Kelly 2016). One high-profile suit Trump is opposing was filed by lawyers on behalf of 21 young Americans (9–20 years of age) on the grounds that his policies (like Obama's) threaten their future.

Regardless of how courts address climate challenges, by attacking Trump's policies and projects, climate justice and other climate activists can find unprecedented unity. Trump's plan is to build filthy Keynesian infrastructure (fossil-fuel pipelines, airports, roads, and bridges), cancel international climate obligations, retract shale gas restrictions and the ban on the Keystone oil pipeline, encourage drilling, defund renewable energy and public transport, as well as to destroy the Environmental Protection Agency (EPA) (Trump 2016). His choices for the main climate-related Cabinet positions left no room for doubt: former ExxonMobil chief executive Rex Tillerson as Secretary of State, Scott Pruitt as EPA Director (based on his Oklahoma career attacking the EPA), and former Texas governor Rick Perry as Secretary of Energy. As the leader of ExxonMobil, Tillerson was not only a major contributor to climate policy inertia for several decades, his recent contract for a massive $500 billion Siberian oil drill earned him the Russian "Order of Friendship" from Vladimir Putin in 2013. A year later, the deal was postponed due to sanctions following Putin's invasion of the Crimea, and tightened sanctions in mid-2017 make the project's revival unlikely.

The climate critique of Trump is also the basis for divestment, for example, of firms associated with Trump's cabinet and top officials (Goldman Sachs bank, ExxonMobil oil, Koch Industries oil, Lockheed Martin military, Pfizer drugs, General Dynamics military, Wells Fargo bank, Amway beauty, and Breitbart media). A broader world divestment movement would build on conceptual tools that have been around for years and that immediately came to life after Trump's election (Bond 2017):

- A decade earlier, Joseph Stiglitz argued that "unless producers in America face the full cost of their emissions, Europe, Japan, and all the countries of the world should impose trade sanctions against the U.S."
- Journalist Naomi Klein reacted to Trump's election: "We need to start demanding economic sanctions in the face of this treaty-shedding lawlessness."
- Representing French business, conservative ex-president Nicolas Sarkozy threatened, "I will demand that Europe put in place a carbon tax at its border, a tax of 1–3% for all products coming from the U.S. if the U.S. doesn't apply environmental rules that we are imposing on our companies."
- The *New York Times* quoted a leading Mexican official at the UNFCCC COP22 summit in Marrakesh: "A carbon tariff against the U.S. is an option for us," a stance echoed by a Canadian official.

Some INGOs are already playing a major role in these crucial battles. Even as it became obvious that the carbon trading strategy countenanced by Greenpeace had failed, the impact of the group's attacks on Shell Oil was formidable in 2015, far outweighing the failed EU ETS reforms in strategic importance. Any institutional cost-benefit analysis of the INGOs' emissions market advocacy (e.g., the astonishing $200 million spent during 2009–2010 on U.S. congressional lobbying for cap-and-trade legislation) would logically place Blockadia strategies far ahead in the benefits category although not without considerable costs (Shell's legal threats against Greenpeace plus the Portland court's fines for blocking its bridge access in mid-2015, for instance). Similarly, 350.org's commitments to direct action grow more vibrant the more the frustrations rise about the slow pace of state and corporate decarbonization. In late 2016, this was evident at the Standing Rock showdown where several INGOs assisted Native Americans in fighting (and initially defeating) the DAPL in a manner suffused with respect and local ownership. (Partly as a result, the framing of "water protectors" rather than climate warriors was emphasized.)

Some INGO visionaries are aware of the limitations of their structural location. For example, African anti-extractive activists, ranging from faith-movement progressives to ActionAid, have responded vigorously to challenges made by Farai Maguwu and Christelle Terreblanche to the "Alternative Mining Indaba" (AMI), held every February in Cape Town to coincide with the African Mining Indaba of major corporate and state attendees. Instead of being resolutely committed to fighting mining—especially coal, which is increasingly destructive across a range of constituencies—AMI tends toward mild-mannered reforms. The dispute recorded in Maguwu and Terreblanche (2016)—including several dueling op-ed articles in March 2016 in the main African ezine, *Pambazuka*—is one example of the INGO-climate justice tensions noted above.

Another example is a reform network of capital and the state, the Extractive Industries Transparency Initiative (EITI). In 2016, EITI witnessed a legitimacy challenge from the (INGO) "Publish What You Pay" movement of Soros-funded NGOs (some of which have grassroots climate justice connections) when EITI imposed a "civil society" representative in their decision-making processes through a dubious process.

In other words, more connections between these differently located and philosophically divergent types of civil society—INGOs and climate justice activists—may unearth further frictions. I believe it is incumbent now upon the better-resourced INGOs to take up the challenge made by CIVICUS (2014) to provide not just auto-critique but new modes of operation sensitive to the (often more radical) grassroots agenda.

This means a complementary move by climate justice groups might be considered, both to scale up their critique so they can offer concrete *global scale* analysis and start networking properly, and to gather sufficient confidence to take on INGO rhetoric, much of which was learned in struggles within the system. This is the argument made by one of the world's leading contemporary historical materialists, David Harvey (1996:400–401), who insists that climate justice activists must become more forward-looking and

> deal in the material and institutional issues of how to organize production and
> distribution in general, how to confront the realities of global power politics
> and how to displace the hegemonic powers of capitalism not simply with
> dispersed, autonomous, localized, and essentially communitarian solutions
> (apologists for which can be found on both right and left ends of the political

spectrum), but with a rather more complex politics that recognizes how environmental and social justice must be sought by a rational ordering of activities at different scales.

In turn, I believe, the climate justice movement organizations, which often suffer from excessive localism (or expressed more positively, "militant particularism," as Harvey calls it), should attempt to link up more decisively with each other, take the broadest terrain as their mandate (including cultural and spiritual features of ecological and social life), and seek to rationally reorder the space economy in a way that directly confronts capitalism's neoliberal discourses. In addition, Harvey (1996:401) suggests:

> The reinsertion of "rational ordering" indicates that such a movement will have no option, as it broadens out from its militant particularist base, but to reclaim for itself a nonco-opted and non-perverted version of the theses of ecological modernization. On the one hand that means subsuming the highly geographically differentiated desire for cultural autonomy and dispersion, for the proliferation of tradition and difference within a more global politics, but on the other hand making the quest for environmental and social justice central rather than peripheral concerns. For that to happen, the environmental justice movement has to radicalize the ecological modernization discourse.

To radicalize ecological modernization, climate justice groups should not boycott the neoliberal nature thesis but instead engage and search out ways to avoid a "post-political" quagmire (Swyngedouw 2010) when it comes to combating climate change and the corporations and states behind it. Both Bryant (2016) and Bracking (2015) cite the critique of climate governance in Swyngedouw (2010), in which carbon markets and the Consumer Goods Forum represent "the predominance of a managerial logic in all aspects of life [and] the reduction of the political to administration where decision making is increasingly considered to be a question of expert knowledge and not of political position."

To return to Andrew Jamison, a typology of the dichotomy between green business and critical ecology leads to a third option (Table 8.1) that transcends even environmental justice: ecosocialism (a term that I have inserted into Jamison's rubric, but it is only a semantic intervention). In the first row, Jamison concedes that green business can sometimes, perhaps often, co-opt environmentalism into the nexus of capital accumulation, using concepts of sustainable development, a problem observed above. The critical ecology movements (including climate justice) resist, drawing upon concepts of environmental justice. The battle of environmentalists and green NGOs against transnational corporations, states, and global agencies will not succeed without a dialectical advance to the next stage: hybrid red-green networks. As for emblematic forms of action, the commercial, brokerage functions of green business—often with INGO legitimization (such as in the carbon trading and climate justice examples)—come into direct cultural conflict with the repertoire of resistance

Table 8.1 Dialectics of environmentalisms and ecosocialism (adapted from Jamison 2001).

Terrain	Green business	Critical ecologies	Ecosocialism
Type of agency	Transnational corporations, states, and global agencies	Environmentalists and green NGOs	Hybrid red-green networks
Forms of action	Commercial, brokerage	Popularization, resistance	Exemplary mobilization
Ideal of "science"	Theoretical, expert	Factual, lay	Situated, contextual
Knowledge sources	Disciplines	Traditions	Experiences
Competencies	Professional	Personal	Synthetic

tactics utilized by climate justice activists. The ecosocialist project, in contrast, has to advance to the stage of what Jamison terms "exemplary mobilization."

Intellectual buttressing remains crucial; hence the ideal articulation of "science" is also worth pursuing. The "theoretical expert" inputs, no matter how flawed in reality, that are used by ecological-modernization promoters working from a green business standpoint, contrast with the factual and lay languages of activists. Can we build on the second by defying the first to achieve a situated, contextual science, such as in the natural capital controversy? The knowledge sources that undergird such efforts are typically divided into the technical disciplines of green business, the political traditions of ecosocial justice, and the transcendental experiences of the ecosocialist project. As for the terrain of competencies, the green-business suits claim professionalism; the critical ecologists invoke personal commitment; and ecosocialists strive for a synthetic understanding of the personal, professional, and, above all, political.

Ultimately this dialectic tension will allow us to draw out "*differences in framing environmental problems*" because they derive from quite substantive "*differences in normative and theoretical positions.*" Exploring the tensions in positionality between INGOs and climate justice, it is extremely difficult, yet perhaps not impossible, to identify "*ways in which more inclusive framings might enable more societally relevant and impactful research and more concerted action/practice*" (Lele et al., this volume). This will surely be a matter of debate through praxis in the months and years ahead.

References

Accion Ecologica Colectivo Miradas críticas del Territorio desde el Feminismo. 2014. La Vida en el Centro y el Crudo Bajo Tierra: El Yasuní en Clave Feminist. http://www.accionecologica.org/component/content/article/1754 (accessed Sept., 18, 2017).

180   *P. Bond*

Biggs, S. 2011. The Rights of Nature: The Case for a Universal Declaration of the Rights of Mother Earth. Ottawa: Council of Canadians, Fundación Pachamama and Global Exchange.

Bond, P. 2012a. Durban's Conference of Polluters, Market Failure and Critic Failure. *ephemera* **12**:42–69.

———. 2012b. Politics of Climate Justice: Paralysis above, Movement below. Pietermaritzburg: Univ. of KwaZulu-Natal Press.

———. 2014. Can Natural Capital Accounting Come of Age in Africa? TripleCrisis: Global Perspectives on Finance, Development, and Environment, Part 2. http://triplecrisis.com/can-natural-capital-accounting-come-of-age-in-africa-part-2/. (accessed Sept. 7, 2017).

———. 2016. Who Wins from "Climate Apartheid"? African Climate Justice Narratives About the Paris COP21. *New Politics* **60**:122–129.

———. 2017. Tripping up Trumpism through Global Boycott Divestment Sanctions. *Counterpunch*, January 20. https://www.counterpunch.org/2017/01/20/tripping-up-trumpism-through-global-boycott-divestment-sanctions/. (accessed Sept. 7, 2017).

Bond, P., K. Sharife, F. Allen, et al. 2012. The CDM Cannot Deliver the Money to Africa: Why the Clean Development Mechanism Won't Save the Planet from Climate Change, and How African Civil Society Is Resisting. *EJOLT Report* **2**:1–124.

Bracking, S. 2015. The Anti-Politics of Climate Finance: The Creation and Performativity of the Green Climate Fund. *Antipode* **47**:281–302.

Brecher, J. 2017. Social Self-Defense: Protecting People and Planet against Trump and Trumpism. In: Labor Network for Sustainability: Making a Living on a Living Planet. http://www.labor4sustainability.org/uncategorized/social-self-defense-protecting-people-and-planet-against-trump-and-trumpism/. (accessed Sept. 7, 2017).

Brundtland Report. 1987. Our Common Future. Oxford: Oxford Univ. Press.

Bryant, G. 2016. The Politics of Carbon Market Design: Rethinking the Techno-Politics and Post-Politics of Climate Change. *Antipode* **48**:877–898.

Bullard, R. D. 2000. Dumping in Dixie: Race, Class and Environmental Quality. New York: Westview Press.

Büscher, B., W. Dressler, and R. Fletcher, eds. 2014. Nature™ Inc.: Environmental Conservation in the Neoliberal Age. Tuscon: Univ. of Arizona Press.

CIVICUS. 2014. An Open Letter to Our Fellow Activists across the Globe: Building from Below and Beyond Borders, Aug. 6. Rustler's Valley, South Africa. http://blogs.civicus.org/civicus/2014/08/06/an-open-letter-to-our-fellow-activists-across-the-globe-building-from-below-and-beyond-borders/. (accessed Sept. 7, 2017).

Club of Rome. 1972. The Limits to Growth: A Report for the Club of Rome's Project on the Predicament of Mankind, ed. D. H. Meadows et al. New York: Universe Books.

Daly, H. E. 1996. Beyond Growth: The Economics of Sustainable Development. Boston: Beacon Press.

ETC Group, EcoNexus, The African Biodiversity Network, and Gaia and Biofuelwatch. 2010. Biofuels, Bioenergy and Biochar: False Solutions Lead to Land-Grabbing. Press Statement, Dec. 10, Cancun, Mexico. http://archive.li/x2jQR. (accessed Sept. 18, 2017).

Fioramonti, L. 2014. How Numbers Rule the World: The Use and Abuse of Statistics in Global Politics (Economic Controversies). London: Zed Books.

FOEI. 2015. Call to Action! Mobilising around COP21: What, Where, When? Email from Dipti Bhatnagar. Maputo. 12 August, Reproduced By Friends of the Earth International. http://jwsr.pitt.edu/ojs/index.php/jwsr/article/viewFile/578/651. (accessed Nov. 28, 2017).

Climate Justice during the Decline of Global Governance   181

Gore, A. 2009. Our Choice: A Plan to Solve the Climate Crisis. New York: Rodale Books.
Harvey, D. 1996. Justice, Nature and the Geography of Difference. Oxford: Basil Blackwell.
Hawken, P., A. Lovins, and L. H. Lovins. 1999. Natural Capitalism: Creating the Next Industrial Revolution. Boston: Little, Brown and Co.
Herbert, T., Sir. 1634. A Relation of Some Yeares Travaile Begunne Anno 1626 in Afrique and the Greater Asia. London: William Stansby and Iacob Bloome.
Jamison, A. 2001. The Making of Green Knowledge: Environmental Politics and Cultural Transformation. Cambridge: Cambridge Univ. Press.
Klein, N. 2013. Conversation. Earth Island Journal: News of the World Environment Autumn. http://www.earthisland.org/journal/index.php/eij/article/naomi_klein/. (accessed Sept. 7, 2017).
———. 2014. This Changes Everything: Capitalism vs. The Climate. New York: Simon & Schuster.
Kovel, J. 2007. The Enemy of Nature: The End of Capitalism or the End of the World. London: Zed Books.
Lander, E. 2010. Reflections on the Cochabamba Climate Summit. Amsterdam, Transnational Institute, 29 April. https://www.tni.org/es/art%C3%ADculo/reflexiones-sobre-la-cumbre-del-clima-en-cochabamba?content_language=en. (accessed Sept. 7, 2017).
Latouche, S. 2004. Why Less Should Be So Much More: Degrowth Economics. Le Monde diplomatique. http://mondediplo.com/2004/11/14latouche. (accessed Nov. 28, 2017).
Leiserowitz, A., E. Maibach, C. Roser-Renouf, S. Rosenthal, and M. Cutler. 2017. Climate Change in the American Mind: November 2016. New Haven: Yale Program on Climate Change Communication.
Leonard, A. 2009. Story of Cap and Trade. The Story of Stuff Project and Climate Justice Now! with Free Range Studios. http://www.storyofcapandtrade.org. (accessed Nov. 28, 2017).
Leonard, A., L. Fox, and J. Sachs. 2007. Story of Stuff. Free Range Studios. https://storyofstuff.org/movies/story-of-stuff/. (accessed Nov. 28, 2017).
Linebaugh, P. 2008. The Magna Carta Manifesto: Liberties and Commons for All. Berkeley: Univ. of California Press.
Lohmann, L. 2006. Carbon Trading: A Critical Conversation on Climate Change, Privatisation and Power. Develop. Dialogue 48:1–362.
Mace, G. M., B. Reyers, R. Alkemade, et al. 2014. Approaches to Defining a Planetary Boundary for Biodiversity. Global Environ. Change 28:289–297.
Magdoff, F., and J. B. Foster. 2011. What Every Environmentalist Needs to Know About Capitalism: A Citizen's Guide to Capitalism and the Environment. New York: Monthly Review Press.
Maguwu, F., and C. Terreblanche. 2016. Alternative Mining Indaba Debate. Pambazuka News. https://www.pambazuka.org/global-south/we-need-real-%E2%80%9Calternatives-mining%E2%80%9D-indaba. (accessed Nov. 28, 2017).
Martinez-Alier, J. 2014. Leave the Oil in the Soil, Leave the Coal in the Hole, Leave the Gas under the Grass. Seminar at Grantham Institute: Climate Change and the Environment. http://www3.imperial.ac.uk/newsandeventspggrp/imperialcollege/naturalsciences/climatechange/eventssummary/event_2-10-2014-15-31-40. (accessed Sept. 12, 2017).
Martinez-Alier, J., and J. Spangenberg. 2012. Green Growth. Ejolt. http://www.ejolt.org/2015/09/green-growth/. (accessed Sept. 7, 2017).

182 *P. Bond*

McAfee, K. 2012. Nature in the Market-World: Ecosystem Services and Inequality. *Development* **55**:25–33.

Moore, J. 2013. Anthropocene or Capitalocene? On the Origins of Our Crisis. Part 1: Excerpt from Ecology and the Accumulation of Capital. https://jasonwmoore.wordpress.com/2013/05/13/anthropocene-or-capitalocene/. (accessed Sept. 7, 2017).

Penzenstadler, N., and J. Kelly. 2016. How 75 Pending Lawsuits Could Distract a Donald Trump Presidency, USA Today, 25 October. https://www.usatoday.com/story/news/politics/elections/2016/10/25/pending-lawsuits-donald-trump-presidency/92666382/. (accessed Sept. 7, 2017).

Reyes, O. 2015. Leaders' Declaration G7 Summit (Climate Section) Annotated. Washington, D.C.: Institute for Policy Studies. 9 June. https://genius.com/G7-leaders-declaration-g7-summit-climate-section-annotated. (accessed Sept. 7, 2017).

Robbins, P., and S. Moore. 2015. Love Your Symptoms: A Sympathetic Diagnosis of the Ecomodernist Manifesto. Entitlecollective, June 19. https://entitleblog.org/2015/06/19/love-your-symptoms-a-sympathetic-diagnosis-of-the-ecomodernist-manifesto/. (accessed Sept. 7, 2017).

Salleh, A. 2010. From Metabolic Rift to Metabolic Value: Reflections on Environmental Sociology and the Alternative Globalization Movement. *Org. Environ.* **23**:205–219.

Scoones, I., M. Leach, and P. Newell. 2015. The Politics of Green Transformation. London: Routledge.

Steffen, W., K. Richardson, J. Rockström, et al. 2015. Planetary Boundaries: Guiding Human Development on a Changing Planet. *Science* **347**:1259855.

Stiglitz, J., and N. Stern. 2017. Report of the High-Level Commission on Carbon Prices. World Bank Carbon Pricing Leadership Coalition, Washington, Dc, 29 May. https://static1.squarespace.com/static/54ff9c5ce4b0a53decccfb4c/t/59244eed17bffc0ac256cf16/1495551740633/CarbonPricing_Final_May29.pdf. (accessed Sept. 7, 2017).

Sullivan, S. 2013. Should Nature Have to Prove Its Value? Green Economy Coalition, London, July 8. https://www.greeneconomycoalition.org/news-analysis/should-nature-have-prove-its-value. (accessed Sept. 7, 2017).

Swilling, M., and E. Annecke. 2012. Just Transitions: Explorations of Sustainability in an Unfair World. Cape Town: Univ. of Cape Town Press.

Swyngedouw, E. 2010. Apocalypse Forever? *Theory Cult. Soc.* **27**:213–232.

Tanuro, D. 2014. Green Capitalism: Why It Can't Work. Toronto: Fernwood Publishing.

TheRules.org. 2015. SDGS. London. https://therules.org/tag/sdgs/ (accessed Sept. 7, 2017).

Trump, D. 2016. An America First Energy Plan. New York. 27 May. https://www.donaldjtrump.com/press-releases/an-america-first-energy-plan.

World Bank. 2014. The Little Green Data Book 2014. Washington, D.C.: World Bank. https://elibrary.worldbank.org/doi/epub/10.1596/978-1-4648-0175-4. (accessed Nov. 28, 2017).

———. 2015. The Little Green Data Book 2015. World Development Indicators. Washington, D.C.: World Bank. http://documents.worldbank.org/curated/en/443931468189562382/The-little-green-data-book-2015. (accessed Nov. 28, 2017).

———. 2017. Carbon Pricing Watch 2017. Washington, D.C.: World Bank. https://openknowledge.worldbank.org/handle/10986/26565. (accessed Sept. 23, 2017).

9

Reflections on the State of Climate Change Policy

From COP21 to Cities

Manfred Fischedick, John Byrne, Lukas Hermwille,
Job Taminiau, Hans-Jochen Luhmann,
Franziska Stelzer, and Daniel Vallentin

Abstract

This Ernst Strüngmann Forum seeks to link justice, sustainability, and diversity agendas. In support, this chapter discusses how linkages between these three concepts have formed and changed in the climate change discourse, particularly in light of the recent Paris Agreement. As the latest addition to the portfolio of international climate change agreements, the Paris Agreement establishes a landscape in which nation-states, subnational actors, and transnational networks will be able to reconfigure existing linkages between sustainability, diversity, and justice, and perhaps improve upon them.

Here, three possible developments are identified which may substantially influence the reconfiguration process. Recognition is given to the sustainability and justice deficits that have plagued the "top-down" character of the international climate change discourse, and it is hypothesized that the Paris Agreement opens the door for "bottom-up" movements to claim a larger segment of climate change policy decision making and design. In turn, the "polycentric" landscape created by such "movement from below" appears to emphasize concepts such as inclusivity and transparency perhaps allowing for explicit climate justice commitments. Finally, to advance societal transformation and embrace diversity, it is hypothesized that the scientific endeavor needs to be transformed from a purely analytical pursuit to an effort that makes use of the wide range of scientific competences and provides support for transformative innovations to change unsustainable sociotechnical systems.

Introduction

Climate change is transforming global society. Will humanity respond by initiating social transformation through cooperative and reflexive change to remove future risks and protect vulnerable communities from present and mounting harm? A comprehensive vision of a truly sustainable and zero- or low-carbon society, associated with a clear and convincing implementation strategy, does not yet exist. Sustainability science can tell us much about the components of such a society and potential implementation pathways, but the true challenge is to attract political backing for functional models that bring different stakeholders together in support of chosen approaches. In brief, a social transformation would appear to require envisioning and enacting sustainability simultaneously. If we are to succeed in limiting global warming "well below 2°C" as stipulated in the Paris Agreement (UNFCCC 2016a, Art. 2), the global community needs to confront this challenge and take action immediately.

The Paris Agreement has sent a strong signal for societal transformation. It ingrains a new collective understanding of the challenge ahead and captures a strong normative obligation for nation-state action. Still, the agreement also suffers from a significant ambition gap and lacks enforceable mandated action (Civil Society Review 2015; Rogelj et al. 2016). Civil movements will continue to contest the continued dominance of the constituents of unsustainable sociotechnical systems worldwide (Hermwille et al. 2015). The policy landscape emerging from the Paris Agreement challenges nation-states, subnational actors, and civil society networks (from local to transnational scales) to reconfigure existing linkages between sustainability, diversity, and justice, as well as significantly improve upon them. The Paris Agreement forcefully reconsiders previously accepted global emission management strategies and opens the door for new ideas and experimentation.

In this chapter, we situate climate change policy and science in this post-Paris landscape, exploring three hypotheses that potentially provide for simultaneous envisioning and enacting of the needed societal transformation. First, cities and regions can and must play an important role in leveraging climate action through a multilevel governance system (Hypothesis 1). Showing signs of innovative experimentation and diffusion, the inclusion of these subnational actors, often articulated through regional, national, and transnational networks, departs from the overarching and singular story perpetuated throughout two decades of climate change negotiations. In addition it ties with the post-Paris multinarrative, nonlinear, and versatile landscape. Second, the climate justice deficit experienced throughout the world can be confronted by "polycentric" design and implementation that expressly incorporates civil society strategies to halt the use of unburned fuels and to empower local energy development (Hypothesis 2). Shared learning, adaptive management, and democratic legitimacy are key components of such strategies. Finally, a new line of inquiry needs to reflect more fully a social purpose of realizing a sustainable and equitable future (Hypothesis 3). Such "transformative inquiry" (i.e., science that makes use of the wide range of scientific competencies to support broad societal change) is required to address the complexity and diversity of our climate and societal challenge. Transformative inquiry can help to provide innovations that have the potential and power to change proactively unsustainable socio-technical systems, to gain a better understanding of system behavior and the decision-making processes, and to help serve the dual aims of a just and sustainable future.

Three Hypotheses

Changing Governance Architecture: Cities and Regions as Pioneers of Climate Governance

Hypothesis 1: A shift is already underway from the nation-state basis of policy design to more regional and locally based policy design. Local and regional levels of policy design are, or can be, motivated by very different political drivers. The concrete measures and investments they promote can shape a climate policy landscape better focused on sustainability and justice.

Subnational policy design efforts have increasingly positioned themselves as drivers of appropriate climate policy formulation (Jordan et al. 2015). Drivers of this emerging and accelerating activity include the "painfully slow" and seemingly paralytic character of global climate change negotiations (Neslen 2015). This movement "from below" embodies a pluralism that extends beyond previously introduced ideas of "regime complexes" that remained international, top-down, and state-centric (Biermann 2014). Captured under the designation of "polycentricity," this movement places the thrust of innovative climate governance within diverse, experiential, and multilevel engagements (Ostrom 2014). Polycentricity is positioned as a new means to organize political space, enable societal response and action, and importantly, govern climate change in the presence of "governance gaps" (Abbott 2014). Indeed, intensifying action at the local level has, for many, advanced the notion that polycentricity is a credible alternative strategy for climate change (e.g., Martinez-Alier et al. 2016; Ostrom 2014).

Cities and regions, in particular, have openly argued for their relevance in a polycentric system of governance (Castán Broto and Bulkeley 2013). With urbanization as a defining feature of modernity, cities are a key source of leadership potential grounded in features such as population (by 2030, about 60% of the global population will live in cities), capital (cities produce about 85% of global GDP), and energy use with its associated greenhouse gas emissions (cities generate 71–76% of energy-related greenhouse gas emissions). In addition, the high-density urban morphologies of cities face vulnerability to climate impacts, especially in coastal regions and along rivers. Proponents of city action furthermore posit the economic advantage (Colenbrander et al. 2015; Gouldson et al. 2015; Sudmant et al. 2015): a recent report by the New Climate Economy, for instance, put the global economic opportunity for low-carbon urban actions at USD 16.6 trillion over 2015–2050 (New Climate Economy

2015). Advantages of such urban climate strategies include economic benefits (such as energy cost savings and mitigation of energy price volatility), environmental improvements (e.g., local air-pollution abatement), and social advancement (e.g., green job growth).

Following a "governance by diffusion" strategy, urban climate change action and experimentation is gaining significant momentum (Hakelberg 2014). The Urban Climate Change Governance Survey, based on results from 350 cities worldwide, underscores the widespread diffusion of climate change action as 75% report activity in both mitigation and adaptation (Aylett 2014). Moreover, city commitments to climate change frequently outstrip their national counterparts in terms of ambition and coverage (e.g., Lombardi et al. 2014; Reckien et al. 2014) inspiring "hope that climate governance *in toto* is more active than critics transfixed by UNFCCC-related meetings have assumed" (Jordan et al. 2015). Often driven by co-benefits (e.g., improvement of local air quality, regional economic impulses, job effects) and backed by their citizens, cities play an active role, even in those cases where an enforcing legislative climate policy framework is missing.

Estimates of potential performance use these commitments to arrive at impressive totals (e.g., ARUP 2014). Empirical insight into the comprehensive performance of the strategy, however, remains challenging in part due to the strategy's strong reliance on "soft results" such as awareness raising, iterative learning, trust building, and democratic legitimacy as key distinguishing parameters (Jordan et al. 2015). Examples of potentially transformative efforts are listed in Table 9.1.

A common expression of urban climate change planning is the formation of (transnational) municipal networks (e.g., C40, ICLEI, Covenant of Mayors, European Green Capital). Such networks can deliver at least three strategies to facilitate effective governance: (a) an information and communication function promotes knowledge exchange, "best practices" diffusion, or improves local-technical expertise; (b) a project funding and cooperation function assists in the rollout of urban projects; and (c) a recognition, benchmarking, and certification function encourages experimentation and rewards leadership (Kern and Bulkeley 2009). Moreover, bottom-up activity can provide political pressure to raise ambition at other levels of government or drive deeper levels of cooperation (Keohane and Victor 2016).

The finding that (transnational) municipal networks (e.g., C40, ICLEI, Covenant of Mayors) themselves promote diffusion and implementation (Hakelberg 2014) combined with evidence of horizontal diffusion (i.e., where communities adjacent to other communities that participate in climate protection efforts are more likely to do so themselves) suggests a possibility of (urban) climate governance outgrowth. Additionally, evidence that "orchestration" of cooperation in climate change can be facilitated by states and intergovernmental organizations further allows for the possibility of governance outgrowth—meaning that these traditional actors often guide, broaden, and

Table 9.1 Overview of a selection of urban climate change activities and experiments (Bartlett and Satterthwaite 2016; Bond 2012; Bulkeley and Schroeder 2012; Francesch-Huidobro 2016; Reckien et al. 2014).

Region	City	Actions
Global	100% renewable energy movement	Dynamic movement of so-called 100% renewable energy cities/regions. Throughout the world, cities or regions pursue 100% renewable energy supply.
Africa	Cape Town	The "One Million Climate Jobs!" social movement brings together labor officials, community activists, and environmentalists to stimulate green job creation on a large scale.
Asia	Shanghai, China	While dominated by hierarchical governance, Shanghai has positioned itself as a leader in experimentation with low- or no-emission public transportation, particularly buses.
East Asia	Seoul, South Korea	One Less Nuclear Power Plant Initiative: innovative energy solution-searching effort seeking to avoid or sustainably generate energy within Seoul's jurisdiction equal to replacing one nuclear power plant, eliminating energy risks, reducing transmission losses, and enhancing citywide resiliency and sustainability.
Europe	Collection of Dutch, Danish, and Belgian cities	Cities pursue aggressive carbon-neutral emission profile by 2020 or 2025.
Middle America	Mexico City, Mexico	Actively exchanges best practices with other cities.
North America	Portland, OR, U.S.A.	Full-scale reconceptualization of the urban form.
	Los Angeles, CA, U.S.A.	World's largest LED streetlight retrofit program.
	New York, NY, U.S.A.	Climate justice agenda includes actions on health-care access, high-speed internet access, and addressing racial and ethnic disparities.
South America	Manizales, Colombia	Multiple risk exposure conditions accelerate shift to prevention, resilience, and vulnerability reduction (as opposed to emergency response).
	Quito, Bogotá, Curitiba	Pioneering cities for bus rapid-transit systems.

strengthen transnational and subnational climate governance (Hale and Roger 2014). The process can be self-intensifying: communities currently dissuaded from climate action due to perceived costs of greenhouse gas reduction efforts could be convinced by the mounting initiatives of neighboring jurisdictions and by the growing stock of learned good practices and demonstrable action benefits. The "driver's seat" for appropriate climate policy making, therefore, could be shifting to regional and locally based policy design.

Climate Justice in a Polycentric World

Hypothesis 2: Polycentric design and implementation can enhance climate justice prospects by moving decision making closer to those who are impacted. Open question: Can transformative change from the bottom up work globally only if an overarching set of justice principles are imposed?

A fracture of monocentric, top-down governance strategy into a polycentric movement of policy design and implementation (Hypothesis 1) will likely be accompanied with broad repercussions for climate justice (Hypothesis 2). While the top-down climate change approach employed throughout much of the negotiations raised questions of justice and equity in relation to operational principles (Okereke 2010), the approach has received frequent criticism for its failure to bring about climate justice (Barrett 2014; Ciplet et al. 2013; Shue 1993). Polycentric governance structures, captured in local movements, domestic networks, and transnational "bottom-up" agreements to share information, lessons learned, and political strategies reorient the subject of governance from the administrative state and technical experts to the "chaotic" landscape of cities, NGOs, communities, and so forth. (Bäckstrand and Lövbrand 2016). This landscape could be a source of transformative change through processes of critique, resistance, and new action. Divestment movements, for instance, enforce change by redirecting financial support away from conventional energy actors while local policy design and innovation opens new areas for community-prioritized and designed investment.

Polycentric policy design efforts could expand from the dominant reform process of environmental commodification and equally consider previously neglected dimensions of sustainability and justice (Bond 2012, 2015). For instance, an evaluation of four polycentric transnational networks found that such action can produce benefits for equity, inclusivity, information, accountability, organizational multiplicity, and adaptability (Sovacool 2011). Global measurement, reporting, and disclosure databases, capturing commitments and plans for thousands of cities, further stress the apparent desire for (global) accountability and responsibility in urban climate governance (Gordon 2016).

Experimentation by a diverse group of subnational actors has produced climate policy actions with impacts beyond the Kyoto Protocol mechanisms, especially of carbon trading (e.g., Hoffmann 2011). Similarly, polycentric response strategies deployed by subnational actors exhibit features such as "individual as coauthor" and other participatory characteristics. Thus, a recent study found that 34 out of the 627 urban climate change experiments surveyed were led by community-based organizations such as grassroots movements and that 296 of 627 experiments were performed by a partnership involving a multitude of diverse actors often including community-based organizations or citizens directly (Castán Broto and Bulkeley 2013). Community-based organizations, nongovernmental organizations, and the general population are often considered key supporting partners for local climate change action (Aylett 2014; Castán Broto and Bulkeley 2013; Hoffmann 2011).

Furthermore, public involvement can bring new knowledge and goals to the fore and can contribute to the determination of what is considered a "good outcome" depending on context. A comparative case study of the multistress environment of Delhi, Bogotá, and Santiago de Chile shows, for instance, that policy makers emphasize policy directions that advance local adaptive capacity and realization of co-benefits (Heinrichs et al. 2013). Participants in polycentric climate change action also emphasize and value the social and technical learning components provided by their membership in the effort (Galaz et al. 2012). Moreover, learning establishes a trust-building process as participants engage each other through successive rounds of experimentation and problem resolution (Ostrom 2014; Wilkinson 2010).

At a provincial scale, the process of developing a "climate protection plan" for North Rhine-Westphalia is an important example for broader stakeholder participation aiming (among other things) for introduction of additional competences and implementation culture for the proposed measures (Fischedick 2015). Actually, more than 400 different stakeholders have been involved in this process (including energy utilities, energy intensive industries, consumer associations, labor unions, NGOs, city networks) which has included common efforts to identify robust mitigation strategies for the state level, to bundle them in consistent scenarios, to discuss results of a comprehensive impact analysis of different future pathways (including socioeconomic aspects), and to propose and assess suitable policy instruments for the implementation process.

In terms of enhancing climate justice, polycentric policy can be seen to underscore the following:

1. Inclusivity or active involvement from a diverse number of stakeholders is important.
2. Internal and external accountability as project sponsors of polycentric policy designs have community ties and disclose intent and progress in (inter)national databases.
3. "One-way street" thinking prevalent in technocratic solutions (such as fiscal regulation, subsidies, and technical efficiency) is inadequate and inappropriate; the search for solutions is time and space contingent.
4. Adaptive management "demands constant revisability of ends as these are rethought and adjusted or altered in the course of experimentation and mutual learning" (Wilkinson 2010) or when placed in different contexts.
5. Shared learning and, through successive rounds of successful experimentation and problem resolution, trust building is necessary.
6. Democratic legitimacy of community representation is key.

Transformative Inquiry on the Rise?

Hypothesis 3: Successful implementation of ambitious greenhouse gas (GHG) mitigation strategies requires increased participatory approaches and a more proactive role of science (e.g., more societally relevant and impactful research, greater concerted action/practice).

In the face of persistent justice and sustainability problems challenging economic development, transformational changes are crucial. Proposed changes can include, for instance, large-scale transitions of practices, infrastructures, as well as values and priorities. The problems we are facing are "wicked" (Jahn et al. 2012), meaning that they are global, complex, and urgent. As such, science needs to move past its descriptive analytic role and should orient toward the society/nonscientific public (Lele and Norgaard 2005).

Recent approaches to this topic include Responsible Research and Innovation—a solution-oriented sustainability research strategy which embodies societal impact assessment frameworks. Sustainability research, in particular, has developed methods and approaches for transdisciplinary (stakeholder) participation (Clark 2007; Clark and Dickson 2003; Kates et al. 2001; Lele and Norgaard 2005; Miller 2013), knowledge integration (Jahn et al. 2012; Lang et al. 2012; Scholz and Tietje 2002), and for strengthening the science–society interfaces (Schäpke et al. 2015; Schneidewind and Scheck 2013) aiming to understand and contribute to transformations.

The latter, in particular, implies a reorientation toward experimental approaches by developing a new generation of experimental settings, such as living laboratories (e.g., Voytenko et al. 2016), urban (sustainability) transition labs (e.g., Loorbach and Rotmans 2010; Wiek and Kay 2015; Wittmayer et al. 2014), and "real-world" laboratories (e.g., Schäpke et al. 2015). Despite their differences, the settings share a focus on interventions in actual political-economic contexts undertaken by stakeholders in transdisciplinary collaborations with scientists and researchers. Furthermore, they share a double aim of understanding and at the same time contributing to societal change toward sustainability (see Schneidewind 2013). Accordingly, they are research endeavors, meaning they produce evidence regarding possible solutions to given sustainability problems (Wiek and Kay 2015) and at the same time pursue a transformational mission and therefore apply solutions with explicit climate justice objectives (Voytenko et al. 2016).

Another common goal of these approaches is to change the relationship of science and society from clearly disconnected to closely intertwined by (a) focusing on societally relevant problems, (b) enabling mutual learning processes among researchers from different disciplines (from within academia and from other research institutions) as well as actors from outside academia, and (c) aiming to create knowledge that is solution oriented, socially robust (see, e.g., Gibbons 1999), and transferable to both scientific and societal practice.

In light of these challenges, the role of science in society has been critically scrutinized, leading to calls for a new "social contract" (Funtowicz and Ravetz 1993; Gibbons 1999; Lubchenco 1998). Such a contract holds science accountable for its role in fostering or hindering progress toward sustainability around the world. This position endorses research in pursuit of effective solutions to complex societal problems. In return, it demands a fundamental shift in research design (Miller et al. 2014; Sarewitz et al. 2012; Wiek et al. 2012) and of the institutional science and funding system.

The Paris Agreement: Harbinger of a
New Climate Policy Paradigm?

Over time, the perception of the climate change problem has shifted significantly. The scope of the problem definition gradually increased from a rather narrowly defined environmental problem to include the developmental perspective, and ultimately to a fundamental transformation of global societies (Hermwille 2016). The "collective action" paradigm shaped decision making as the global character of the issue and was argued to necessitate international consensus as well as, more importantly, planetary carbon-emission management. When assessing the Paris Agreement, it is this paradigm that commonly guides the analysis. Can the Paris Agreement provide a basis for international governance of the great transformation? Is it inclusive enough to overcome the current confrontational style of climate policy and establish a sense of meaningful cooperation? Will it help to establish a common understanding that this transformation needs political guidance? And, ultimately, will it help global leaders to take the right decisions?

A separate assessment of the Paris Agreement seeks answers to a different set of questions. Testimony to the confrontation and disagreement among global leaders can be found in the realization that the pursuit of such a global collective action solution has been over two decades in the making. To date, global commitments have been unable to slow the pulse of the key identifier of climate change: the climbing pattern of atmospheric greenhouse gas concentrations first measured in Mauna Loa. Further, negotiation failures—most prominently in 2009 at the Copenhagen Bella Center—underline the sustainability and justice deficit which led to many openly questioning the viability and desirability of the collective action paradigm (e.g., Ostrom 2012). Can the Paris Agreement motivate and facilitate subnational and local creativity, innovation, and leadership to a globally meaningful level that can overcome the deficit left by 20 years of top-down negotiation impasse? Do communities embrace climate action due to global enforcement or due to the local climate, economic, and justice benefits of intervention? Does the Paris Agreement allow for local stakeholders to be decision makers?

The Paris Agreement provides six elements which, in our view, and in spite of all the remaining shortcomings of the international negotiation process and format, offer a promising foundation for successful governance of the required transformation:

1. An arena in which stakeholders can engage in a spirit of trust and cooperation
2. A shared transformational vision, not necessarily in terms of a clear picture about a mutual way forward, but at least as a common sense of direction
3. Sufficient resources, though not enough to finance the transformation, but at least to get the implementation process started
4. Transparency to provide the required information to further build trust and to allow for reflexivity
5. A mode to address undesired effects of the transformation
6. A process with a shared agenda and schedule

The following analysis is based on the Wuppertal Institute's more extensive analysis of COP21 and the Paris Agreement (Obergassel et al. 2016).

The Return of Environmental Multilateralism

After the diplomatic disaster of Copenhagen, confidence in the multilateral negotiation process had declined dramatically. As the French COP President Laurent Fabius stated in his speech before the final draft was tabled: "[I]f, today, we were so misfortunate as to fail, how could we rebuild hope? Confidence in the very ability of the concert of nations to make progress on climate issues would be forever shaken" (Fabius 2015).

To avoid the breakdown of environmental multilateralism, three elements were crucial: First, diligent preparation and outstanding leadership was provided by the French COP Presidency and the UNFCCC Secretariat. Second, a "high ambition coalition" emerged, comprised of Small Island States, Least Developed Countries, and the EU, and ultimately even included countries like Japan, Brazil, and the United States. This coalition helped to push the outcome toward the upper end of what seemed politically possible. Third, by allowing national determination of intended effort, Parties were able to reach an agreement which envisages climate action by all nations, partly repairing the deep schism between developed and developing countries within the UNFCCC. The success of the Paris conference, therefore, restored some of the confidence that had been lost over the last decade (cf. Bodle et al. 2016).

Normative Vision: The Long-Term Goal

The agreement's ambition of limiting global warming to "well below 2°C above preindustrial levels and to pursue efforts to limit temperature increase to 1.5°C above preindustrial levels, recognizing that this would significantly reduce the risks and impacts of climate change" (UNFCCC 2016a, Art. 2), represents a quantitative increase compared to the previous wording, but it is also a qualitative modification.[1] The 2°C threshold of the Copenhagen Accord and consequently the Cancún Agreements has been widely interpreted as a goal to be "achieved." This implies an economic cost-benefit calculation in which the 2°C threshold marks the point at or around which the cost of abatement of GHG emissions and the expected benefits of avoided cost through climate change impacts are deemed to break even (for comprehensive discussion, see Grubb et al. 2014). The sense of urgency of the 2°C goal was never beyond question to those familiar with the matter, but it may still have linguistically created a "comfort zone" and a sense of remaining flexibility that was never justified. The ultimate objective of the Convention is to avoid dangerous climate change. The long-term goal of the Paris Agreement can only be understood one way: any global warming is dangerous.

Furthermore, countries agreed that the temperature limit is to be reached by achieving "a balance between anthropogenic emissions by sources and removals by sinks of greenhouse gases in the second half of this century." Other formulations would probably have worked better as a norm to guide the behavior of actors. For example, a goal of full decarbonization would have provided a much less ambiguous mandate. However, from a climate science point of view, the actual formulation is even more inclusive, as it also encompasses greenhouse gases other than CO_2 and particularly the land-use sector. The opportunity becomes even more pressing when open questions and various risks are considered that are associated with "net negative emissions." If action is delayed, many climate protection scenarios do require negative emissions (most likely after 2050) to limit global warming to just 2°C. From the current perspective, it is hard to imagine, for instance, that the combination of the usage of huge amounts of biomass and carbon capture and storage, that is, the combination of two problematic strategies, could do the job and help extend the fossil fuels era. The Paris Agreement, thus, provides a strong mandate to accelerate the global societal transformation away from the age of fossil fuels. However, as elaborated above, the Paris Agreement insufficiently considers the justice consequences of its provisions, relegating the active pursuit of climate justice to the actions of polycentric actors.

Legal Construction

Unlike the Kyoto Protocol, the Paris Agreement does not take the shape of a protocol as per Article 17 of the UNFCCC. It follows an innovative legal

---

[1] Due to unequal distribution of warming effect, we note that by limiting the average *global* temperature increase to 1.5°C, some regions may experience temperature increase well above 1.5°C accompanied by more severe impact and threats.

approach: it is neither an amendment to the Convention nor a protocol. The Paris Agreement constitutes a new form of a (dependent) treaty under international law (Obergassel et al. 2016). This somewhat peculiar legal construction was chosen owing to the domestic politics of the United States as this construction would allow the ratification per executive order by the President of the United States as opposed to ratification through the Senate.

The domestic politics of the United States' constituted a secondary condition to the international negotiations in even more ways. For the same reasons, the U.S. delegation was not prepared to accept any legally binding obligations that go beyond what was already ratified in preceding international agreements or what is already reflected under current national legislation (i.e., predominantly the existing Clean Air Act). Hence, the Paris Agreement obliges Parties to communicate nationally determined contributions (NDCs) and to implement policies accordingly, but not to actually achieve them. Pledges are voluntary, but the process is compulsory (Clémençon 2016). Since a system of accountability through formal obligations was not possible under the current political realities, Parties reverted to a compromise: an accounting and transparency system paired with periodic moments of concentrated political attention.

Transparency

The Paris Agreement establishes, for the first time, a universal transparency system. While previously there had been separate reporting and review systems for industrialized and developing countries, now there will be only one system. This could substantially increase the transparency requirements for mitigation actions by developing countries. At the same time, the new system meets the demands of developing countries by also including adaptation and requiring developed countries to increase transparency on their provision of (financial) support. The details of the transparency framework will be part of the fine print in the Paris Agreement to be drafted in the coming years.

Climate Finance

The finance section of the Paris Agreement is weak. It does not contain any compulsion to scale-up climate finance (cf. Clémençon 2016). Only the accompanying decision text reiterates that the goal of mobilizing an annual USD 100 billion of North–South financial flows in 2020 and beyond, promised in Copenhagen, is still valid. What is more, industrialized countries were not prepared to provide a clear road map for how this goal could be achieved. The only step forward in Paris, albeit a small one, was that Parties agreed to set a new collective financing target by 2025. In this context, the USD 100 billion figure is now considered the bottom floor of financial contributions, rather than the ceiling as it was before Paris.

Addressing the Downsides: Loss and Damage, and Adaptation

One reason the Paris Agreement won the support of developing countries was its recognition of two decade-long demands: First, it elevates the standing of adaptation in the international climate regime. Crucially, action on adaptation is to be reviewed and accelerated every five years in parallel to the contribution cycles for mitigation.

Second, the Paris Agreement recognizes that there are adverse climate impacts that cannot be adapted to, but can only be dealt with. This was a crunch issue until the very end because developed countries feared that the inclusion of the concept of "loss and damage" in the agreement could be used to justify compensation and liability claims. The final outcome acknowledges both positions. The Paris Agreement features a separate article on loss and damage, while the decision text contains a clause that excludes compensation and liability claims. Treating "loss and damage" as a distinct issue, as opposed to a subcategory of adaptation, focuses and legitimizes ongoing international discussion on specifics such as possible appropriate response options and responsibility to act to assist recovery from damages and losses.

The Post-Paris Landscape

The Paris Agreement clearly does not "resolve" climate change as an environmental problem. The agreement imposes legal obligations on signatories to formulate and communicate climate policy objectives, the so-called NDCs. However, it does not obligate them to achieve those contributions.

The emissions reductions pledged by countries are out of line with its global target. Assuming these pledges are fully implemented, the global mean temperature would most likely still increase substantially (see Figure 9.1). Yet it is important to observe that this shortfall of ambition has been explicitly highlighted in the decisions accompanying the Agreement (UNFCCC 2016b, para. 17).

The Paris Agreement aims to address the lack of legal compulsion by creating a reputational risk through the establishment of mandatory transparency framework and review provisions. Parties agreed that "successive nationally determined contribution will represent a progression beyond the Part[ies'] then current nationally determined contribution" (UNFCCC 2016a, Art. 4.3), ensuring that the policy cycles induced by the Agreement resemble a ratchet mechanism. Reneging on earlier pledges is prevented. Starting in 2018, these mandated "stocktakes" will create moments of concentrated political attention every five years that may be used to foster political pressure on governments and corporations, strengthening a growing critique of the global political economy. Operating at all scales, civil society movements and networks are expected to provide ammunition for critique of nation-state and corporate action by revealing any shortcomings in goal fulfillment while simultaneously broadcasting promising results of experimentation and local leadership.

196

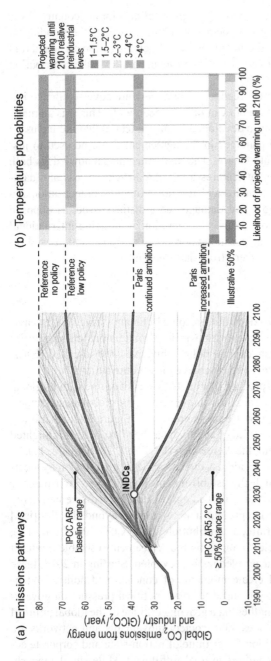

Figure 9.1 Global CO$_2$ emissions and probabilistic temperature outcomes of Paris. (a) Global CO$_2$ emissions from energy and industry (includes CO$_2$ emissions from all fossil fuel production and use, and industrial processes such as cement manufacture that also produce CO$_2$ as a by-product) for the four emissions scenarios explored in this study. The IPCC AR5 emissions ranges are from the IPCC scenario database. The IPCC AR5 baseline range comprises scenarios that do not include new explicit GHG mitigation policies throughout the century. The IPCC AR5 2°C 50% range comprises scenarios that limit global warming until 2100 to less than 2°C with at least a 50% chance. The faint lines within the IPCC ranges represent the actual emissions trajectories that determine the range. (b) Likelihoods of different levels of increase in global mean surface temperature change during the twenty-first century relative to preindustrial levels for the four scenarios. Although (a) shows only CO$_2$ emissions from energy and industry, temperature outcomes are based on the full suite of GHG, aerosol, and short-lived species emissions generated by the GCAM simulations. The illustrative 50% scenario in (b) corresponds to an emissions pathway that achieves a 50% chance of maintaining temperature change below 2°C until 2100. Other 50% pathways could lead to a range of temperature distributions depending on cumulative CO$_2$ emissions and representations of other GHGs. Reprinted with permission from AAAS (Fawcett et al. 2015).

Furthermore, the Paris Agreement has no termination date. This fact should promote a long-term outlook for the development of national policies and investment decisions in line with long-term goals. Also, Parties are urged to develop long-term low greenhouse gas emission-development strategies. This exercise could further facilitate the anchoring of climate protection in all government decisions.

As such, the Paris Agreement now deploys a pacemaker that stimulates and synchronizes the climate policy making on national and international levels. It creates periodic political moments, each of which can move us closer to a sustainable, carbon-free future. The intensity and effectiveness of these moments will depend, in part, on whether the spirit of cooperation experienced in Paris can be continued and transferred to a number of challenges that are not directly linked to the core of the Paris Agreement (see, e.g., Spencer et al. 2015). The prospect of transformation could grow if cooperation or even coordination can be achieved on issues such as

- exchanging competences (e.g., in terms of providing long-term strategies or scenarios[2]) (DDPP 2015);
- expanding and improving effectiveness of public-private R&D for the most promising technologies and mutual efforts to expedite the deployment of low-carbon technologies;
- avoiding frictions and incoherent market signals through uncoordinated or even contradictory (national) policies;
- addressing concerns of carbon leakage through concerted policy interventions among competing states;
- aligning global financial flows and the financial sector with the long-term goal of the Paris Agreement (see also UNFCCC 2016a, Art. 2.1c); and
- increasingly managing "exnovation"—deliberate divestment and phase-out (Kimberly 1981)—of unsustainable, high-carbon technologies, and industries, including the mitigation of social and economic disruptions that this may entail.

The pledge-and-review architecture provides an opportunity for strategic experimentation that was largely unavailable during the past two decades of "top-down" climate change negotiations. The multilevel character of climate policy could drive much of the national improvements in each successive "stocktake": "Diplomacy does not happen in a vacuum. The positions countries take internationally are determined by their domestic political situations" (Obergassel et al. 2016:39). The opening of such a "bottom-up" avenue for climate action could seek to fill sustainability and justice deficits that have been a part of the

---

2 Scenarios play an important role for sketching plausible future pathways and giving an orientation about room for maneuver if pursuing a specific mitigation goal. However, very often scenarios focus mainly on technological effects and do not fully reflect the role of behavior. As such, more advanced scenarios and underlying modeling instruments are needed in the future.

international governance effort. Climate justice, with its emphasis on unequal distributions across local communities, has an inherent bottom-up feature that has remained largely illegible to the international governance structure evidenced by, for example, the observation that financial flows to stem injustice are not being delivered to the communities that are most at risk and least responsible (e.g., Barrett 2014; Ciplet et al. 2013).

This "bottom-up" opportunity could drive the emergence of a "polycentric" paradigm that distinguishes itself from the technocratic strategy of planetary emissions control and focuses, instead, on creativity, experimentation, and innovation. The capacity for success in a polycentric system depends partly on (a) the successful roll-out of climate change action by a diversity of actors, (b) an enhanced prospect for climate justice, and (c) a societally relevant and impactful role for the scientific community. These three elements, discussed above in the three hypotheses, are critical questions that will shape the narrative and future evaluation of the Paris Agreement.

Conclusions

The Paris Agreement marks a new stage in the long history of climate change negotiations as it entrenches a trajectory first embarked upon in the wake of the Copenhagen COP failure. Ultimately, how this will play out remains uncertain. The provisions of the Paris Agreement offer promising conditions for renewed and vigorous climate change action despite its apparent shortcomings. Moving forward, the "heartbeat" of the Paris Agreement architecture will be shaped by national, subnational, and transnational efforts and innovation.

The official commitments by nation-states will likely fuel much of the discussion on the international stage at each successive point of stocktaking. Of particular interest to the international arena will be the efforts and improvements made by those nation-states heretofore unwilling to lock themselves into international agreement. However, concerns about sustainability and justice deficits linger as early evidence evaluating commitments suggests that nation-states will have to substantially raise ambition levels in order to meet their self-imposed objectives and avoid dangerous climate change, and that more needs to be done to advance fair burden sharing.

Leadership on climate change action might emerge from a new angle. The experimentation in subnational and transnational networks could form a distributed policy design pathway capable of overtaking the observed glacial speed of nation-state decision making. Community-based trial and error, negotiation, and creativity could produce sustainability and justice models available for rapid diffusion and adaptation to other contexts. A possible path forward for the Paris Agreement and the UNFCCC, in this light, is to act as a facilitator of such diffusion and recombination.

New sustainability and justice ideas could also emerge from a redirection of scientific effort. Past analytical endeavors have firmly established the importance of climate change and the need for societal transformation. A collaborative, solution-oriented scientific approach could reinvigorate the production of societal transformation options and help clarify to all participants its actual implementation.

The questions addressed in this chapter, in the form of three hypotheses, do not presume to establish the answers to existing sustainability and justice concerns. Rather, our intent was to highlight key trends and developments that could have substantial effects on the overall direction and level of change generated by the Paris Agreement, in an effort to advance understanding and stimulate further conversation.

Acknowledgments

The analysis of the Paris Agreement draws heavily on discussions with colleagues at the Wuppertal Institute. The text contains contributions from Wolfgang Obergassel, Christof Arens, Nicolas Kreibich, Florian Mersmann, Hermann E. Ott, and Hanna Wang-Helmreich. The Foundation for Renewable Energy & Environment (FREE) wishes to acknowledge the contributions from FREE colleagues Joohee Lee, Joseph Nyangon, and Jeongseok Seo.

References

Abbott, K. W. 2014. Strengthening the Transnational Regime Complex for Climate Change. *Transnat. Env. Law* **3**:57–88.

ARUP. 2014. Working Together: Global Aggregation of City Climate Commitments. http://publications.arup.com/publications/g/global_aggregation_of_city_climate_commitments. (accessed Oct. 27, 2016).

Aylett, A. 2014. Progress and Challenges in the Urban Governance of Climate Change: Results of a Global Survey. Cambridge, MA: MIT Press.

Bäckstrand, K., and E. Lövbrand. 2016. The Road to Paris: Contending Climate Governance Discourses in the Post-Copenhagen Era. *J. Environ. Pol. Plan.* DOI: 10 .1080/1523908X.1522016.1150777.

Barrett, S. 2014. Subnational Climate Justice? Adaptation Finance Distribution and Climate Vulnerability. *World Dev.* **58**:130–142.

Bartlett, S., and D. Satterthwaite. 2016. Cities on a Finite Planet: Towards Transformative Responses to Climate Change. New York: Earthscan.

Biermann, F. 2014. Earth System Governance: World Politics in the Anthropocene Earth System Governance: World Politics in the Anthropocene. Cambridge, MA: MIT Press.

Bodle, R., L. Donat, and M. Duwe. 2016. The Paris Agreement: Analysis, Assessment and Outlook. Dessau-Roßlau: German Federal Environment Agency (UBA) Research Paper.

Bond, P. 2012. Politics of Climate Justice: Paralysis above, Movement below. Scotsville, South Africa: Univ. of KwaZulu Natal Press.

Bond, P. 2015. Can Climate Activists "Movement below" Transcend Negotiators' "Paralysis above"? *J. World-Sys. Res.* **21**:250–270.

Bulkeley, H., and H. Schroeder. 2012. Global Cities and the Politics of Climate Change. In: Handbook of Global Environmental Politics Handbook of Global Environmental Politics, ed. P. Dauvergne, p. 560. Cheltenham: Edward Elgar.

Castán Broto, V., and H. Bulkeley. 2013. A Survey of Urban Climate Change Experiments in 100 Cities. *Global Environ. Change* **23**:92–102.

Ciplet, D., J. T. Roberts, and M. Khan. 2013. The Politics of International Climate Adaptation Funding: Justice and Divisions in the Greenhouse. *Global Environ. Politics* **13**:49–68.

Civil Society Review. 2015. Fair Shares: A Civil Society Equity Review of INDCs. http://civilsocietyreview.org/. (accessed July 18, 2017).

Clark, W. C. 2007. Sustainability Science: A Room of Its Own. *PNAS* **104**:1737–1738.

Clark, W. C., and N. M. Dickson. 2003. Sustainability Science: The Emerging Research Program. *PNAS* **100**:8059–8061.

Clémençon, R. 2016. The Two Sides of the Paris Climate Agreement: Dismal Failure or Historic Breakthrough? *J. Env. Develop.* **25**:3–24.

Colenbrander, S., A. Gouldson, A. H. Sudmant, and E. Papargyropoulou. 2015. The Economic Case for Low-Carbon Development in Rapidly Growing Developing World Cities: A Case Study of Palembang, Indonesia. *Energy Policy* **80**:24–35.

DDPP. 2015. Pathways to Deep Decarbonization 2015 Report. Deep Decarbonization Pathways Project. http://deepdecarbonization.org/wp-content/uploads/2016/03/DDPP_2015_REPORT.pdf. (accessed Oct. 6,2017).

Fabius, L. 2015. COP21: Plenary Session for the Submission of the Final Draft Text, Presented at the UNFCCC. http://www.diplomatie.gouv.fr/en/french-foreign-policy/climate/events/article/cop21-plenary-session-for-the-submission-of-the-final-draft-text-speech-by. (accessed July 18, 2017).

Fawcett, A. A., G. C. Iyer, L. E. Clarke, et al. 2015. Can Paris Pledges Avert Severe Climate Change? *Science* **350**:1168.

Fischedick, M. 2015. The Participatory Process to a Low-Carbon Economy in the German State of NRW. Transition and Global Challenges Towards Low Carbon Societies. *EAI Speciale* **I-2015**:34–37.

Francesch-Huidobro, M. 2016. Climate Change and Energy Policies in Shanghai: A Multilevel Governance Perspective. *Appl. Energy* **164**:45–56.

Funtowicz, S. O., and J. R. Ravetz. 1993. Science for the Post-Normal Age. *Futures* **25**:739–755.

Galaz, V., B. Crona, H. Österblom, P. Olsson, and C. Folke. 2012. Polycentric Systems and Interacting Planetary Boundaries: Emerging Governance of Climate Change–Ocean Acidification–Marine Biodiversity. *Ecol. Econ.* **81**:21–32.

Gibbons, M. 1999. Science's New Social Contract with Society. *Nature* **402**:C81–C84.

Gordon, D. J. 2016. The Politics of Accountability in Networked Urban Climate Governance. *Global Environ. Politics* **16**:82–100.

Gouldson, A., S. Colenbrander, A. Sudmant, et al. 2015. Exploring the Economic Case for Climate Action in Cities. *Global Environ. Change* **35**:93–105.

Grubb, M., J.-C. Hourcade, and K. Neuhoff. 2014. Trapped? In: Planetary Economics: Energy, Climate Change and the Three Domains of Sustainable Development, ed. M. Grubb et al., pp. 1–45. Abingdon, NY: Routledge.

Hakelberg, L. 2014. Governance by Diffusion: Transnational Municipal Networks and the Spread of Local Climate Strategies in Europe. *Global Environ. Politics* **14**:107–129.

Hale, T., and C. Roger. 2014. Orchestration and Transnational Climate Governance. *Rev. Int. Organ.* 9:59–82.

Heinrichs, D., K. Krellenberg, and M. Fragkias. 2013. Urban Responses to Climate Change: Theories and Governance Practice in Cities of the Global South. *Int. J. Urban Regional Res.* 37:1865–1878.

Hermwille, L. 2016. Climate Change as a Transformation Challenge. A New Climate Policy Paradigm? *GAIA* 25:19–22.

Hermwille, L., W. Obergassel, H. E. Ott, and C. Beuermann. 2015. UNFCCC before and after Paris: What's Necessary for an Effective Climate Regime? *Climate Policy* 17:150-170.

Hoffmann, M. J. 2011. Climate Governance at the Crossroads: Experimenting with a Global Response after Kyoto. Oxford: Oxford Univ. Press.

Jahn, T., M. Bergmann, and F. Keil. 2012. Transdisciplinarity: Between Mainstreaming and Marginalization. *Ecol. Econ.* 79:1–10.

Jordan, A. J., D. Huitema, M. Hilden, et al. 2015. Emergence of Polycentric Climate Governance and Its Future Prospects. *Nature Clim. Change* 5:977–982.

Kates, R. W., W. C. Clark, R. Corell, et al. 2001. Sustainability Science. *Science* 292:641.

Keohane, R. O., and D. G. Victor. 2016. Cooperation and Discord in Global Climate Policy. *Nature Clim. Change* 6:570–575.

Kern, K., and H. Bulkeley. 2009. Cities, Europeanization and Multi-Level Governance: Governing Climate Change through Transnational Municipal Networks. *JCMS* 47:309–332.

Kimberly, J. R. 1981. Managerial Innovation. In: Handbook of Organizational Design, ed. P. C. Nystrom and W. H. Starbuck, pp. 84–104. Oxford: Oxford Univ. Press.

Lang, D. J., A. Wiek, M. Bergmann, et al. 2012. Transdisciplinary Research in Sustainability Science: Practice, Principles, and Challenges. *Sustain. Sci.* 7:25–43.

Lele, S., and R. Norgaard. 2005. Practicing Interdisciplinarity. *Bioscience* 55:967–975.

Lombardi, M., R. Rana, P. Pazienza, and C. Tricase. 2014. The European Policy for the Sustainability of Urban Areas and the "Covenant of Mayors" Initiative: A Case Study. In: Pathways to Environmental Sustainability: Methodologies and Experiences, ed. R. Salomone and G. Saija, pp. 183–192. Cham: Springer.

Loorbach, D., and J. Rotmans. 2010. The Practice of Transition Management: Examples and Lessons from Four Distinct Cases. *Futures* 42:237–246.

Lubchenco, J. 1998. Entering the Century of the Environment: A New Social Contract for Science. *Science* 279:491.

Martinez-Alier, J., L. Temper, D. Del Bene, and A.-. Scheidel. 2016. Is There a Global Environmental Justice Movement? *J. Peasant Stud.* 43:731–755.

Miller, T. R. 2013. Constructing Sustainability Science: Emerging Perspectives and Research Trajectories. *Sustain. Sci.* 8:279–293.

Miller, T. R., A. Wiek, D. Sarewitz, et al. 2014. The Future of Sustainability Science: A Solutions-Oriented Research Agenda. *Sustain. Sci.* 9:239–246.

Neslen, A. 2015. EU Calls for Urgency in "Seriously Lagging" Paris Climate Talks. *Guardian*, August 20, 2015.

New Climate Economy. 2015. Seizing the Global Opportunity: Partnerships for Better Growth and a Better Climate. Washington, D.C.: New Climate Economy.

Obergassel, W., C. Arens, L. Hermwille, et al. 2016. Phoenix from the Ashes: An Analysis of the Paris Agreement to the UN Framework Convention on Climate Change. Wuppertal: Wuppertal Institute.

Okereke, C. 2010. Climate Justice and the International Regime. *Climate Change* 1:462–474.

Ostrom, E. 2012. Nested Externalities and Polycentric Institutions: Must We Wait for Global Solutions to Climate Change before Taking Actions at Other Scales? *J. Econ. Theory* 49:353–369.

———. 2014. A Polycentric Approach for Coping with Climate Change. *Ann. Econ. Fin.* 15:97–134.

Reckien, D., J. Flacke, R. J. Dawson, et al. 2014. Climate Change Response in Europe: What's the Reality? Analysis of Adaptation and Mitigation Plans from 200 Urban Areas in 11 Countries. *Clim. Change* 122:331–340.

Rogelj, J., M. den Elzen, N. Höhne, et al. 2016. Paris Agreement Climate Proposals Need a Boost to Keep Warming Well Below 2°C. *Nature* 534:631–639.

Sarewitz, D., R. Clapp, C. Crumbley, D. Kriebel, and J. Tickner. 2012. The Sustainable Solutions Agenda. *New Solutions* 22:139–151.

Schäpke, N., M. Singer-Brodowski, F. Stelzer, M. Bergmann, and D. J. Lang. 2015. Creating Space for Change: Sustainability Transformations: The Case of Baden-Württemberg. *GAIA* 24:281–283.

Schneidewind, U. 2013. Transformative Literacy: Gesellschaftliche Veränderungsprozesse Verstehen und Gestalten. *GAIA* 22:82–86.

Schneidewind, U., and H. Scheck. 2013. Die Stadt Als „Reallabor" für Systeminnovationen. In: Soziale Innovation und Nachhaltigkeit: Perspektiven Sozialen Wandels, ed. J. Rückert-John, pp. 229–248. Wiesbaden: Springer.

Scholz, R. W., and O. Tietje. 2002. Embedded Case Study Methods: Integrating Quantitative and Qualitative Knowledge. Thousand Oaks, CA: Sage.

Shue, H. 1993. Subsistence Emissions and Luxury Emissions. *Law Policy* 15:39–60.

Sovacool, B. K. 2011. An International Comparison of Four Polycentric Approaches to Climate and Energy Governance. *Energy Policy* 39:3832–3844.

Spencer, T., R. Baron, M. Colombier, and O. Sartor. 2015. Reframing Climate and Competitiveness: Is There a Need for Cooperation on National Climate Change Policies? Paris: OECD.

Sudmant, A. H., A. Gouldson, S. Colenbrander, et al. 2015. Understanding the Case for Low-Carbon Investment through Bottom-up Assessments of City-Scale Opportunities. *Climate Policy* 17:299–313.

UNFCCC. 2016a. Paris Agreement, United Nations Framework Convention on Climate Change. http://unfccc.int/files/meetings/paris_nov_2015/application/pdf/paris_agreement_english_.pdf.

———. 2016b. Report of the Conference of the Parties on Its Twenty-First Session, United Nations Framework Convention on Climate Change. http://unfccc.int/resource/docs/2015/cop21/eng/10a01.pdf (accessed July 17, 2017).

Voytenko, Y., K. McCormick, J. Evans, and G. Schliwa. 2016. Urban Living Labs for Sustainability and Low Carbon Cities in Europe: Towards a Research Agenda. *J. Cleaner Prod.* 123:45–54.

Wiek, A., and B. Kay. 2015. Learning While Transforming: Solution-Oriented Learning for Urban Sustainability in Phoenix, Arizona. *Curr. Opin. Environ. Sustain.* 16:29–36.

Wiek, A., B. Ness, P. Schweizer-Ries, F. S. Brand, and F. Farioli. 2012. From Complex Systems Analysis to Transformational Change: A Comparative Appraisal of Sustainability Science Projects. *Sustain. Sci.* 7:5–24.

Wilkinson, M. 2010. Three Conceptions of Law: Towards a Jurisprudence of Democratic Experimentalism. *Wisconsin Law Rev.* 673:673–718.

Wittmayer, J. M., N. Schäpke, F. van Steenbergen, and I. Omann. 2014. Making Sense of Sustainability Transitions Locally: How Action Research Contributes to Addressing Societal Challenges. *Critical Policy Studies* 8:465–485.

10

Energy and Climate Change

Sun-Jin Yun, John Byrne, Lucy Baker, Patrick Bond,
Götz Kaufmann, Hans-Jochen Luhmann,
Peter D. Lund, Joan Martinez-Alier, and Fuqiang Yang

Abstract

This chapter summarizes a series of discussions at the 23rd Ernst Strüngmann Forum,
which aimed at understanding how differences in framing environmental problems
in the area of energy and climate change are driven by differences in normative and
theoretical positions. Utilizing the diverse expertise of individual group members,
twelve framings were identified that shape the energy and climate debate. These
framings are used to explore how more inclusive engagement of these framings might
contribute to more societally relevant and impactful research.

Background

Over a century ago, scientists provided evidence that the burning of coal in
industrial countries causes CO_2 concentrations to increase in the atmosphere
(e.g., Arrhenius 1896; Callendar 1938). Since then, through increasingly sophisticated scientific models, early evidence of human impact has been upheld
with greater accuracy. Findings reported by the Intergovernmental Panel on
Climate Change (IPCC) are unequivocal: between 1850 and the present there
has been a rapid increase in atmospheric CO_2 concentrations, from 280–400
ppm, and this increase is due to human activity (IPCC 2014:3, 5). The principal impacts of human-induced change in atmospheric chemistry are (a) rising
average surface temperatures and (b) global mean sea-level rise (IPCC 2014:9,
11). Scientists at NASA's Jet Propulsion Laboratory have concluded that the
minimum atmospheric CO_2 concentration will remain above 400 ppm for the
next several decades "unless something dramatic happens with humans and the

Group photos (top left to bottom right) Sun-Jin Yun, John Byrne, Lucy Baker,
Patrick Bond, Peter Lund, Hans-Jochen Luhmann, Joan Martinez-Alier, Fuqiang Yang,
Götz Kaufmann, Lucy Baker, Hans-Jochen Luhmann, Peter Lund, Götz Kaufmann,
Sun-Jin Yun, Fuqiang Yang, Patrick Bond, Joan Martinez-Alier, John Byrne, Sun-Jin
Yun, John Byrne, Lucy Baker

planet" (Schmidt 2017). Climate change, as a result of human activity, is now accepted as fact by the scientific community, and its principal consequences— a warming of the planet due to rising concentrations of CO_2 and increased risks of coastal inundation due to sea-level rise—are likewise regarded as beyond scientific doubt. Only a dramatic shift in human use of energy to low- and no-carbon sources (IPCC 2014:28) will avert the worst effect of climate change that humans now face: irreversibility of the effects of climate change (IPCC 2014:16).

In light of human-induced change in atmospheric chemistry, the consequent increase in average surface temperature and global mean sea-level rise, and the recent concern that without a fundamental change in our reliance on carbon-intensive economic development we as a species face the threat of irreversibility, one must ask: Why did it take so long for climate change to become an issue of global importance?

One explanation is that climate change itself is the product of the modern energy–society relationship. The drive to increase capitalist industrial economies required an extraordinarily rapid use of energy, initially supported by a coal-mining regime and "urban exudation" (Mumford 1936/2010: 169). This created a carbon-intensive ideology of progress that went unquestioned for nearly a century, in part, because it contributed to a spectacular increase in economic growth (Maddison 2001:29):

> From 1000 to 1820 the upward movement in per capita income was a slow crawl—for the world as a whole the rise was about 50 per cent. Growth was largely "extensive" in character. Most of it went to accommodate a fourfold increase in population. Since 1820, world development has been much more dynamic, and more "intensive." Per capita income rose faster than population; by 1998 it was 8.5 times as high as in 1820; population rose 5.6-fold.

The wealth boom of the last 150 years, however, has been deeply unfair in its distribution. So much so that social inequality is at risk of becoming embedded in nature itself, as the resources and services of ecosystems are managed by economic and technological forces that largely serve the ambitions of a small percentage of the human population (Byrne et al. 2002; Bond 2012).

The positive belief in the link between fossil energy use and economic growth became so entrenched socially and politically that an observer of the period noted that seen through the lens of this ideology, "a clear sky" would be taken as evidence of a labor strike or lockout rather than an environmental goal to ensure human well-being (Mumford 1936/2010). The consequences of this ideology and its global impact are well known. For society to question the validity of pursuing this ideology, let alone address what might be necessary, seems only possible under the threat of a catastrophe.

Thus, the global energy and climate debate is about the looming catastrophe and the extraordinary social change needed to avert it. As seen in Figure 10.1, the projection of the world's energy future continues to rely on the presumption

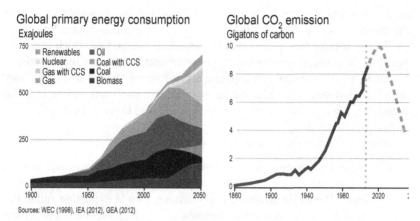

Sources: WEC (1998), IEA (2012), GEA (2012)

Figure 10.1 Global primary energy consumption and global CO_2 emission (GRID-Arendal 2015).

that our destiny is to be energy and carbon intensive. Only a spectacular shift in social commitment to the current development formula can bring about the needed reduction path in CO_2 emissions to challenge the prospect of irreversibility. Even so, the change discussed during 23 years of UNFCCC negotiations is modest compared to the change that science has forecast as necessary. In brief, humanity faces a crisis that derives from its economic success, the injustice that accompanies this success, and the need to change dramatically society's relation to living nature.

Framings

In our discussions, we identified twelve different framings. We searched at length for an appropriate language to characterize these framings and their political and/or analytical affinities. In some cases, the broad focus is on societal organization and operations (e.g., ecological modernization), whereas others champion specific strategies to address climate change and energy transformation. Several framings reflect long-standing conflicts over societal organization and operations (e.g., political economy vs. neoliberalism). In addition, some framings specifically inform the discourse on climate action and energy change (e.g., climate justice, energy sovereignty, and green economy).

We categorized the framings according to their political, economic, and/or analytical affinities and provide examples of key proponents. It is important to stress, however, that these framings are not inclusive: they are intended to provide an overview of normative and theoretical/conceptual positions that affect current discourse. Importantly, these framings do not agree on how society should characterize or address climate change problems or energy-related transformative responses. Continued conflicts in this area should be anticipated, and we view this conflict to be essential. We strongly recommend that the emerging conflict be treated as a means to expand understanding of the challenges society faces, as well as the alternative responses which might be possible: from social movements, to governmental or business sector actions/ inactions, to international (dis)agreements. While we are not sanguine about the likelihood that such fundamental conflict will end soon, we believe the engagement in issues raised by the framings will lead to more socially relevant and impactful research.

Four of these framings focus on the market:

1. *Neoliberalism* espouses a policy philosophy that limits public actions to those that are consistent with market logic. Currently, fossil fuel markets still provide profitable opportunities, both in extraction and in the correction of negative environmental effects (e.g., carbon sequestration and geoengineering), and neoliberal policy prefers that markets decide the extent and terms of use for fossil fuels. Policies based on increasing carbon prices or emissions permits (European Union Emission Trading Scheme, EU ETS) or grassroots activities against "unburnable fuels" are treated as naive. Insofar as corporations and the interests of invest- ment and finance drive politics, climate change policies will be imple- mented only to the extent that they guarantee capital accumulation.

2. *Ecological modernization* changes the energy mix with new technol- ogies and economic instruments (e.g., carbon pricing, taxes, REDD, markets) in emissions permits. Actions are undertaken by governments or middle-level institutions, such as cities and regions. Here the empha- sis is on technological change and economic instruments.

3. *Sustainable development* has its roots in the 1987 Brundtland Commission by the United Nations: "Sustainable development is de- velopment that meets the needs of the present without compromising the ability of future generations to meet their own needs" (Brundtland Report 1987). To meet these "needs" (in particular, the essential needs of the world's poor), economic alternatives to the current quantitative growth-driven economy should be prioritized. This perspective in- cludes critiques of the neoliberal path of development, including the "steady-state economy" (Daly 1991), "limits to growth" (Meadows et al. 1972), and the idea of limitations imposed by the state of technology and social organization on the environment's ability to meet present and future needs.

4. *Green economy* aims to "put a price" on nature for the sake of maxi- mum efficiency and rationality, for example, in carbon markets and other forms of emissions trading and virtual water sales that are in- creasingly packaged in exotic investment instruments. The economics of ecosystems and biodiversity (TEEB) within the UN Environment Program aims to "make nature's values visible" and thus "help decision

makers recognize the wide range of benefits provided by ecosystems and biodiversity, demonstrate their values in economic terms, and, where appropriate, capture those values in decision-making." The payment for ecosystem services (PES) approach is being pursued vigorously in many terrains, for example, natural capital accounting.

Three framings are analytically oriented:

1. *Sociotechnical systems analysis:* This perspective guides the discourse of climate change from unilateral "single-source, single-country" thinking into a much broader "global, no-boundaries" frame. This provides an important objective, evidence-based input to discussions with stronger elements in climate justice, and to some extent also on sustainability.

2. *Cost-benefit analysis*: Regardless of whether framings recommend action or inaction, they share a common methodology of valuation in money units to compare present and future costs and benefits or losses. The results depend on arbitrary discount rates chosen. Despite its simplicity, it has been a successful frame for the political discussions because of the social prominence of economics in politics. It is a state-driven approach.

3. *Common-pool resource management*: Climate change is an issue that can be approached in terms of the theory of common-pool resources. Access to the atmosphere and the oceans as sinks exhibits rivalry and non-excludability: there is rivalry, but so far no mechanism for excludability. Effective instruments for management of the commons (Ostrom 1991) must be established in the next conference of the parties, or elsewhere.

Five framings address aspects of the postmarket economy:

1. *Political economy*: This perspective sees energy as embedded within broader social, economic, and political forces and processes, and asserts that, if a just transition is to be achieved and inequality of access *between* and *within* countries and generations are to be addressed, a reconfiguration of infrastructures, institutions, technology and ownership; and modes of production and consumption are needed.

2. *Political ecology*: Regardless of whether they recommend action or inaction, political ecology framings share a common methodology of valuation that compares present and future costs and benefits (or losses) in ecological terms. The results depend on discount rates which political ecologists often see as arbitrarily chosen. This framing has been used by adherents to criticize the social prominence of economics in politics.

3. *Ecosocialism*: Proponents of this perspective respect the merits of valuing nature (though not counting it for the sake of marketization), at the

same time confirming the role of anti-market social movements, including those of indigenous people and ecofeminists, in nature's stewardship. It is a state-driven approach.

4. *Climate justice*: Draws on "critical ecology movements" which invoke environmental justice, demand stronger laws and enforcement, and engage in campaigns against corporations and states which exploit the environment. This approach regards one of its purposes as supporting "strong sustainability." A common feature of this approach is to distinguish livelihood emissions and luxury emissions, to discuss who should reduce emissions, and how to represent liability for past excessive emissions, that is, the idea of ecological debt (see Agarwal and Narain 1991).

5. *Energy sovereignty*: This is a recent framing, inspired partly by new technological opportunities, such as distributed electricity generation, as well as by approaches which include environmental justice and concerns over the protection of diversity. Energy sovereignty is similar to concepts of "food sovereignty" in promoting the ability of small regions to determine their own plans for energy use and production, the elimination of "energy poverty," and the reduction of greenhouse gas emissions. Similar to the hope for a municipal "hydroelectric socialism" around 1900 (small dams of municipal property), publicly or privately owned energy sources would be made locally available according to needs and taken out of the sphere of capital accumulation.

Table 10.1 captures the key ideas, values, and concerns that emerged from our examination of these framings.

New Methods

What type of methods might best help researchers and communities assess the challenges of energy (and more broadly, social) transformation in the face of a rapidly warming world? Are new analytical methods required, or will current ones suffice? From our discussions, four methods emerged as potential candidates to improve understanding of the challenges we face, as well as to promote dialogue among proponents of the different framings.

Justice-Based Transformation Pathways

The IPCC has rightly earned praise for its efforts to synthesize available research and evidence in climate change. Through each of its five assessments and several special reports, the IPCC has provided the human community with an evidence-based understanding of the phenomenon and the risks that all forms of life face.

Table 10.1 Response of framings to the core issues of sustainability, justice, and diversity.

Framings	Sustainability	Justice	Diversity
Market-focused framings			
Neoliberalism	Sustainability of capitalism (accumulation of capital); sustainability of growth and new investment opportunity	Promotes cornucopian approaches—expands wealth and corrects environmental problems with the new wealth	Depletion or expansion of biodiversity depending on efficient management
Ecological modernization	Regulation of the economic system to maintain serviceable ecosystems	Pursues win-win situations of improved economy and environment	Adopts Environmental Kuznets curve thinking (internalizing externalities)
Sustainable development	Economic sustainability/weak sustainability	How to balance the present and the future by using cost-benefit analysis	Concerned with human diversity, both intra- and intergenerational needs
Green economy	More concern about economic sustainability than sustainable development	Weaker concern on justice than sustainable development	Concerned with economic profit and values bio- or other forms of diversity only when they are profitable
Analysis-focused framings			
Sociotechnical systems analysis	Analytical, nonterritorial approach to environmental sustainability	Context-based assessment of intra- and intergenerational equity	Methodologically possible to include diversity linked to production and value chains
Cost-benefit analysis	Interested in instrumental sustainability	Discounts future values to prioritize the "needs of the present"	Interested in pricing diversity
Common-pool resource management	Effective instruments for management of the commons	Adopts cooperation principle within communities to manage access on behalf of all members equitably	Biological diversity (diversified ways of protecting diversity through place-based management)

Table 10.1 (continued)

Framings	Sustainability	Justice	Diversity
Postmarket economy framings			
Political economy	Challenges the sustainability of the orthodox economic growth model	Concerned with the distribution of, access to, and ownership of resources and technology	Challenges nature's commodification; in this way, it assists efforts to protect biodiversity and livelihoods
Political ecology	Concerned with political, economic, environmental, and social relations	Objection to inequitable enclosure of atmospheric commons and unequal distribution of the costs of environmental change	Emphasizes the need to understand peoples' relationship to their environment
Ecosocialism	Collective ownership of the means of production, use of democratic planning that makes it possible to pursue ecological rationality	Democratic control, social equality, and the predominance of human need over profit making	Ecosystem planning at the large national scale
Climate justice	Emission constraint within carrying capacity, concerned about sustainable consumption	Distributed justice; luxurious emission vs. livelihood emission; ecological debt	Concerned with cultural and livelihoods diversity
Energy sovereignty	Community-based sustainability (local energy)	Self-governance principle; the right of individuals, communities, and peoples to make their own decisions on energy generation, distribution, and consumption	Can support relevant diversity efforts based on local energy sources appropriate to their ecological, social, economic and cultural circumstances

To construct the much-needed assessments of justice-based pathways that will lead to a change in course, it is vital that IPCC's network of researchers be connected to the world's most vulnerable communities. Admittedly, this would take the IPCC beyond its mission of creating synthetic knowledge. However, these assessments are urgently needed now, and they must be based on the best available information. Although the IPCC, as an organization, has achieved a certain level of interdisciplinary knowledge, an even more robust commitment to interdisciplinarity is needed. Linking research institutions (such as the IPCC and others) with community networks that are at greatest risk is necessary if we are to address the challenge of climate change and energy transformation.

Climate Life-Cycle Analysis

In most legal frameworks for carbon emission reduction, penalization of carbon emissions is addressed solely from a production-based viewpoint. For example, the EU ETS is based on point-emissions of facilities whereas the United Nations Framework Convention on Climate Change (UNFCCC) Paris climate agreement relates to nations as carbon sources.

Systems that account for production-based emissions do not address the embedded emissions in products. They overlook the role of consumption in emissions (including the roles of value and production chains) and neglect key issues such as carbon leakage.

By focusing solely on singular, time- and space-constrained carbon source—without understanding that all products carry a CO_2 (or a greenhouse gas) history with them—key international agreements on climate and national mitigation measures may lose effectiveness and turn out to be costly exercises. In addition, sustainability and justice may also be jeopardized. For example, cheap goods produced in China, which Western consumers require and enjoy, cause high emissions locally and are accounted for solely in the national inventory. Alternatively, if a country adds unilaterally a carbon tax to its otherwise resource-effective production to reduce greenhouse gas emissions, this might transfer that industry to a third country which does not have any CO_2 restrictions and end up causing much higher overall emissions.

Thus, it is important to pay acute attention to the emissions of a product or service over the whole life cycle ("from cradle to grave"); this could be realized through a spatiotemporal type of life-cycle analysis (LCA), termed here as the climate LCA. From the climate policy side, this would mean putting more emphasis on consumption rather than production. We recognize that introducing this as a new basis for climate agreements may be complicated and involve methodological hurdles. However, in terms of justice, sustainability, and diversity, it could fill a major gap. For future science, this could be highly motivating.

We recognize that the idea of accounting for embedded emissions over value chains is not a new idea. Already in the late 1970s, in the aftermath of the oil crises, net energy analyses (e.g., based on Leontief's input–output model) were proposed to guide energy investments. In the late 1980s, research incorporated embedded CO_2 (Lund 1989) and later environmental analysis and LCA were introduced. Peters (2008) elaborated models for a consumption-based carbon inventory.

There are multiple challenges associated with a climate LCA system. For example, there are particular demands on quality of knowledge (e.g., data across nations and territories), politics (e.g., transparency), and monitoring, in which different areas of science need to be strongly engaged. This research could also bring about new knowledge with serious political implications. For this reason, such research needs to address both the theoretical framework as well as strong sociopolitical-economic linkages, for example, with issues related to climate justice and sustainability. Such research requires collaboration with other disciplines. We also see here an analogy to strategic environmental assessments, where science may provide important knowledge to local or regional communities and groupings on impacts such as diversity that may affect their habitat and living conditions (e.g., large-scale hydropower schemes).

In summary, climate LCA offers the potential to open up new avenues in global mechanisms on climate change mitigation and local energy solutions, or at least better and more objectively understand how the carbon emissions originate and how the value and production chains affect emissions. This could guide the discourse of climate change away from a unilateral "single-source, single-country" thought process to a much broader "global, no-boundaries" frame. It remains to be seen how this will affect the creation of new global mechanisms (e.g., a global CO_2 tax, equity-based burden distributions), but it would certainly provide objective evidence-based input to that discussion, with strong elements in climate justice and, to some extent, sustainability.

China Coal Cap Initiative

The coal cap initiative in China provides an example of an ongoing LCA. China plays a key role in climate change mitigation as it is the world's largest energy consumer and carbon emitter. China's energy production is heavily based on coal, which in 2015 accounted for 64% of its electricity. Nearly 30% of global energy-related CO_2 emissions emanate from China. The strong economic growth from which the global economy has benefited has driven up China's emissions. A considerable share of China's emissions originates from products intended for exports to Western economies, but the national carbon inventories do not recognize carbon export. Moving from a production- to consumption-based carbon accounting, putting more emphasis on carbon intensity in production and restricting carbon leakage could change this.

Meanwhile, China has intensified its efforts to reduce CO_2 emissions, in particular by decreasing the use of coal. A three-year effort, the "China Coal Cap Project," was launched in 2013 to provide government authorities with

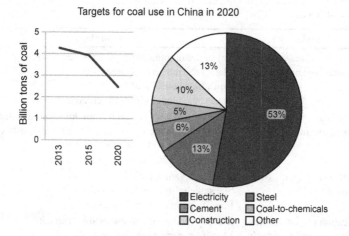

Figure 10.2   Progress and prospects for coal use by sector in China by 2020, provided by the National Defense Research Council of China.

recommendations on policies and their implementation. The first phase (2014) included detailed background analysis on health, climate, environmental, and other impacts vis-à-vis coal use, but it also proposed a coal cap to reduce the share of coal in energy from 64% to 17% by 2050. In the second phase (2015), a coal cap strategy was submitted to the 13th Five-Year Plan 2016–2020: the cap was broken down to regional, provincial, and municipal levels as well as to major coal-using industries. The third phase (2016) focuses on preparing a coal cap action and monitoring plan while laying the foundations for a long-term energy transition and development strategy. Pilot projects in three provinces will be launched to promote implementation of coal caps in practice, in particular in coal-intensive industries. In addition, extensive analysis of impacts and policy effectiveness is planned as well as spreading best practices to accelerate reduction of coal use.

Through these efforts, coal use has dropped by 7% in two years. The goal for 2020 is a 20% reduction from the peak in 2013 (Figure 10.2). The coal cap initiative advocates for China's leadership in global green governance, but it also presents a great opportunity for international cooperation. China is now in the position to be able to develop its green leadership to protect the environment and cope with climate change. China may also take the lead in achieving the 2030 UN Agenda for Sustainable Development.

Critical Policy Analysis

Critical policy analysis has emerged to oppose what Sachs (2002:33) characterized as "conventional development thinking"; namely, the assumption that the market economy should decide the direction and value of development.

216

S.-J. Yun et al.

Table 10.2   Tasks needed to conduct critical environmental justice research (Dryzek 2008; see also Kaufmann 2013).

Tasks	Methodology
Criticism of technocratic and accommodative analysis	Reveal assumptions behind technocratic description Clarify the particular frame Don't be a consultant! Reveal the tunnel vision of research
Explication of dominant and suppressed meanings	Differentiate your research from dominant discourse Contribute to redefinition of environmental justice Collect and provide the relevant data Reflect the spatial scale you are choosing
Identification of agents of impairment	Name relevant stakeholders See stakeholders' polyrationality Identify people's tacit beliefs
Identification of communicative capacities and standards	Identify factors of communicative capacity Identify communicative standards

This is tantamount to accepting a universal ideal of development "irrespective of the fact that [the] world is already dehumanized and dehumanizing" (Irigaray 2003:167) and embracing the view that "equity is a problem of the poor." Instead, critical policy analysis adopts the view that "justice is about changing the rich and not about changing the poor" (Sachs 2002:33).

Methodologically, critical policy analysis contests the assumptions and moral standards of existing power elites, which the social system (Luhmann 1995) of (social and natural) sciences does not examine. This also entails consideration of "climate change" as a discourse–regime (Foucault 1981; Costa 2011) dichotomy, as it has already been characterized (Vlassopoulos 2012). Dryzek (2008:200) offers guidance to researchers on how this method can be applied to pursue environmental justice research that avoids the pitfalls of conventional policy analysis. Table 10.2 addresses the tasks identified by Dryzek (2008).

Action Research

Action research is a method that combines "knowledge" from both academics and activists. To see the potential for issues related to climate change, consider the following examples.

Local communities are directly impacted by decisions to extract fossil fuels in terms of when and/or where to leave them in the ground as well as associated political opportunities. Utilizing their own criteria, local communities could benefit from the use of formalized multiple criteria methods. Similarly, as communities and lay persons become generally aware of climate change, they might react to locally perceived impacts (e.g., sea-level rise, retreating glaciers, and changes in vegetation). A concerted effort to build an action–research network would allow researchers to learn about the local impacts that are of greatest concern to these communities. This would support a synthetic learning process between science and society.

From a technical perspective, it can be argued that coal (particularly brown coal) should be the main focus of reduction because it produces more CO_2 per unit of energy than oil or natural gas. An action–research perspective could, however, show that while there are indeed grassroots actions to "leave the coal in the hole" (e.g., Fuleni in South Africa, Sompeta in Andhra Pradesh in India, Laubnitz in Germany), actions elsewhere oppose natural gas fracking or favor "leaving oil in the soil." Rigorous accounting of avoided carbon emissions, discussions of "leakage," and so forth by academics would be inspired through, and accompany, such actions.

Policy and Strategy Discussions in the Discourse: Highlighted Tensions

In our deliberations, we discussed the typical management interventions suggested by several framings to address the challenge of climate change and the need for energy transformation, and considered policies that go beyond management intervention. Here we offer a synopsis of that debate to highlight tensions in the policy component of the energy and climate change discourse. Again, the framings identified above inform our effort to understand key tensions surrounding policy. We did not attempt to discuss the full range of policy proposals and practices, as this is beyond the scope of this report.

Carbon Pricing

Significant political and analytical effort has been expended, in different ways, to price carbon. International initiatives (e.g., the Kyoto Protocol) and national policies (e.g., China's recent creation of markets to facilitate in-country carbon trading) typify attempts to use this policy tool.

In either of its two main forms—carbon trading and carbon taxation—carbon pricing establishes a right to pollute through the purchase of emission commodities in a market. Well-documented problems of fraud have been associated with European and U.S. carbon markets. The approach is criticized for being distributionally regressive, allowing the rich, for example, to maintain energy-intensive lifestyles while shifting the burden of social change to poorer countries.

This policy tool can lock in existing power relationships by encouraging change at the margins. Taxation simply raises the cost of the next unit of output rather than entailing the full-fledged restructuring that many industries require.

Generally, carbon pricing has not raised the cost of pollution sufficiently to provide an incentive for wealthy societies to decarbonize. The size of the world's carbon market peaked in 2008 at $140 billion, and by 2014 had dipped to $50 billion. Several analyses suggest that prices above $150–500 per ton are needed to instigate the dramatic change inherent in the science forecasts that we must achieve (Ackerman and Stanton 2012). In 2006, the European Union price peaked at $35/ton and the current price has fallen to about $6/ton; in California it hovered around $12/ton, Korea around $9/ton, and China at $3–7/ton, depending on the city.

Some now propose to redesign the tool of carbon pricing as a "cap and dividend" scheme (i.e., carbon tax plus redistribution). To date, however, this approach has attracted only modest interest.

Notwithstanding lackluster performance (this is putting it mildly), proponents of the following framings continue to champion the use of carbon pricing: neoliberalism, sustainable development, green economics, and ecological modernization. This underscores a real tension in the case of climate policy: despite a record that fails tests of sustainability, justice, and diversity, carbon pricing is still accorded a powerful role in discussions about societal action. This tension can be traced to deeper concerns of the viability of the market economy in a warming world.

Nonmarket Policy Strategy

In the Paris Agreement of December 2015 (UNFCCC 2016), Article 6 calls for the adoption of "cooperative mechanisms." It reflects the widespread critique of market mechanisms pursued under the Kyoto Protocol. The Article specifically adopts a mandate of "nonmarket cooperative mechanisms" (Paris Agreement Art. 6.8–6.9), but lacking full definition, it remains up to the Parties to elaborate proposals for defining the mechanism, allowing cooperation without revealing market features. Scientists should have a major role in elaborating such proposals.

We do not assume that Article 6 will lead to transformative change underscoring as it does the support for market-based mechanisms and, in this regard, may reflect increasing pressure on framings which are market-focused. The conflicts in framings focused on the market and/or the postmarket economy in the energy and climate change sphere, illustrate the creative value of framing conflict to rethink environmentalism.

Municipal and Citizen-Led Policy

Recently, citizen movements and local governments have played significant roles in recasting the energy and climate change debate. The emergence of activities at this scale can serve as a catalyst for change and a source of some of the most aggressive inventive strategies emanating from it. One representative example is Seoul.

Seoul's civil movements and metropolitan government have worked together to reduce grid electricity use rapidly via a campaign called "One Less Nuclear Power Plant (OLNPP)." Initiated in April 2012, OLNPP's initial goal was to "retire" one Korean nuclear reactor by December 2014 through city-wide conservation and local renewable energy supply strategy that would cut grid consumption. By June 2014, Seoul had met its goal, lowering the country's need for nuclear generation by 6%. In August 2014, OLNPP launched its next initiative to "retire" a second reactor—a step that directly challenged the national government's nuclear expansion policy. Despite several efforts by the national government to interfere in the campaign and a negative media coverage (which falsely accused OLNPP of threatening an economic slowdown and eventually higher electricity prices), civil support remains high.

The actions in Seoul led to four provincial governors signing a "Joint Declaration for Local Energy Transition" in November 2015. Together, these governors and the municipal government of Seoul represent 49.2% of the country's population.

Some argue that nuclear power is a so-called "clean energy" option. The movement launched in Seoul, however, regards nuclear power as a key driver for Korea increasing the energy intensity of its economy and pursuing unchecked economic growth. These features undermine decarbonization by promoting production and consumption of goods from materials that are carbon-based (e.g., steel, cements, plastics) and interrupting carbon stores that are provided by forests, prairies, and undeveloped land and replacing them with buildings, streets, and so forth. For this reason, the campaign measures its progress in tons of oil equivalent.[1] Moreover, the politics of nuclear power are seen as antidemocratic, fostering the consolidation of energy policy making at the national level by technocratic, corporate, and military elites, ignoring the desires of communities and local governments. Finally, the primary motive of the OLNPP campaign is to reduce energy use in any form, singling out nuclear power for its current dominance in the national energy mix and politics, but aiming to shift society and the economy away from "more is better" to "enough is enough."

The recently launched Solar City Seoul initiative[2] clarifies the political and social underpinnings of OLNPP. The city has adopted a 1 GWp solar power target for installation by 2022 on building rooftops. A part of the project (currently the largest urban solar initiative in the world) is dedicated to reducing electricity bills for low- and moderate-income families. Moreover, the mayor has cited the initiative as a means to end the use of coal-powered electrical generation that is associated with city pollution problems. This distributed solar

---

[1] To obtain the One Less Nuclear Power Plant report, see http://energy.seoul.go.kr/en/olnpp.jsp#none as well as the Seoul Metropolitan Government website: http://english.seoul.go.kr/policy-information/policy-focus-2017/one-less-nuclear-power-plant/

[2] See http://english.hani.co.kr/arti/english_edition/e_national/820207.html (accessed Feb. 7, 2018).

power plant reflects the principles of sustainability, democracy, and justice
guiding Seoul's civil society-led OLNPP program.

New Policies for "Unburnable Fossil-Fuel Reserves"

To explore new policies with respect to "unburnable fuels," reserves were calculated using the IPCC's Fifth Assessment Report database (Jakob and Hilaire
2015). The amount of carbon embodied in fossil fuels yet to be released (and
consequently still to be burnt) was found to be 1000 Gt of CO_2 under the conditions that (a) the 2°C target will be met by humankind and (b) other emitting
sectors (e.g., forests, agriculture) will keep to the predefined limits delineated
in the IPCC Fifth Assessment Report (IPCC 2014). "Unburnable fossil-fuel
reserves" is the complement to this; their exact volume is not known but is projected to be about ten times what is still allowed to be burned. Where it occurs,
which national states are stakeholders of these resources, and to what extent
is not exactly known (McGlade and Ekins 2015). The breakdown by fuel type
(i.e., coal, oil, and natural gas, which have quite different carbon content) has
only been roughly established.

   Who are the actors of social and technical changes in the field of energy and
climate change? The typical policy discussions and analysis leave aside movements for climate and environmental justice. These movements have provided
the impetus for a "Blockadia" strategy of leaving fossil fuels in the ground.
This strategy, however, could prove to be the most important effort to date to
act on climate change by transformative energy action.

The UNFCCC and the Paris Agreement

There are contrasting evaluations of the achievements made by the UNFCCC.
The Paris Agreement is no exception. Some celebrate the Paris Agreement because it is viewed as having established a landscape in which nation-states,
subnational actors, and transnational networks will be able to reconfigure existing linkages between sustainability, diversity, and justice, and, perhaps, improve upon them. In turn, this could open up opportunities for "bottom-up"
movements to claim a larger segment of the decision-making and design processes involved in climate change policy. Many, however, criticize its nonbinding approach to nationally determined contributions in the volunteer emission
reduction target without clear consideration of ecological debts of the North.
Still others regard the Paris Agreement as "a fraud or a fake, unless greenhouse
gas emissions are taxed across the board" (Hansen 2015).

   Binding targets, absent from the Paris Agreement, are preferred by the
postmarket economy framings and in many instances by the analysis-focused
framings. Market-focused framings in the energy and climate space support
what others see as a notable failure. This testifies to the continued power of

market thinking, but it also underscores the increasing isolation of these framings politically and analytically. Again, we wish to stress that framing conflict should be seen as instigating inventive rethinking of environmentalism.

One concrete example is the rising importance of civil movements. In spite of many problems acknowledged by even its supporters, the Paris Agreement is being used by civil movements to demand that national governments respond to climate change by transforming economic structures. When governments are reluctant to act, civil society demands action based on international agreements. When industries are reluctant to act, governments as well as civil society can demand action: they can work together to assert energy sovereignty and self-designed climate policies. A recent assessment of the Paris Agreement by several civil society organizations suggests that it can be used to mobilize political challenge to governments and industries that fail to meet the objectives of a just and sustainable response to climate change (Climate Equity Reference Project 2015).

Civil society has been the main source of criticism of the Paris Agreement for its lack of commitment to environmental justice. As we have noted, this critique is fast becoming the dominant source of challenge to inaction on energy and climate problems. A key example in this regard is the focus on the international system's exclusion of indigenous groups from the negotiations generally, and specifically the lack of reimbursement negotiations about the already existing damage caused by the former colonial powers. The international apparatus built to address the problem of climate change is far from answering this criticism.

The U.S. secession from the Paris Agreement in June 2017 must be viewed as a major setback for the UNFCCC. The departure of the largest per capita emitter of greenhouse gases from the Agreement underscores again the environmental justice failings of the process and structure. In addition, it contests the efficacy of the international approach. However, internationally and in the United States, actions by the private, public, and nongovernmental sectors to curb emissions are growing and can be attributed to the commitment by civil society to demand action. The strong rebuke of the U.S. decision by European, Asian, African, and Latin American government leaders as well as by corporate leaders may indicate that the obligation to act now is being felt in these quarters.

The search continues to realize a thorough rethinking of environmentalism to address the problem of energy and climate. Hopefully the framings presented and applied here will assist efforts to find a suitable environmentalism that can meet our urgent challenge.

References

Ackerman, F., and E. A. Stanton. 2012. Climate Risks and Carbon Prices: Revising the Social Cost of Carbon. *Econ. Polit.* 6:1–25.

Agarwal, A., and S. Narain. 1991. Global Warming in an Unequal World: A Case of Environmental Colonialism, Centre for Science and Environment. http://cseindia.org/ challenge_balance/readings/GlobalWarming%20Book.pdf (accessed Jan. 20, 2017).

Arrhenius, S. 1896. On the Influence of Carbonic Acid in the Air Upon the Temperature of the Ground. *Philos. Mag. J. Sci.* **5**:237–276.

Bond, P. 2012. Politics of Climate Justice: Paralysis above, Movement below. Pietermaritzburg: Univ. of KwaZulu-Natal Press.

Brundtland Report. 1987. Our Common Future. Oxford: Oxford Univ. Press.

Byrne, J., L. Glover, and C. Martinez. 2002. The Production of Unequal Nature. In: Environmental Justice: Discourses in International Political Economy. Energy and Environmental Policy, vol. 8. New York: Routledge.

Callendar, G. S. 1938. The Artificial Production of Carbon Dioxide and Its Influence on Temperature. *Q. J. Meterol. Soc.* **64**:223–240.

Climate Equity Reference Project. 2015. Fair Shares: A Civil Society Equity Review of INDCs. OXFAM. https://oxf.am/2t8lAqa. (accessed Jan. 20, 2017).

Costa, S. 2011. Researching Entangled Inequalities in Latin America: The Role of Historical, Social, and Transregional Interdependencies. Working Paper No. 9. desiguALdades.net

Dryzek, J. S. 2008. Policy Analysis as Critique. In: The Oxford Handbook of Public Policy, ed. M. Moran et al. New York: Oxford Univ. Press.

Foucault, M. 1981. The Order of Discourse. In: Untying the Text: A Post-Structuralist Reader, ed. R. Young, pp. 48–78. Boston: Routledge and Kegan Paul Ltd.

GRID-Arendal. 2015. Frozen Heat: A Global Outlook on Methane Gas Hydrates. Publ. No. 164. Arendal: GRID.

Hansen, J. 2015. James Hansen, Father of Climate Change Awareness, Calls Paris Talks "a Fraud." *Guardian* Dec. 14, 2015.

IPCC. 2014. Contribution of Working Groups I, II and III to the Fifth Assessment Report of the Intergovernmental Panel on Climate Change. In: Climate Change 2014: Synthesis Report, ed. R. K. Pachauri and L. A. Meyer. Geneva: IPCC.

Irigaray, L. 2003. Between East and West: From Singularity to Community. New York: Colombia Univ. Press.

Jakob, M., and J. Hilaire. 2015. Climate Science: Unburnable Fossil-Fuel Reserves. *Nature* **517**:150–152.

Kaufmann, G. F. 2013. Environmental Justice and Sustainable Development: With a Case Study in Brazil's Amazon Using Q Methodology. PhD Dissertation, Freie Universität Berlin. Düsseldorf: Südwestdeutscher Verlag für Hochschulschriften.

Luhmann, N. 1995. Social Systems. Stanford: Stanford Univ. Press.

Lund, P. D. 1989. Assessment of the Effectiveness of Renewable and Advanced Technologies in Reducing Greenhouse Gases Based on Net Energy Analysis: The Energy Breeder Concept. Paris: Proc. of Energy Technologies for Reducing Emissions of Greenhouse Gases, OECD/IEA.

Maddison, A. 2001. The World Economy: A Millennial Perspective, OECD Development Centre. http://www.oecd.org/dev/developmentcentrestudiestheworldeconomy-amillennialperspective.htm (accessed July 31, 2017).

McGlade, C., and P. Ekins. 2015. The Geographical Distribution of Fossil Fuels Unused When Limiting Global Warming to 2°C. *Nature* **517**:187–190.

Meadows, D. H., D. L. Meadows, J. Randers, and W. W. Behrens, III. 1972. The Limits to Growth. New York: Universe Books.

Mumford, L. 1936/2010. Technics and Civilization. Chicago: Univ. of Chicago Press.

Ostrom, E. 1991. Governing the Commons: The Evolution of Institutions for Collective Action. New York: Cambridge Univ. Press.

Peters, G. P. 2008. From Production-Based to Consumption-Based National Emission Inventories. *Ecol. Econ.* **65**:13–23.

Sachs, W. 2002. Ecology, Justice, and the End of Development. In: Environmental Justice: Discourses in International Political Economy. Energy and Environmental Policy, ed. J. Byrne et al., pp. 19–36, vol. 8. New York: Routledge.

Schmidt, L. J. 2017. Satellite Data Confirm Annual Carbon Dioxide Minimum above 400 ppm. *Global Climate Change* Jan. 30, 2017.

UNFCCC. 2016. Paris Agreement, United Nations Framework Convention on Climate Change. http://unfccc.int/files/meetings/paris_nov_2015/application/pdf/paris_agreement_english_.pdf (accessed Jan. 20, 2017).

Vlassopoulos, C. A. 2012. Competing Definition of Climate Change and the Post-Kyoto Negotiations. *Int. J. Climate Change Strat. Manag.* **4**:104–118.

Water

11

Accounting for Water

Questions of Environmental Representation in a Nonmodern World

Margreet Zwarteveen, Hermen Smit,
Carolina Domínguez Guzmán, Emanuele Fantini,
Edwin Rap, Pieter van der Zaag, and Rutgerd Boelens

Abstract

This chapter discusses new science policy initiatives that involve water in terms of environmental representation: who represents, what is represented, for what purpose (equity, sustainability). One such initiative, Water Accounting, adheres, at least on paper, to a modernist treatment of science and politics (and nature–society) as being distinct and distinguishable. The effect of this is that particular water "facts" (e.g., the water consumed per unit of land, or the crops produced per unit of water) appear as the (only or most) objective starting points for improving water management and governance; other "facts" and other possible ways of representing water come to be seen as less true or important. Typically modernist claims of neutrality and independence also make it difficult to recognize how water accounting representations, measurements, and calculations derive from specific epistemic and policy communities, whose members have specific concerns, are part of specific knowledge traditions, and pursue specific societal projects of betterment. Inspired by social studies of science, it is suggested that water accounting and other attempts to speak for water or represent the environment become more useful and honest when efforts are made to address, more explicitly, the entanglements between science and politics. This abandonment or relaxation of modernism includes embracing diversity, plurality, or multiplicity as well as acknowledging, accepting, and reconciling the existence of many different ways to engage with, relate to, and account for water—many different versions of water. It also includes replacing the aspirations of transcendence, integration (or universality, commensuration), and inclusion (consensus) with equally difficult to attain, but more modest and pragmatic, ideals of situatedness, translation (mediation), and contestation (or dissent).

228   *M. Zwarteveen et al.*

Introduction

On the face of it, the reframing of water as an environmental question has been very successful. The water resources literature is replete with alarmist references to the closure of river basins, the depletion of aquifers, or the disappearance of wetlands. Most agree that water can no longer be simply considered as something to be captured and made available for societal purposes, arguing that it should instead be treated with caution. The widespread recognition of water as a global environmental concern coincides with a surge in international initiatives to speak differently—more ecologically or environmentally green—about water. These often combine large-scale attempts to count, measure, map, and predict changes in the globe's water resources with proposals for new ways to appreciate, valuate, and account for them. Focusing on water used in agriculture, initiatives such as the World Water Assessment Program, the Water Foot Print, and Water Accounting share the ambition of making science's abilities of accurate representation available and legible to water decision makers to support them in their efforts to use and allocate water in more environmentally conscious ways. Water Accounting, for instance, as explained below, aspires to become "the standard reporting and planning system" for water, helping "managers to manage water consumption more tightly (following certain well-defined targets) and to understand which flows to manipulate by means of retention, withdrawals and land-use change" (Bastiaanssen et al. 2015). Initiatives such as Water Accounting and the Water Footprint share the objective to measure and calculate water consumption (of a crop, but also of a consumer, farmer, company, chain, community, or nation) to allow (economic) valuation and comparison.

The enthusiasm with which these initiatives are embraced, developed, and sponsored by a wide variety of actors suggests that they offer an attractive and powerful way to express environmental concerns in water. In this chapter, we discuss whether this enthusiasm is justified. We do this by examining which, and whose, concerns are addressed by these new science–policy initiatives and by exploring how they are addressed.[1] Engaging with the larger aim of this book, we are particularly interested in their ability to address and combine concerns of sustainability with those of equity and diversity. Speaking for/about water is a question of environmental representation in two senses of the word:

1. Representation as an operative term within political processes that seek to extend visibility and legitimacy to the environment as a societal concern, thus creating political and societal support for institutional or technological interventions to help conserve or protect it

---

[1] Politics is as much about who has the authority to speak as about what is spoken and how. For elaborations of this for water, see Boelens and Zwarteveen (2005) as well as Zegwaard et al. (2015). For a more general discussion of so-called ontological politics, see Mol (1999, 2002).

2. Representation as the normative function of language, which either reveals or distorts that which is assumed to be true about the environment

Hugely oversimplifying, and following Latour (2004), one could say that modern societies, at least ideally, have delegated these two forms of representation to specific societal realms: political representation happens (or is supposed to happen) in the realm of political decision making, whereas representation in terms of revealing what is real belongs (or is supposed to belong) to the realm of science or academia. The assumption of a fundamental divide between nature and society both mirrors and, importantly, feeds this separation.

Here we show that new international policy initiatives in water, at least on paper, seem faithful to this modernism in how they present, promote, and discuss the knowledge they produce. One clear effect of this is that the cultures and identities of the producers of knowledge and their particular social projects disappear from the water facts presented, as do many of the assumptions, stories, and labor (the semiotic and physical work) used to measure, calculate, and compare waters across times and places. In addition, because of their categorization of water as something natural to anchor their truth claims, initiatives such as Water Accounting and the Water Footprint make waters appear the same irrespective of where and how they are accessed or used, by whom or for what purpose, and irrespective of the social relations (of ownership and labor) and meanings of which they form a part. These waters thus come to appear as similar, or even as the only possible versions of water. These "natural order" versions of water are those to which—by modernist agreement and tradition—only scientific experts have access.

By granting some waters universal status, this distinctly modern way of knowing water may foreground some concerns while neglecting (or making it more difficult to see and address) others. We also fear that by implicitly delegating contentious allocation decisions to scientists, these initiatives may short-circuit processes of political and democratic decision making (cf. Blok and Jensen 2011). Thus, we make a plea for the project of environmental representation to abandon or relax modernist claims and languages. Inspired by a rapidly expanding body of scholarly work in the social studies of science, science and technology studies, and feminist technoscience studies, which calls into question the divides between politics and science (and by implication between nature and society),[2] our arguments here entail an invitation to all who are engaged in environmental representation to reflect on their practices. We suggest that attempts to represent water will be more successful if they (a) acknowledge that science and politics are, and will always be, deeply

---

[2] We have been particularly inspired by Donna Haraway (1991, 1997, 2008), Bruno Latour (2004), and David Harvey (e.g., Haraway and Harvey 1995). For water, Erik Swyngedouw's various writings (e.g., Swyngedouw 1999, 2009, 2013) are noteworthy and influential. In his writings on "modern water," Jamie Linton also provides important ingredients to this discussion (e.g., Linton 2010; Linton and Budds 2014). See also Zwarteveen (2015).

intertwined and (b) abandon the modernist idea that there is only one correct way to know and represent water.[3] We thus propose to replace or perhaps complement attempts to commensurate very different waters in one overarching language, or set of values, with efforts to explicitly accept and support different ways to engage with, relate to, and account for water—different versions of water. This implies replacing transcendence, integration (or universality), and inclusion (consensus) with the equally difficult to attain but more modest and pragmatic principles of situatedness, translation (mediation), and contestation (or dissent) as guiding values. Water accounting initiatives do not have to be given up in the process. However, rather than assuming and asserting that many waters (or watery realities) can and should be expressed and calculated in water accounting terms, it becomes important to specify more clearly and delineate the types of questions that water accounting representation can answer, and for what (political or societal) purposes. As well as toning down universalist claims and ambitions, this includes efforts to accept and acknowledge that there are other ways of imagining and speaking for water, including nonscientific ones.

Three notions emerging from the social studies of science are particularly important for our argument. The first concerns the treatment of knowledge as performative; that is, the idea that knowledge—in our case knowledge about agricultural water—does not simply *describe* but also tends to *enact* realities into being (Law 2009). Considering (scientific) data as "facts" is based on the widely held assumption that reality has a definite form that is independent of the tools that are used to measure or count it. We propose a different and less accepted view of scientific knowledge in treating it as performative: rather than unambiguously emanating from some preexisting reality, knowledge actively helps produce reality.[4] Concepts (scientific) divide, map, and categorize; they are a way to help make sense of complexities by creating order. They demarcate, define, delineate, and indeed proactively establish and produce the boundaries between what matters and what can be ignored. Considering science and knowledge practices as performative also means that

---

[3]   We are not the first to make this argument. Recent pleas for nonmodern ways of knowing and engaging with nature include Castree (2014) and Haraway (2008). Scholars who have made this argument for water include Linton (2010), Linton and Budds (2014), and Swyngedouw (1999, 2009, 2013).

[4]   This view is not incompatible with philosophical realism or the preservation of an empiricist-realist belief in a world that is independent of the knower. In fact, we, along with others (e.g., Latour and Haraway), remain committed to accurate accounts of reality. Haraway argues for situated "knowledges" which maintain a strong commitment to objectivity (learning to see well) while denying that everyone will see in precisely the same way. For Haraway, "seeing well" is not just a matter of having good eyesight: it is a located activity, cognizant of its particularity and of the accountability requirements that are specific to its location (Haraway 1991:191). In situated knowledge-making projects, embodied knowers engage with active objects of knowledge whose agency and unpredictability unsettle any hope for perfect knowledge and control.

differences between research findings—accounts of realities—produced by different methods or in different research traditions can no longer be treated as different *perspectives* of a single reality: they become, instead, the enactment of different *realities*. This implies a move from assuming that the world is one—a uni-verse—to accepting the existence of multiple worlds that are produced through diverse social and material practices and relations. These worlds need not necessarily be disconnected. There may be resonances or overlaps between them (Law 2009).

The second notion follows from the idea of Bruno Latour that realities (and knowledge of realities) "depend on practices that include or relate to a hinterland of other relevant practices" (Law 2009:241). Sustainable knowledge rests in, and reproduces, more or less stable networks of relevant instruments, representations, and the realities that these describe. This is what makes realities—together with the techniques and representations that enact them—seem stable, durable, and reliable (Law 2009:241–242). This also means, and is the third notion, that realities are only real within particular networks (of people, devices, funds) or systems of circulation. Truths, therefore, are not universal (Law and Mol 2001): they are only "realized" in definite form within the networks of practices that perform them. The question of how this happens then becomes important: this is a question of power, interests, traditions, and culture.

By exposing how water accounting initiatives can fall into the trap of modernism, our aim is not to expose these initiatives as wrong, or to accuse their developers. To the contrary, some of us directly collaborate in water accounting efforts and subscribe to its aspirations of using water in wiser and greener ways. We also find the idea of using freely available remote-sensing data for improving the democracy of water decisions attractive. However, we are concerned about the close affinity that water accounting has to particular policy networks and societal projects of betterment, and fear that if the "facts" produced by water accounting come to be seen as the only possible truth, this will lead to the privileging of some waters (and associated uses and users) over others. In particular, as we discuss below, water accounting foregrounds efficiency and productivity as concerns, thereby risking to eclipse concerns of equity or diversity. Hence, by showing that water accounting is not that modern after all, our aim is to improve sensibility to how the production of water accounting measurements, maps, and sheets may help structure relations of dependence and power.

Below, we first establish how water accounting can be characterized as modern. Its implicit appeal to a natural and global order as the foundation of knowledge—an order that can be exposed or discovered through the rigorous methods of science—is distinctly modern. In addition, its active cherishing of a specific form of objectivity—one that is anchored in the idea that it is possible to see (and know) from an undetached and unmediated position—belongs to a modernist repertoire. After this discussion, we focus on whether, or to what extent, water accounting meets or can ever meet its own implicit modernist claims. We demonstrate that water accounting facts are more compatible with some versions of reality (and some political projects of improvement) than others. When using a modernist approach to science, it becomes difficult or even impossible to discuss this. Finally, we end with a discussion of whether and how water accounting tools can be used in less modernist ways. This discussion explores what would be needed for water accounting tools to support not only greater productivity or efficiency, but also goals of equity or diversity.

The Modernism of Water Accounting

Many emerging science policy initiatives concerned with the agricultural use of water utilize the exciting new possibilities of satellite imagery to measure water use changes and water footprints from a distance, in the hopes that such measurements will add rigor to policy discussions about how to best use and allocate water across competing uses and users. Water accounting initiatives, based on the idea that there is much water to be gained (or saved) in agriculture by more efficiently matching water gifts to crop requirements, are a case in point. This idea combines the oft repeated statement that most fresh water is used in agriculture, with the widely held belief[5] that much of this water is used inefficiently (see, e.g., Postel 2000; Gleick 2002). Important objectives of these initiatives include identifying where water can be used more efficiently, finding ways to hold users accountable for their irrigation practices, and making consumers aware of how much water went into the production of the fruits and vegetables they buy in the supermarket.

A foundational premise of water accounting is that the availability of accurate knowledge about sources and uses of water is a crucial precondition for achieving these objectives. Water Accounting (2016), for instance, states:

> Water problems around the world are increasing; however, information useful for decision makers within the water sector and related to the water sector seems to be decreasing. Solving water problems requires information from many disciplines, and the physical accounts (describing sources and uses of water) are the most important foundation.

The idea is that water is in need of representation by science in ways that are usable in policy and decision-making processes. This reflects a particular, yet implicit, view of the science–policy interface and posits one particular version of water and logic of water use. First, the assumption seems to be that water

---

[5] By calling this a belief, we are not suggesting that water is not used inefficiently. We are referring to how definitions and uses of efficiency terms are not uniform or agreed upon within the community of irrigation scholars, with claims about efficiency often being inappropriately used outside the contexts to which they apply, leading to false estimates of water savings at the basin scale (Perry 2007; Lankford 2012a, b; van Halsema and Vincent 2012).

decision making happens in clearly identifiable spaces by known actors who are rational and accountable to their words and deeds. This corresponds poorly to actual practices. In this neoliberal and globalizing world, water governance is controlled by an ever-increasing number of actors, only few of whom are identifiable or identify themselves as water decision makers. These actors have widely differing perspectives, influence, and interests; operate in different overlapping domains; and draw on different rationalities, values, resources, norms, and legal repertoires to articulate, frame, and defend their positions (Franks and Cleaver 2007). The environments in which they operate and strategize is complex and difficult to predict. Actual decisions about water uses and allocations are only partly informed by what would be considered as scientific knowledge: they occur through often messy, multilayered, and multiple negotiations (Zwarteveen 2015). Second, water accounting focuses on the water used as an input for agricultural production (or economic profits) and stipulates efficiency or productivity (e.g., expressed as units of water per quantity of crop or profit produced) as the main concern. Because the Water Accounting initiative adheres to modernist ideas of objectivity, these waters and the logic behind them risk being perceived as the only possible or true ones, as we will explain.

Like many international science–policy initiatives in water, water accounting speaks to an emergent policy consensus that water problems are (or should be treated as) global in nature. Water Accounting, for instance, refers to the need for assessing planetary boundaries or states that "a system of water accounting has so far been missing as an important element in the emerging system of global water governance" (Bastiaanssen et al. 2015:10). Texts also make use of global projections of population growth, by stating that "producing enough food to meet the demands of a global population of 9.1. billion people by 2050 require levels of food production in 2007 to be increased by approximately 60%, and doubled in sub-Saharan Africa and parts of South and East Asia" (Bastiaanssen et al. 2014:6) or that annual agricultural water use will need to increase from approximately 7,100 km3 globally to between 8,500 and 11,000 km3 in order to meet projected food requirements in 2050 (de Fraiture et al. 2010). The reference to water problems as "global problems" not only draws useful attention to global connections and interdependencies, it also—in our opinion much less usefully—suggests the need for and possibility of one coherent global view, which is also one that only (some) scientists can express and articulate. It is a small step from the articulation of environmental problems and solutions as global problems to the claim that the world is a "universe": that there is only one correct way of knowing this world. In other words, there is a risk that a preference for the global also feeds beliefs in the existence of the universal. Such beliefs erroneously posit the "global" as the larger whole, of which "the local" is a subsystem or a specification. "Globalness" then comes to signify (the existence of) one overarching order, the one that scientists first need to unravel and discover before policy makers can intervene in it.

The belief in such a global universal order underlies the aspiration that water accounting shares with many other science–policy initiatives in water: that of integrating all available knowledge on "water flows, fluxes, stocks, and the services and benefits related to water consumption" in one all-encompassing database or overarching map. Water Accounting (2016), for instance, states:

> Solving water problems requires information from many disciplines....The information has to be coherent and synchronized in order to provide an integrated picture.

In a paper explaining water accounting, Bastiaanssen et al. (2015:2) express the hope that it "should be adopted by environmentalists, agronomists, economists and lawyers alike," thereby indeed functioning as a common, integrative language; a language that commensurates different waters and either reflects or establishes one single water order:

> The working hypothesis is that by having an approved central database on water-land-ecosystems at the negotiation table with standard nomenclature and clear data, confusion becomes minimal, and trust among parties will get to a higher level.

Thus, water accounting appears to be guided by the conviction that there is, or should be, one correct or right way of representing water.

Water accounting makes use of satellite-produced earth observations and aerial photographs to produce maps that cannot be established from the ground: a single image presents (proxies for) the distribution of evapotranspiration, biomass growth or water productivity for large areas. The attractiveness of this way of knowing water flows and stocks, especially when on-the-ground measurement stations are few, is clear. Yet, our concern is with how the use of earth observations tallies with (and further nourishes) ideas of transcendence: ideas that it is possible to see and control water from a detached and unmediated position, a global vantage point or eye in the sky.[6] Such ideas match with a modernist interpretation of objectivity as consisting of being detached and unconnected, with independence as its central feature. Water Accounting (2016) states that it "is a multi-institutional effort from international knowledge centers (IWMI, UNESCO-IHE and FAO) that are neither politically nor geographically connected to a given river basin...[and] provides independent estimates of water flows, fluxes, stocks, consumption and services, that in the near-future becomes certified." It is thus guided by the assumption or desire to produce facts more or less independently from wider policy contexts. The possibility to do this is anchored in the unequivocal placement of water on the nature side of the well-known and typically modernist society–nature dichotomy. Hence, the "white paper" defines water management as an interference in natural flow processes

---

[6] For a critical feminist analysis of Earth Observation Satellites, see Litfin (1997); for a discussion of the dangers of using pixels to represent the commons, see Lele (2001).

and refers to the physical, quantitative understanding of hydrological processes as the basis for such interferences (Bastiaanssen et al. 2015).

In all, the Water Accounting initiative—just like many other international science–policy initiatives in water—is clearly built upon a modernist onto-epistemological scaffolding. The rest of our discussion will reflect on this scaffolding, first in terms of whether it meets its own modernist claims and then in terms of how it addresses questions of power, inequality, and politics in its analysis of water problems. We continue by discussing the possibility of using the tools, methods, and maps of the Water Accounting and other science–policy initiatives in less modernist ways, in support of the production of political-scientific representations of, and interactions with the environment that are more accountable to equitable, diverse, and sustainable forms of living.

Networks, Affinities, and Waters

A first way to test the modernism of water accounting initiatives is to question their independence or neutrality by tracing their origins: Through which networks do they happen and circulate? This is not a difficult exercise as these initiatives bring together old epistemic friends who share a long history of developing and actively promoting and circulating a particular body of water knowledge. This body of knowledge is one that is closely linked to water management interventions in developing countries, funded by development cooperation money and dominated by the disciplines of engineering and economics (see Goldman 2001, 2007). Indeed, water accounting initiatives bring together a very particular group of international water experts—the International Water Management Institute (IWMI), the FAO (World Food Organization), the World Bank, and more recently IHE-Delft—and provides them with the opportunity to use advanced technological means (remote sensing and modeling) and a contemporary grand challenge (global water scarcity) to breathe new life into an old and very familiar project: that of measuring and mapping consumptive uses of water in order to improve its productivity (e.g., IWMI 2007).

The many collaborations and overlaps (in ideas and people) between the international water policy network (development cooperation donors, UN agencies, the World Bank, and regional development banks) and international centers of scientific expertise (FAO, IWMI, IHE-Delft) produce a close circularity between how water is scientifically understood and how it is enacted in policy proposals for regulating and controlling it. It is no coincidence that the very donor organizations (e.g., USAID, the Asian Development Bank, the African Development Bank, and the World Bank itself) that finance development of the new water accounting tools and indicators are also the ones most interested in the river basin scans, maps, and water resources assessments that they facilitate (Bastiaansen et al. 2015). Water accounting initiatives are thus less independent or objective than they hope or claim to be: they exist and circulate (obtain funds, legitimacy, and credibility) thanks to and because of a particular network of scientists, policy makers, and donors. Viewed another way, water accounting initiatives thus far seem to have been more convincing and effective in mobilizing funds for their projects than in supporting actual water decision-making processes with their tools. One could even speculate that much of the modernist language used serves primarily the purpose of mobilizing support, rather than expressing deep epistemological convictions

One possible implication of this is that the maps and measurements produced by water accounting initiatives do not produce a universal water, but rather one very particular version of it.[7] We suspect that this version exists and circulates precisely thanks to its affinity with specific communities of practice, citation circles, projects, and flows of funding. If true, knowledge of water accounting is deeply situated: it is attached to, and depends on, specific people (bodies) with particular institutional, financial, and political affiliations, helping them achieve *their* water ambitions. It may even be that it is because of these attachments and affiliations that knowledge of and tools for water accounting become mobile, durable, and eventually "true" or at least effective.

The Water Accounting Version of Water as Efficient, Quantifiable, Valuable

We have suggested that water accounting belongs to a particular epistemic tradition in agricultural water science, one aimed at efforts to improve the effectiveness of water projects and water policy reforms funded and supported by development cooperation and loans. Inspired by optimistic beliefs that development happens through technological progress and economic growth, the overarching problem diagnosis of this body of work was, and still is, that there is a gap between the potential (determined on the basis of results obtained in virtual or real-life laboratories, experiments, and pilot plots) and actual performance of water delivery systems (variously expressed in terms of poor efficiencies, water productivities or yield gaps). Articulating water problems in this way frames the search for solutions as consisting of ways to close this gap and, in the process, (re-)formulates the water problem as primarily one of efficiency or productivity. Efficient or productive water use, in turn, tends to get defined as the precise matching of water deliveries to crop water demands, or the avoidance of waste or losses of water when conveying it from the source to the root zone of plants.

---

[7]   Sletto (2008) provides another clear example of how environmental knowledge is produced as part of specific institutions. He shows that the production of environmental knowledge is central to the institutional cultures of environmental planning agencies and shaped by political-economic processes, dominant narratives, and particular institutional desires to produce "conservation" landscapes.

This is a very particular framing or enactment of the water problem, one that emphasizes water as an input for the production of crops, or perhaps more accurately profits. The water that counts, and that needs to be accounted for, is the water that produces a particular (quantifiable and marketable) kind of value, for only when this value is produced can returns to investments (in infrastructure, reforms) be expected (Gilmartin 1994, 2003; van Halsema 2002). This version of water is particularly compatible with approaches to the regulation, management, and control of water that are optimistic about the possibility to combine economic growth with efficiency and environmental conservation; that is, modes of resource regulation that aim to deploy markets as the solution to environmental problems, based on the conviction that sustainability depends on maintaining natural capital (Bakker 2004, 2007; Ahlers 2010; Robertson 2012).

This further underscores the point that water accounting is deeply social and political in that it belongs to specific policy networks and is more suitable to support some societal projects of betterment than others. The projects that water accounting seem to favor are those closely aligned with market-based solutions to problems of development and sustainability. As noted, it may well be that it is partly because of this affinity with dominant policy models that the maps and measurements of water accounting obtain authority. It may also be that it is precisely their resonance and alignment with powerful funding, policy networks, and ideologies that enables their exposure and makes them popular, legitimate, and indeed true, as much, or more, as the accuracy with which they represent water.

The Water Accounting Version of Water as Natural

Water Accounting treats, defines, and analyzes water as something that is, in essence, "natural," the behavior of which can be explained through reference to a universal natural order or logic. This has two effects: First, it suggests that water can be rather straightforwardly read off actual realities "out-there," as if these realities unequivocally exist prior to being mapped, measured, or known. This conceals the hard, and often messy, work involved in water accounting science-in-action: producing measurements, maps, and sheets entails engaging in the repetitive, laborious, and painstaking activities of labeling, marking, repeating, cleaning, numbering, noting, interpreting, and controlling. This work is partly performed by different technologies, machines, algorithms, and computer programs through advanced tools of registration, classification, aggregation, measurement, and calculation (for an example of what this entails for water accounting, see Bastiaanssen et al. 2014). Whereas a modernist account would use these tools to help uncover the pre-given order of things, a social studies of science account has it that tools themselves help produce this order. Accordingly, observed phenomena do not simply depend on certain material instrumentation; they are *thoroughly constituted by the machines and apparatuses that make them appear* (cf. Latour and Woolgar 1986; Blok and Jensen 2011:32). The point is that science requires work and an enormous amount of laborious, meticulous, and routine manipulation of artifacts: facts are indeed literally produced. Accepting this prompts the need to rethink what is true, shifting the discussion about objectivity and representation from just one of accuracy to one of the *translations and alliances* needed to mediate (move, displace) a phenomenon or substance into a textual or visual fact.

The second effect of ontologically defining water as natural is that it makes the water accounting version of water appear almost as if it moves and circulates by and of itself: it situates the power to change water flows in the hands of "decision makers," who themselves remain rather opaque and anonymous. The human labor, technologies, institutions, and funds required to (organize water's) access and transport, and indeed "own" it, do not enter into the analysis, nor do the specific social relations through which this labor and funding are mobilized become visible. The particular water use configurations shown on Water Accounting maps thus reveal little or nothing about the historical or contemporary struggles over water's access and control; the social arrangements in place to share, care, or control it; or the multiple ways of engaging with and making sense of water that help explain how it "behaves" and moves.

To give but one example, consider the water accounting maps produced for an irrigation scheme in Sudan (e.g., Figure 11.1). These maps clearly show circular patches of high water productivity next to smaller rectangular spaces where water productivity is much lower and displays greater variation. The highly productive circles represent the areas that were irrigated by the center-pivot irrigation sprinkler systems introduced a decade ago. On these colorful maps, these circular areas are the places where most value per drop of water is produced. The maps suggest, therefore, that these areas should be appreciated and encouraged. The particular abstractions through which this valuation exercise happens, however, leave many things out of the equation. Hidden (and thus eliminated from the public policy deliberations that water accounting hopes to inform) are the highly unequal deals, sharing arrangements, and favors that have produced this seemingly homogenous and highly productive space. The maps, for instance, do not tell how the Sudanese company that owns the sprinkler pivots and exploits the scheme succeeded in cheaply negotiating access to this land, water, and infrastructure in 2006. Prior to this, in 1990, people in the investment area had agreed with Kuwaiti and Sudanese public investors to construct a smallholder irrigation scheme, which would have provided them with opportunities for engaging in the cultivation of irrigated crops as well as water and fodder for their cattle. In 2006, however, the Kuwaiti government sold its share, which consisted of the main irrigation infrastructure, to this Sudanese company. The Sudanese government allowed this company to lease the land of the smallholder scheme for 25 years. The company subsequently fenced off the area and bulldozed away most of the

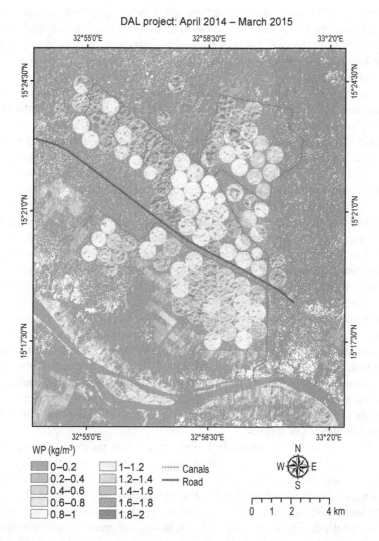

Figure 11.1　Water accounting map that shows water productivities in the area irrigated by a Sudanese company; circles represent the lands irrigated by the efficient center-pivot irrigation schemes.

existing scheme infrastructure to transform it into a highly mechanized pivot irrigation scheme. It started a profitable farming enterprise to produce fodder (alfalfa) for export to dairy farms in Saudi Arabia.

What the maps do also not reveal is what happened to the smallholders who were thrown out of the 1990 irrigation scheme to make place for the company farm. Many were no longer themselves involved in the hard labor on the farms; instead, they hired laborers from South Sudan, Darfur, and Nomadic tribes

240 *M. Zwarteveen et al.*

in the area. Yet, they protested fiercely when their ability to farm was taken away. Security forces had to be called in to quash their revolt. Village leaders benefited from the deal that was reached and played a key role in allocating the 4,000 *feddan* left for smallholders. Laborers left the area after the company began operations.

None of this is visible when observing the pivots quietly moving around to complete 12- or 24-hour irrigation cycles. These imported pivots make perfectly round wetted areas to produce a solid 18 tons of alfalfa per hectare. Water accounting maps support the story of the company scheme as a highly productive and efficient one, by positively contrasting the circular company "islands of high water productivity" to the less productive rectangles of smallholder farms. What would be the effect of this representation if it were actually used to inform water allocation decisions? The existence and mapping of productivity differences might suggest that those who are less productive are less deserving of support and encouragement. It could also provide arguments in favor of more pivot irrigation schemes. Water accounting representations may thus help convince those responsible for water allocation decisions to sell or rent out other smallholder schemes to private investors. At another level, the narrow focus of the maps on plot-level productivities may divert policy and political attention away from the question of whether it is better to use water to grow food for the Sudanese market rather than to use it to produce fodder for mega-dairy farms in Saudi Arabia. While both the Ministry of Agriculture and the business plans of the company, since the 1970s, have repeatedly referred to Sudan as a bread basket of the world to justify their actions, an ironic effect of the maps may be that they mark the smallholder farms instrumental in producing this "bread" as inefficient, thus creating the risk of cutting them off from new investments or support.

This example demonstrates how dangerous it is to present water accounting measurements and maps as the only possible reality, or as unequivocally representing (speaking for) "nature" or "the environment." To do so shelters them from political debate (thereby directly clashing with the stated ambition of water accounting initiatives to contribute to water democracy) and renders invisible the deeply social and political relations and processes through which this "nature" has come into being. The situation of irrigation schemes along the Blue Nile in Sudan is not unique in that it is one where all available water has been allocated. This means that interventions that intend to change (improve) uses or users inevitably entail reallocations and often dispossessions.

Dealing with Difference: From Commensuration to Connections, Translations, and Contestations

As discussed, the science of water accounting, just as all of science, is replete with culture and politics: it comes from particular epistemic communities, emerges from distinct successor science projects, and becomes true and obtains legitimacy through (associations with) particular policy and funding networks. The maps and measurements produced by water accounting do not describe the world as it is, but instead produce one particular version of it. This version is easier to align with some societal and political projects of betterment than with others. To our knowledge, water accounting maps and assessments have not yet been directly translated into, or informed , interventions or water allocation decisions, nor have they been used to legitimize or promote specific investments. We fear, however, that their use of a modernist objectivity allows them to be captured, interpreted, and used by those who promote a certain logic of calculation and efficiency as the only, preferred, or most important one in talking and deciding about agricultural water uses or about wise water uses in general. Although presented as a new and green logic, this water accounting logic is the same logic that has long dominated agricultural water science, particularly in debates about irrigation efficiencies.[8] It is a logic that makes some uses and users appear productive and/or efficient, thus qualifying them for support and approval. Others, in contrast, appear as wasteful or environmentally destructive, which sets them up to become candidates for sanctions or for projects of improvement, training, or awareness raising. It is also a logic that can easily be made compatible with larger efforts to use pricing or valuation for the regulation, management, and control of water, such as payments for ecosystem services.

The high water use and delivery efficiencies or water productivities favored by water accounting appear (are made or become real) on farms where farmers have the desire and (technological and financial) means to indeed optimize the drops of water used against yields or incomes obtained. Perhaps there are industrialized farms or commercial plantations in desert areas for which this is the case. However, as the example of the Sudanese company farm illustrates, the crop per drop or income language does not show how rights to water or land were obtained in the first place (through which negotiations and deals), thereby eclipsing historical investments in technologies or infrastructures for accessing water. Thus, we maintain that the language used in water accounting is not the only possible, best, or true language, nor is the water this language enacts (i.e., water as an input to the cultivation of marketable crops) the only possible or true version of water. Farmers and others concerned about water may be interested in many more things than just optimizing crop or income per drop. Farmers may, for instance, aspire to use water for improving the ease of farming operations or to spread risks. They may take pride in their own particular variety of crops, nurturing it partly for its own sake and cultivating it

---

[8] This logic is far from straightforward, as the many ongoing discussions within irrigation expert communities about how to best define and operationalize efficiency demonstrate (see, e.g., Lankford 2012a, b; van der Kooij et al. 2013). The familiarity of this logic in this particular epistemic community may explain why water accounting initiatives generate enthusiasm, support, and funds here.

not just to maximize profits, but also for reasons of taste, beauty, and heritage to mention just a few.[9] There may be trade-offs between water and labor (e.g., when farmers over-irrigate to reduce weed growth) or between water and fertility (e.g., when farmers sequence their water gifts to allow the roots of their fruit trees to grow deeper, thus allowing them to make better use of the soil's fertility) (Domínguez Guzmán et al. 2017).

Moreover, water accounting measurements—like all representations of water—are deeply scale sensitive. There are places in the world where the optimization of water use does not occur at the level of the farm or of a single crop over a single season, but involves sophisticated social and political mechanisms—which may have become more or less fixed in infrastructure— of sharing available waters across larger areas (watersheds or river basins), or over time, among different users. These mechanisms may have evolved historically as part of the collective investments in the construction or maintenance of infrastructure, to form wider social fabrics that govern the organization of socionatural relations. Indeed, living with water in specific places for generations often has yielded intricate ways of looking after and caring for or protecting it, or the infrastructure constructed to transport it, for future generations. Such forms of living with and caring for water come with their own ways of expressing or enacting water: with their own ontological definitions and logics, and their own repertoires for making water real (Boelens and Hoogendam 2002; Boelens and Gelles 2005; Domínguez Guzmán et al. 2017). In other words, there are many ways of "doing" water that differ from how emerging policy–science initiatives like water accounting "do" it. Different waters have their own specific bodies of knowledge and communities of knowers. They are also associated with their own words and systems of meanings, and embedded in their own practices (e.g., of accessing and transporting water, of assessing and classifying quantities and qualities, of irrigating crops, of sharing and distributing it, of making decisions about it, and of conserving and protecting it).

The modern way to address the simultaneous existence of multiple waters is to try and commensurate and integrate them. Can the waters embodied in the above-mentioned practices, however, be (more or less accurately) expressed in water accounting terms and somehow be shown on maps? Can concerns (e.g., of equity and diversity) that these waters highlight be integrated in calculations, so that the decision makers they are supposed to support can also be held accountable to values other than productivity and efficiency? We have tried to grapple with this latter question in a recent project, the Nile Water Lab (nilewaterlab.org) in which we explored how water accounting maps compare with more ethnographic attempts to understand changing water uses and flows. One tentative conclusion is that there are indeed many values and measurements

---

[9] Jan Douwe van der Ploeg's long list of publications about farming styles and the persistence of family and smallholder types of farms is insightful here.

beyond those of productivity and efficiency which can be expressed in water accounting terms and that can be shown on maps. Whether, how, and to what extent the information and data required to produce these values and measurements can be (just) obtained from remote-sensing data is, however, doubtful. Water security, for instance, would require detailed information about (changing) property relations, information which may be difficult to gauge from aerial photographs. Another conclusion is that the grid-like Euclidian spatialities of water accounting waters are less suitable for expressing the mutual interactions (and indeed co-constitution) of water flows and people, and thus for explaining why differences between watery places and people came about in the first place. For generating such explanations other forms of representation are more suitable, such as historic narratives or pictures.

The question of how and whether remote-sensing tools can be used to address a range of different realities is one that has also been addressed by others. One suggestion that originates from their attempts is using and answering such questions playing around with a variety of temporal and spatial scales and categories (e.g., of land cover or use) to inform the production of maps (Guyer and Lambin 1993; Jiang 2003; Walker and Peters 2007). They propose that it is necessary to compare and contrast explicitly the measurements, classifications, and values used by different scientists with those used by irrigators or water users, to make the resulting differences the topic of conversation and discussion. What happens when water is *not only* a natural resource but is, for instance, also a relative or a goddess? Most exercises of this kind conclude that the modernist approach for dealing with the existence of multiple ontologies (i.e., multiple versions of reality) tend to result in positing some ontologies as superior (see, e.g., Robbins 2003; Zubrow 2003; Comber et al. 2005; Turnbull 2007). This is because a modernist treatment would usually consist of attempts to make different waters fit one overarching scientific logic. The history of irrigation development is replete with examples of the effects of such a modernist treatment, examples of the sometimes violent destruction and erasure of existing ways of living and ways of using water in favor of more scientific and therefore supposedly more efficient, rational, and productive modern irrigation schemes (Boelens 2015). "Local" and "other" waters then get treated as informed by, or merely consisting of, tradition and culture, placing them in contradiction to the universal and global modern waters of scientifically informed policy experts (see Robbins 2003; Bonelli et al. 2017).

Our proposal for addressing multiple waters is different: rather than trying to find a singular currency that allows the commensuration of all differences, or devising a singular grid on which all different waters can be mapped, we suggest that there is merit in finding ways to acknowledge and live with many different orders, repertoires, registers, languages, choreographies, and idioms that express or do water. Importantly, this requires acknowledging that culture and politics are inherent to *all* forms of knowledge, including scientific ones. Robbins described one attempt to do this for an environmental mapping exercise in Rajasthan, India. Although not specifically focusing on water, his conclusion is insightful: he argues that the simultaneous expression of a variety of knowledges—those from "experts" as well as local "laymen"—creates a level playing field that allows comparing and contrasting them to see whether and how they resonate or clash with each other (Robbins 2003:239). Robbins's example illustrates how all "categorical imaginations" (based on forestry typologies, ecological classifications or hydrological units) are inevitably partial, preliminary and situated. They may clash, but they may also usefully converge.[10] Robbins suggests that exploring such clashes and convergences is useful to elicit legitimate disagreements; this can draw attention to how proposals which change uses or allocations of environmental resources may benefit some but not others, creating contestation and sometimes conflict. The simultaneous existence of different waters, in other words, reflects not just how disparate groups of people engage with, depend on, or interfere with water, it also shows that there are likely to be disagreements and clashes over access, rights, and futures that specific water decisions and investments help enable.

In agreement with Robbins, we propose that the question of how to address different waters is not mainly one of integration, standardization, or commensuration. It also concerns travels, connections,[11] translations, or networks between many different, sometimes contingently, emerging orders and forms of patterning. Rather than explaining away the waters that do not fit dominant patterns or orders, or forcefully reshaping and normalizing them to make them fit, we propose to combine the water accounting methodology with other ways of knowing and representing water. In this way, water accounting can become a useful starting point to compare and contrast different waters, as well as to help elicit the logic, values, and futures that inform them. Rather than only mapping and accounting for the waters that fit one universal logic, this would also highlight those that do not fit, and provide a potentially useful way to bring accepted normalcy into relief as an entry point for discussing or perhaps to challenge them (for further discussion, see Leigh Starr 1990; Law and Mol 2002).

To reach practical decisions about how to most wisely deal with and allocate water in specific situations, it is clear that some procedure is needed to handle, combine or merge multiple ideas, interests, and opinions. We suggest

---

[10] This proposal is not new, but derives directly from what many of the science and technology scholars cited here (e.g., Law, Mol, Haraway, and Latour) suggest. For elaborations and specifications of the argument for the particular case of mapping, see Turnbull (2007) and Zubrow (2003).

[11] Turnbull (2007:140) makes a similar argument to address multiplicity in the production of maps by "reconceiving mapping and knowing performatively and hodologically...through focusing on the encounters, tensions and cooperations between traditions and utilizing the concept of cognitive trails—the creation of knowledge by movement through the natural and intellectual environment....differing modes of spatially organized knowledges can then be held in dialogical tension that enables emergent mapping."

that a pragmatic discussion of what works, and for whom, provides a more interesting guide than modernist beliefs in the possibility to assess water in an objective, nonpolitical way (see also Latour 2004, whose ideas were inspired by Dewey's pragmatism). Rather than integration or commensuration, the task then becomes one of forming strategic alliances; that is, learning to translate from one language to another and communicate in ways that can surpass disciplinary, ethnic, and other such boundaries (Haraway and Harvey 1995).

Conclusions

Emerging science–policy initiatives in water can be used to rethink the question of the political-scientific representation of the environment and the politics of scientific knowledge production. How can these be democratically organized? How can "universals" (durable, mobile, and stable knowledge; insights, tools or technologies that can travel between places and times) be produced without simultaneously causing dangerous elite concentrations of knowing? How can novel and stronger forms of accountability be created for what Haraway calls "livable worlds."

Water accounting initiatives have grounded our discussion, as these epitomize a contemporary surge in science–policy initiatives in water that combine adherence to a modernist conception of science with advanced remote sensing and computing powers to produce the "facts" supposedly needed to govern, manage, and use water in wiser ways. Although we share the enthusiasm of the developers of water accounting and similar initiatives about new possibilities of producing and making available advanced and detailed geo-referenced information about water stocks, flows, and services, we are worried by the distinctly modernist ways in which this information may be captured and used to obtain legitimacy and funding. In particular, we are concerned about how this modern treatment makes the particular versions of water that water accounting produces, appear to be the only possible ones. The distinct social and political origins of these waters disappear, also making the specific societal projects of betterment that they promote appear as the only or best ones.

In this chapter, we have tried to demonstrate that water accounting maps are as much a reflection of the specific networks and communities of those who produce them, as of the hydrosocial features of the areas they represent. In our view, this does not necessarily discredit them as useless or faulty. Instead, we have used the example of the deep situatedness in water accounting to call for a different way of interpreting and treating the maps and measurements produced. We argue that the usefulness or value of water accounting knowledge does not depend on strong (claims of) detachment and universality. To the contrary, water accounting will become better—truer, more useful—when these modernist claims are replaced with nonmodernist ones.

What does this mean? Instead of modernist detachment, we suggest that there is merit in actively cherishing and acknowledging different forms of rootedness or situatedness. Here it helps to reconsider the idea of scientists as engaged in uncovering universal or global orders, replacing it with one that imagines the task of researchers as modestly intervening or carefully tinkering. Rather than being only about representional accuracy, the question of environmental representation then becomes also one of (organizing forms of) permanent critical scrutiny of why and how some knowledge travels (obtain legitimacy, durability) and some does not. This includes asking uncomfortable questions about the implications of geographical distance between centers of knowledge production and those of knowledge application, especially when this distance also marks, and is maintained by, differences in economic and political power. The production of nonmodern forms of knowledge warrants continuous investigation of how produced knowledge helps "order society" by performatively sparking communities, authority, and reality into being, thereby also disciplining and normalizing "others" (creatures, spaces). This must be accompanied by a different organization of accountability in research, one that makes researchers more visible socially, culturally, and politically.

Admittedly, this is rather theoretical and idealistic, and seemingly far removed from the urgent task of producing answers to pressing problems of water scarcity, pollution, or floods. Yet, much of it may be, above all, a *practical* matter related to a rethinking of how research is funded in relation to how accountability for produced research results is organized. It may start rather humbly by simply acknowledging what many researchers actually do in their everyday work with water, instead of focusing on what they say they do. It involves the relatively straightforward task of clearly situating knowledge production efforts in the specific political decision-making efforts or imaginations of futures of which they always form a part. Pragmatic proof of effectiveness, which works to help achieve a particular goal, then becomes necessarily part of the evidence needed to establish the value of produced knowledge.

Perhaps more difficult, the reconstruction of water policy–science initiatives on a less modernist scaffolding requires that researchers (and by extension their financial and political supporters) move away from modernist truth claims and hopes for total information, transparency, and control (to know and steer from a relatively invisible, global and detached position) and instead learn to accept that knowledge will always be local, tentative, preliminary, and partial, and that solutions are clumsy and temporary.

Acknowledgments

Much of the research in this chapter was financed by the CGIAR Water, Land and Ecosystems Program, as part of the project "Accounting for Nile Waters. Connecting

Investments in Large Scale Irrigation to Gendered Reallocations of Water and Labor in the Eastern Nile Basin." We thank the two formal reviewers for their insightful reflections, and the participants to the Ernst Strüngmann Forum for their ideas and comments.

References

Ahlers, R. 2010. Fixing and Nixing: The Politics of Water Privatization. *Rev. Radical Pol. Econ.* **42**:213–230.

Bakker, K. 2004. An Uncooperative Commodity: Privatizing Water in England and Wales. Oxford: Oxford Univ. Press.

———. 2007. The Commons versus the Commodity: Alterglobalization, Antiprivatization and the Human Right to Water in the Global South. *Antipode* **39**:430–455.

Bastiaanssen, W., L. T. Ha, and M. Fenn. 2015. Water Accounting Plus (WA+) for Reporting Water Resources Conditions and Management: A Case Study in the Ca River Basin, Vietnam. Winrock International. http://wateraccounting.org/files/White_Paper_Water_Accounting_Winrock.pdf. (accessed Dec. 15, 2016).

Bastiaanssen, W. G. M., P. Karimi, L.-M. Rebelo, et al. 2014. Earth Observation Based Assessment of the Water Production and Water Consumption of Nile Basin Agro-Ecosystems. *Remote Sens.* **6**:10306–10334.

Blok, A., and C. Jensen. 2011. Bruno Latour: Hybrid Thoughts in a Hybrid World. London: Routledge.

Boelens, R. 2015. Water, Power and Identity. The Cultural Politics of Water in the Andes. London: Routledge.

Boelens, R., and P. Gelles. 2005. Cultural Politics, Communal Resistance, and Identity in Andean Irrigation Development. *Bull. Latin Am. Stud.* **24**:311–327.

Boelens, R., and P. Hoogendam, eds. 2002. Water Rights and Empowerment. Assen, NL: Van Gorcum.

Boelens, R., and M. Zwarteveen. 2005. Prices and Politics in Andean Water Reforms. *Dev. Change* **36**:735–758.

Bonelli, C., D. Rocca Servat, and M. Bueno de Mesquita. 2017. The Many Natures of Water in Latinoamerican Neo-Extractivist Conflicts. Flows, IHE Water Governance Group. https://flows.hypotheses.org/478. (accessed July 30, 2017).

Castree, N. 2014. Geography and Global Change Science: Relationships Necessary, Absent and Possible. *Geogr. Res.* **53**: 1–15.

Comber, A., P. Fisher, and R. Wadsworth. 2005. You Know What Land Cover Is but Does Anyone Else? An Investigation into Semantic and Ontological Confusion. *Int. J. Remote Sens.* **26**:223–228.

de Fraiture, C., D. Molden, and D. Wichelns. 2010. Investing in Water for Food, Ecosystems, and Livelihoods: An Overview of the Comprehensive Assessment of Water Management in Agriculture. *Agric. Water Manage.* **97**:495–501.

Domínguez Guzmán, C., A. Verzijl, and M. Zwarteveen. 2017. Water Footprints and *Pozas:* Conversations About Practices and Knowledges of Water Efficiency. *Water* **9**:16.

Franks, T., and F. Cleaver. 2007. Water Governance and Poverty: A Framework for Analysis. *Prog. Devel. Stud.* **7**:291–306.

Gilmartin, D. 1994. Scientific Empire and Imperial Science: Colonialism and Irrigation Technology in the Indus Basin. *J. Asian Stud.* **53**:1127–1149.

———. 2003. Water and Waste: Nature, Productivity and Colonialism in the Indus Basin. *Econ. Polit. Wkly.* **29**:1–19.

Gleick, P. H. 2002. Soft Water Paths. *Nature* **418**:373.

Goldman, M. 2001. Constructing an Environmental State: Eco-Governmentality and Other Transnational Practices of a Green World Bank. *Social Problems* **48**:499–523.
————. 2007. How Water for All Policy Became Hegemonic: The Power of the World Bank and Its Transnational Policy Networks. *Geoforum* **38**:786–800.
Guyer, J. I., and E. F. Lambin. 1993. Land Use in an Urban Hinterland: Ethnography and Remote Sensing in the Study of African Intensification. *Am. Anthropol.* **95**:839–859.
Haraway, D. 1991. Simians, Cyborgs and Women: The Reinvention of Nature. London: Free Association Books.
————. 1997. Modest_Witness@Second_Millenium.Femaleman©_Meets_Onco-Mouse^tm: Feminism and Technoscience. London: Routledge.
————. 2008. When Species Meet. Minneapolis: Univ. of Minnesota Press.
Haraway, D., and D. Harvey. 1995. Nature, Politics and Possibilities: A Debate and Discussion with David Harvey and Donna Haraway. *Environ. Plann. D* **13**:507–527.
IWMI. 2007. Water for Food, Water for Life, a Comprehensive Assessment of Water Management in Agriculture. London: Earthscan.
Jiang, H. 2003. Stories Remote Sensing Images Can Tell: Integrating Remote Sensing Analysis with Ethnographic Research in the Study of Cultural Landscapes. *Hum. Ecol.* **31**:215–232.
Lankford, B. 2012a. Fictions, Fractions, Factorials and Fractures: on the Framing of Irrigation Efficiency. *Agric. Water Manage.* **108**:27–38.
————. 2012b. Irrigation Efficiency and Productivity: Scales, Systems and Science. *Agric. Water Manage.* **108**:1–96.
Latour, B. 2004. Politics of Nature: How to Bring the Sciences into Democracy. Cambridge, MA: Harvard Univ. Press.
Latour, B., and S. Woolgar. 1986. Laboratory Life: The Construction of Scientific Facts. Princeton: Princeton Univ. Press.
Law, J. 2009. Seeing Like a Survey. *Cult. Sociol.* **3**:239–256.
Law, J., and A. Mol. 2001. Situating Technoscience: An Inquiry into Spatialities. *Society and Space* **19**:609–621.
————. 2002. Complexities: Social Studies of Knowledge Practices. Durham: Duke Univ. Press.
Leigh Starr, S. 1990. Power, Technology and the Phenomenology of Conventions: On Being Allergic to Onions. *Sociol. Rev.* **38**:26–56.
Lele, S. 2001. Pixelising the Commons and Commonising the Pixel: Boon or Bane? *Common Prop. Res. Dig.* **58**:1–3.
Linton, J. 2010. What Is Water? The History of a Modern Abstraction. Vancouver: Univ. of British Columbia Press.
Linton, J., and J. Budds. 2014. The Hydrosocial Cycle: Defining and Mobilizing a Relational-Dialectical Approach to Water. *Geoforum* **57**:170–180.
Litfin, K. T. 1997. The Gendered Eye in the Sky: A Feminist Perspective on Earth Observation Satellites. *Front. J. Women Stud.* **18**:26–47.
Mol, A. 1999. Ontological Politics: A Word and Some Questions. In: Actor-Network Theory and After, ed. J. Law and J. Hassard, pp. 74–89. Oxford: Blackwell.
————. 2002. The Body Multiple: Ontology in Medical Practice. Durham: Duke Univ. Press.
Perry, C. 2007. Efficient Irrigation; Inefficient Communication; Flawed Recommendations. *Irrig. Drain.* **56**:367–378.
Postel, S. L. 2000. Entering an Era of Water Scarcity: The Challenges Ahead. *Ecol. Appl.* **10**:941–948.

Robbins, P. 2003. Beyond Ground Truth: GIS and the Environmental Knowledge of Herders, Professional Foresters, and Other Traditional Communities. *Hum. Ecol.* 31:233–253.

Robertson, M. M. 2012. Measurement and Alienation: Making a World of Ecosystem Services. *Trans. Inst. Br. Geogr. NS* 37:386–401.

Sletto, B. 2008. The Knowledge That Counts: Institutional Identities, Policy Science, and the Conflict over Fire Management in the Gran Sabana, Venezuela. *World Dev.* 36:1938–1955.

Swyngedouw, E. 1999. Modernity and Hybridity: Nature, Regeneracionismo, and the Production of the Spanish Waterscape, 1890–1930. *Ann. Ass. Am. Geogr.* **89**:443–465.

———. 2009. The Political Economy and Political Ecology of the Hydro–Social Cycle. *J. Contemp. Water Res. Educ.* **142**:56–60.

———. 2013. Into the Sea: Desalination as Hydro-Social Fix in Spain. *Ann. Ass. Am. Geogr.* **103**:261–270.

Turnbull, D. 2007. Maps, Narratives and Trails: Performativity, Hodology and Distributed Knowledges in Complex Adaptive Systems: An Approach to Emergent Mapping. *Geogr. Res.* **45**:140–149.

van der Kooij, S., M. Zwarteveen, H. Boesveld, and M. Kuper. 2013. The Efficiency of Drip Irrigation Unpacked. *Agric. Water Manage.* **123**:430–455.

van Halsema, G. 2002. Trial and Re-Trial: The Evolution of Irrigation Modernisation in the NWFP, Pakistan. Doctoral Dissertation, Wageningen Univ., Wageningen, NL.

van Halsema, G., and L. Vincent. 2012. Efficiency and Productivity Terms for Water Management: A Matter of Contextual Relativism versus General Absolutism. *Agric. Water Manage.* **108**:9–15.

Walker, P. A., and P. E. Peters. 2007. Making Sense in Time: Remote Sensing and the Challenges of Temporal Heterogeneity in Social Analysis of Environmental Change, Cases from Malawi. *Hum. Ecol.* **35**:69–80.

Water Accounting. 2016. Background. http://wateraccounting.org/background.html. (accessed Jan. 9, 2018).

Zegwaard, A., A. C. Petersen, and P. Wester. 2015. Climate Change and Ontological Politics in the Dutch Delta. *Clim. Change* **132**:433–444.

Zubrow, A. A. S. 2003. Mapping Tension: Remote Sensing and the Production of a Statewide Land Cover Map. *Hum. Ecol.* **31**:281–307.

Zwarteveen, M. 2015. Regulating Water, Ordering Society: Practices and Politics of Water Governance (Inaugural Lecture, University of Amsterdam). Amsterdam Univ. Press. https://www.unesco-ihe.org/sites/default/files/oratie_margreet_zwarteveen.pdf.

12

Integrating Sustainability, Justice, and Diversity?

Opportunities and Challenges for Inclusively Framing Water Research

Amber Wutich, Juan-Camilo Cardenas, Sharachchandra Lele,
Claudia Pahl-Wostl, Felix Rauschmayer, Christian Schleyer,
Diana Suhardiman, Heather Tallis, and Margreet Zwarteveen

Abstract

The twentieth century has seen a dramatic increase in human uses of and human impacts on water resources, increasing competition over water as well as depleting or deteriorating its availability. Given its importance to human life and livelihoods, water is becoming one of the major foci of environmental research. The coincidence of water scarcity with poverty in many parts of the world makes it a focal point of international development efforts. With engineering thinking dominating over past decades, water management research has embraced more integrative approaches triggered by an increasing awareness of failures that focused on narrow single issues or technical solutions to address the complex challenges of sustainable water management. This chapter explores whether, when, and how more inclusive framings might enable more socially relevant and impactful research, and lead to more effective action. Discussion begins by establishing what a frame is and then defining what is meant by an "inclusive frame" for interdisciplinary research on environmental problems. Seven frames in water research are examined; emphasis is given to how framings are driven by differences in normative and theoretical positions, which yields very different views on progress and how best to achieve it. Next, the use of more inclusive frames in academic or research contexts is explored using two examples which incorporate multiple normative and theoretical positions. Barriers encountered by academics and researchers, as they attempt to use inclusive frames, are then examined. To explore how inclusive frames can be used to address real-world problems, three cases highlight the possibilities and challenges in applying inclusive frames to research with the goal of informing action and practice.

What Are Inclusive Frames and Why Do They Matter?

In a research context, as elaborated in the introductory chapter (Lele et al., this volume), problem frames define and bound what researchers examine and from which perspective. Water is a complex socioenvironmental phenomenon, essential for human survival and well-being and laden with multiple meanings since time immemorial. Water problems are therefore framed in myriad ways, and academic research on water emphasizes different aspects of complex water problems and points to different paths toward solutions. The fragmentation, tension, and conflict within the academic discourse on water are a matter of concern, insofar as it paralyzes action or undermines the possibility of reasonable solutions. Exploring where and how these tensions are located could open up possibilities for more meaningful dialogue on water. The concept of "framing," as discussed by Lele et al. (this volume), provides a way for us to carry out this exploration.

Frames vary widely in terms of how inclusive they are of normative and theoretical positions. At the normative level, we considered a frame to be more inclusive when it addresses more than one of the three broad values or normative concerns within the sphere of environmentalism: sustainability, justice, and diversity. Following on from Lele et al. (this volume), we consider these broad values to have many layers or components: sustainability (ecological, economic, social); justice (distributional, procedural, interactional recognition); and diversity (biological, cultural, linguistic, institutional). We also recognize that apparently "non-environmental" concerns, such as efficiency or productivity, may be used as expressions or indicators of broader values. Finally, we recognize some potential causal connectivity—conceptual or empirical—between these values. For instance, preserving biodiversity may play a role in accomplishing ecological sustainability. Intergenerational justice, which is a core component of sustainability, may be considered a form of distributional justice. Prioritizing cultural, linguistic, or institutional diversity in decision making may help accomplish procedural justice. These areas of connectivity can be important in determining the normative inclusivity of a frame.

At the theoretical level, the inclusivity of frames refers to their ability to incorporate, combine, or reconcile different representations of social and natural reality (often coming from different disciplines or subdisciplines in academia). For instance, a "bucket model" of groundwater is a less accurate and inclusive framing than a model that incorporates surface–groundwater links. Similarly, a frame that accommodates the ways that social structure constrains people's choices—and people's choices can change social structure—would be more inclusive than others that assume only one of these matters. We acknowledge that there may be trade-offs between theoretical inclusivity and analytical tractability: complex representations are harder to translate in unequivocal predictions or courses for action. Also, in discussing the theoretical inclusivity of frames, we recognize that it is not always possible or desirable to combine different representations: recognizing incompatibilities between representations may be a productive starting point for discussing alternative scenarios or intervention pathways.

We believe, nevertheless, that it is important for research to strive for normative and theoretical inclusivity. Normative inclusivity (i.e., being inclusive of or speaking to a broader set of values) may make research more relevant to societal debates. Theoretical inclusivity (i.e., being inclusive of, recognizing, or reconciling different representations of social and natural phenomena) should make research more accurate and, as a result, interventions based on such research may be more effective. In this chapter, we explore the extent of inclusiveness of different frames in the water sector at these two levels and then explore the possibilities of, and challenges to, more inclusive framings in academia. Finally, we examine the link between inclusiveness and effectiveness on the ground.

How Do We Describe and Compare the Inclusivity of Different Water Frames?

A wide range of frames are commonly used in research on water problems. Rather than choose the "best" frames or the "most influential" frames (however defined) for this analysis, we purposively chose a range of frames based on the configurations of normative values (in terms of sustainability, justice, and diversity) that they express or emphasize: integrated water resource management, adaptive water management, common-pool resources, water footprinting, hydrosocial cycle, human right to water, and ecosystem services.

Table 12.1 presents an overview of the seven frames in terms of the problem, causes, and solutions. By "problem," we mean the manner in which the frame identifies the core problem being considered. "Causes" indicate the frame's typical approach to diagnosing the source of the problem. These often include claims about biophysical and social processes. Following from the diagnosis (but perhaps making additional assumptions), many frames identify or suggest solutions to water problems.

Below, for each of these frames, we assess the following key aspects:

1. A brief *intellectual history* is provided to explain where each frame came from and who developed it. We note if it has changed much over time and mention, where possible, who tends to use it now and how it

Table 12.1 Overview of seven frames for water problems: (1) integrated water resources management, (2) adaptive water management, (3) common-pool resources, (4) water footprinting, (5) hydrosocial cycle, (6) human right to water, and (7) ecosystem services.

Problem	Causes	Solutions
1. Increasing scarcity and inefficient allocation of water	Biophysical: Water flows within the river basin boundary, links different users. Social: Administrative fragmentation leads to disconnected decision making and inefficient allocation	Plan at basin scale; introduce participatory approaches; recognize role of women, recognize water as an economic good
2. Existing water management systems are too inflexible to variability in the environment	Biophysical: Engineering assumption that everything is knowable and predictable. Social: All decisions in the hands of technocrats or engineers	Design management systems that adapt to unpredictability, using buffers, polycentric rules, and multiple knowledge
3. Individuals under-provide and over-extract water due to misaligned individual and group incentives	Biophysical: Water is rival and limited by physical cycles. Social: Groundwater and surface water are non-excludable, often open access	Assign property rights to private owners, the state or self-governed organizations, with clear rules on boundaries, rights, and responsibilities. Governments need to facilitate the formation of self-governed institutions
4. Water is being wasted in the production and consumption of goods	Biophysical: Production and consumption of goods involve water as an input. Social: Invisibility or incomplete accounting of the water used in final goods	Estimate and disseminate the correct direct and indirect amounts of water involved in production and consumption of goods
5. Water is being unfairly or unequally distributed	Biophysical: Downstream flow of water is mediated by technology and human use. Social: Those with more power appropriate more water, with effects (feedbacks) on nature and the powerless	No clear solutions, not prescriptive
6. Some humans are excluded from accessing a minimally sufficient quantity of water	Biophysical: Biophysical cycles limit water available. Social: Lack of legal protections to all and overemphasis on commodification of water leading to its privatization	Legislative action pushing access to water as a human right; subsequent actions by governments to ensure such right with a minimum guaranteed amount of water per capita
7. Environmental degradation decreases provision of ecosystem services, including water regulation and purification, which in turn reduces human well-being	Biophysical: Many ecosystem functions and services known, not well understood, not assessed or accounted for. Social: Neglect in recognizing these services leads to market activities and policies that degrade them	Assessment and valuation of water-related ecosystem services. Policy and/or market tools that capture (internalize) ecosystem service values, to induce transfers from beneficiaries to potential agents of water conservation, and to induce conservation practices

is typically used. This background information helps inform our interpretation of the theoretical inclusiveness of the frame.

2. *Basic assumptions* may have been built into the frame which may not be apparent but are fundamental to the causal claims and solutions. Understanding these basic (but often unstated) assumptions is helpful for assessing normative and theoretical inclusiveness.

3. Every frame emphasizes or is focused on certain *values*, which in the environmental context may be some variants and combinations of justice, sustainability, diversity, efficiency, or productivity, as detailed by Lele et al. (this volume). Frames may also vary in terms of how much room they leave (implicitly or explicitly) to explore other values beyond the focal value(s). Understanding the values promoted in the frame is essential to assessing the normative inclusiveness of each frame.

4. *Additional values promoted through implementation* may be present though not essential to the framing. These may be the values that are typically embraced by people who promote the frame or seek to operationalize its ideas in particular contexts. Information about additional values may be useful to evaluate the capacity of the frame to accommodate alternative values in real-world implementation.

5. The frame's *representational accuracy* is assessed by discussing to what extent it produces a faithful and credible account of the described reality. Representational accuracy is relevant to understand the frame's capacity to produce effective interventions.

6. Each frame's *political effectiveness* is assessed in terms of how influential it has been in bringing about political, institutional, or practical change. This may be due to inherent features of the frame or the acceptability of a particular framing among the powerful actors in the water sector. Hence, explicit separation of representational accuracy from political effectiveness allows us to distinguish between a frame's popularity and its ability to represent reality accurately. Political effectiveness bears directly on the ability of the frame to yield workable real-world solutions.

The purpose of our analysis of these seven frames is to showcase how different framings of environmental problems are driven by differences in normative and theoretical positions. Following this evaluation, we compare the frames, discuss the potential for and challenges to the adoption of inclusive frames in academia, and assess the capacity of inclusive frames for bringing about real-world change in water problems.

Integrated Water Resource Management Frame

Integrated water resource management (IWRM) frames water problems that stem from increasing scarcity of water and (economically) inefficient allocation of water. The biophysical causes of these problems, as framed by IWRM, include water flowing within the river basin boundary and linking different users. The social causes of the problem include administrative fragmentation, which leads to disconnected decision making and inefficient allocation across users and sectors. An IWRM framing often leads to solutions such as planning at the basin scale, introducing participatory approaches, recognizing the role of women, and recognizing water as an economic good.

Key Aspects

1. Intellectual history
   - In 1992, the International Conference on Water and the Environment developed and published the Dublin Principles which sets out recommendations for action at local, national. and international levels to reduce water scarcity.
   - In 1996, the Global Water Partnership was established to foster IWRM. This group is an action network open to all organizations involved in water resources: country government institutions, international agencies, NGOs, research institutions, bi- and multilateral development banks as well as the private sector.
2. Basic assumptions
   - Bringing stakeholders together is the first step to resolving/sharing problems.
   - Participation is power neutral.
3. Values emphasized in the frame
   - Sustainability: mentioned but economic efficiency is strongly emphasized.
   - Justice: included to the extent that the frame recognizes and gives voice to multiple stakeholders; gender is also mentioned.
   - Diversity: not emphasized.
4. Additional values promoted through implementation
   - Inclusion of multiple stakeholders (e.g., conservationists) which may increase diversity.
5. Representational accuracy
   - Strengths: reveals linkages created by water flow or movement and the multiple stakeholders and sectors involved.
   - Weaknesses: assumes that participation automatically translates into fair allocation. Water as an economic good contradicts the idea of stakeholder-based allocation.
6. Political effectiveness
   - Aimed at policy and managerial communities (as opposed to activists, NGOs).
   - Politically conservative as it respects established principles of

governance. Because it is not overtly subversive or antagonistic, this can serve to establish a dialogue.

- Change is generally perceived as incremental. No aim for radical systems change, thus it does not overtly address power issues.
- Adopted more in developing countries; adaptive water management is more common in developed countries.

Example: The Cases of Burkina Faso and Nepal

The starting point for IWRM was the recognition that water flows within a river basin that connects users across locations (upstream to downstream) and across sectors, leading to the inference that both research and management decisions must occur at the basin scale. However, from the idea of coordinated river basin management, the concept metamorphosed in the global water policy discourse in the 1990s into a holistic perspective on water resources management (Lenton and Muller 2009) with objectives "to improve efficiency in water use (the economic rationale), promote equity in access to water (the social or developmental rationale), and to achieve sustainability (the environmental rationale)" (Butterworth et al. 2010:69).

Donor agencies have put in enormous resources to support research as well as a large number of implementation projects using this framework, especially in developing countries.

Researchers have applied this frame by integrating basin hydrology with (typically) economics to identify opportunities for water savings, cross-sectoral transfers, and so on (Molden et al. 2001). Such research has demonstrated, for instance, the importance of looking at water savings at the system level instead of the farm level. On the implementation side, the success of IWRM approaches has been limited (Biswas 2008; Medema et al. 2008).

In Burkina Faso, in spite of adopting policies and laws to enable basin-level management, the setting up of basin-wise nested institutions, the rationalization of pricing policies, and major investments in training, a big gap remains between IWRM principles and outcomes on the ground (Suhardiman et al. 2015). Local-level committees have little autonomy, and newly designed institutions failed to take into account the informal and often undocumented nature of water withdrawals and the complexity of existing land-tenure arrangements (Petit and Baron 2009). In Nepal, legislation that would enable intersectoral coordination could never be passed as existing ministries perceived it as a loss of power and not as a benefit, other than a way of attracting donor funds for individual projects (Suhardiman et al. 2015). Individual project-level implementation seemed to make some headway at the local level but lacked authority to scale up to the basin. Indeed, whether cross-sectoral coordination is really needed to achieve what objectives, for whom, at which (operational) level, and how key government stakeholders could benefit from

IWRM policy formulation and implementation are questions that have not yet
been fully answered.

Adaptive Water Management Frame

The adaptive water management (AWM) frame identifies as its core problem
the inflexibility of water management structures and procedures, making them
too inflexible to address uncertainties such as climate change. The biophysical
causes identified in the AWM frame include the engineering assumption that
everything is knowable and predictable. Social causes identified by the frame
point to decision-making control being placed almost exclusively in the hands
of technocrats or engineers. The main solution suggested by the AWM frame
is to design systems that can be managed adaptively, using techniques such as
emphasizing unpredictability, creating buffers or safe margins, building polycentricity in governance, and drawing from multiple knowledge systems.

Key Aspects

1. Intellectual history
   - Developed in the 1980s by ecologists.
   - Moved beyond the command-and-control approach in IWRM to
     management as a process with built-in feedback loops.
2. Basic assumptions
   - Ecosystems are complex and must be evaluated in conjunction
     with participatory processes.
   - Climate change necessitates more flexible approaches to address
     uncertainties.
3. Values emphasized in the frame
   - Sustainability: resilience and flexibility.
   - Justice: not emphasized.
   - Diversity: Biological and knowledge diversity is not emphasized
     because it is not a goal; it is a means to achieve adaptability.
4. Additional values promoted through implementation
   - Can link to sustainability, intergenerational justice
   - Can be made participatory
   - Can link to diversity by recognizing aquatic life as a stakeholder
5. Representational accuracy
   - Strengths: Can address complex resource management problems
     and uncertain contextual conditions.
   - Weaknesses: Adaptation and learning require particular manage-
     ment frames and cultures. May not sufficiently address path de-
     pendency, low management or decision-making capacity, power

imbalances or inequities, and so forth. In systems where the uncertainty is not high, this may not be an efficient frame.

6. Political effectiveness
   - More common in developed countries.
   - Addresses policy, manager, and planner communities (as opposed to activists, NGOs); generally enhances adaptive capacity and resilience.
   - Good fit for flood management, agricultural settings, contexts where some infrastructure is already established but can be used in initial design for communities receiving new infrastructure.
   - Not a good fit for contexts where infrastructure is insufficient (discussed further below) or too fixed (e.g., German wastewater systems).
   - Tends to be adopted by people who aim to preserve ecosystems; difficult for engineers to adopt because they are trained in command-and-control thinking; cultural, linguistic, and institutional diversity is an instrument, not a goal.
   - Politically conservative in the sense that it respects established principles of governance. It is not overtly subversive or antagonistic, thus it can function as a "space-opener" for conversation. Change is generally perceived as incremental; it does not prompt radical systems change nor does it overtly address power issues.
   - May be compromised by powerful groups if processes of adapting management decisions are not transparent.

Example: The Case for a Transition toward Adaptive Flood Management in the Tisza River in Hungary

In flood management, the shift to adaptive management is aptly captured by the move from "controlling water" to "living with water" (Pahl-Wostl 2015). The EU-funded project NeWater (New Approaches to Adaptive Water Management under Uncertainty) adopted a broad framing of AWM, defining it as "a systematic process for improving management policies and practices by systemic learning from the outcomes of implemented management strategies and by taking into account changes in external factors" (Sendzimir et al. 2010:573).

An example of successful adaptive and integrated management is the paradigm shift in flood management in the Tisza River Basin in Central Europe. Traditional flood management largely focused on keeping the water out of the landscape by using structural measures (e.g., dikes or reservoirs). This was a reactive approach that protected human lives and assets exposed to increasing flood risk because the settlements are on the former river floodplains. Despite a shrinking population density in a region of chronic poverty, rising trends of flood damage from major floods have increasingly challenged the conventional engineering paradigm (Sendzimir et al. 2007). Recent years saw the slow infiltration of more advanced practices, such as polders, used as flood volume retention areas. A more radical change in approach was promoted by an informal network of actors (a "shadow" network) from the government, academia, and NGOs. Their proposal involved a more inclusive framing and a more participatory process, including the involvement of marginalized groups. A major environmental disaster, a cyanide spill, generated increasing political pressure and increased public awareness of environmental problems, facilitating more ecological considerations in flood management policy. Pilot experiments with floodplain restoration and traditional agriculture were initiated, and a combination of change in leadership and a further severe flooding facilitated the scaling-up of innovative programs (Werners et al. 2009; Sendzimir et al. 2010).

Research identified a number of potential measures (soft and hard)—mainly at the local level—that could be used to promote adaptive flood management to build resilience against floods, increase the portfolio of ecosystem services delivered or provided by a more healthy riverine ecosystem, and develop livelihoods for the more marginalized groups.

AWM, thus, highlights the critical role played by social learning processes in transitions to more resilient management in the face of uncertainty. However, learning can be costly and risky—one of the key obstacles to the acceptance of AWM (Medema et al. 2008). Moreover, AWM does not pay sufficient attention to addressing the presence of vested interests and asymmetries of power; these same problems also plague IWRM.

Common-Pool Resource Frame

The common-pool resource frame points to the inadequate provisioning of surface water and/or the depletion of the groundwater resource as major water problems. Biophysical causes for this are that groundwater (and water provisioning) is rival (subtractable) and non-excludable at the scale of the individual user.

Social causes are that groundwater and surface water are often open access. Given the above properties, free riding then happens and the resource depletes or under-provisioning takes place.

For solutions, this frame suggests converting open-access water resources to state, private, or community control. Specifically, the frame recommends creating or enabling collective action institutions at a scale that matches the resource boundary, and creating rules for monitoring, regulation, and sanctioning. The government's role should be to facilitate the formation of such institutions.

Key Aspects

1. Intellectual history
   • Increased in popularity around the 1950s.

- Emerged out of bioeconomics, game theory, and new institutional economics in conversation with natural scientists.
2. Basic assumptions
   - Actors are motivated by self-interest (methodological individualism) and behave rationally.
   - Actors have similar interests and abilities in the resource to exploit it; there is no variation in social, economic, or political power between users.
3. Values emphasized in the frame
   - Sustainability: Avoidance of groundwater depletion, framed as efficiency and maximizing provisioning of the resource.
   - Justice: Fair results assumed to emerge automatically, but no explicit analysis of the causes of inequalities or exclusions.
   - Diversity: Biodiversity not included in the basic framework.
4. Additional values promoted through implementation
   - Focus is on promoting community control (as against state or private control), believed to produce a better balance of power between state and community (but not within the community). In addition, within-community fairness is considered.
   - In some cases, biodiversity has been incorporated into the goals, when the user group itself values biodiversity.
5. Representational accuracy
   - Strengths: Predicts the emergence and continuation of successful collective action in certain situations and has been proven accurate.
   - Biophysical weaknesses: Surface water has upstream–downstream asymmetries that prevent cooperation. In groundwater systems, the assumption of aquifers having a closed boundary is incorrect, as it ignores natural discharge that feeds into downstream flows or reservoirs.
   - Social weaknesses: Cooperation can be fragile, but presence of altruistic individuals can keep cooperation going. Forced cooperation (by powerful individuals) exacts a high cost to weak parties.
   - When analyzing cooperation, a too narrow focus on the natural resource can overlook the deeply embedded nature of local institutions and organizations, networks, interdependencies, and relations of collaboration that encompass many more areas.
   - Lack of attention to political dynamics and power asymmetries results in overestimating the emergence of cooperation, and the acceptance of social hierarchies as long as these are functional for effective resource management.
6. Political effectiveness
   - Represents a potentially attractive alternative to fully public or private forms of natural resource management. Aligns well with the

agenda of powerful actors pushing for state withdrawal as well as those representing communities that demand more control.

• Its popularity has led to widespread promotion of "principles of self-governance" in canal irrigation management, watershed development, and some programs for groundwater management.

Example: The Case of the Acequias de la Vega de Valencia and Its Tribunal

The Acequias de la Vega de Valencia irrigation system is comprised of seven water canals or ditches and a main canal that irrigates small farms in Valencia, Spain. Today, it covers an area of 17,000 ha, and its Tribunal has existed for centuries. There is no exact date of when the Tribunal was created, but it is believed to have started during the Moorish rule in the eighth century. Its mandate is to resolve claims of mismanagement of water by any of the water users. Disputes are discussed every Thursday by its members (*síndicos*) who are irrigators themselves, and who receive the complaints and resolve them orally in front of the rest of the community, with no room for appeal. The *síndico*, who is a member of the ditch involved, does not participate in the deliberation and assignment of the penalty. The Tribunal does not use written documents or lawyers, and over centuries has successfully processed all claims without the need for public intervention.

The success of the Tribunal can be accurately explained through a common-pool resource frame, which defines the problem of water management as a collective action problem where individual and collective incentives may not be aligned since water users may benefit from the contributions of others while not contributing themselves. To use the language of public economics, since water is a rival (or subtractable) but non-excludable good for those water users in the irrigation system, the group needs to solve this social dilemma via institutions. The self-governance solution for this common-pool resource requires that water users endogenously develop, monitor, and enforce the rules that govern the provision and distribution of the water resource. The maintenance of these rules and social norms is a second-order social dilemma in which costly actions by each water user are required for the rules to produce the collective benefits, namely, efficient provision of the water and a fair distribution of the resource.

The research within this frame assumes that each water user has a utility function that maps actions and water benefits onto individual well-being. The social dilemma emerging from the divergence between the individual interest and the group interest requires that self-governed institutions are set in motion to restrict free riding by group members. The research conducted over this landmark case concludes that the Tribunal has successfully created a system of rules and norms that is enforced by the members who have monitoring responsibilities, and that the ease by which disputes are solved weekly, in an open and expeditious manner, has maintained the overall efficiency and fairness of the water management system over centuries (Ostrom 1990; Ortega-Reig et al. 2014).

Many of the usual solutions suggested by this framework, the Valencia Acequias Tribunal being a clear case, imply a prescription for higher government levels not to intervene with top-down regulations as these may erode the capacity of the water user system to self-govern. The common policy prescription, therefore, is to let the group of users, according to their particular conditions, devise rules and norms that are the best fit for the management of their own common pool.

This framing, however, usually omits other ecological functions of water management systems, such as the coexistence of other components associated with water (e.g., biological diversity, soil conservation). The upstream–downstream hydrological dynamics are often ignored in this framework. The common-pool resources approach usually concentrates on one resource: water in this case. Nonetheless, in this and most cases, there are other ecological functions associated with the containing ecosystems that can be of critical importance for the sustaining of life for the water users. An implication is that biodiversity, as a value in itself in many water management systems, plays a minor or nonexistent role. Likewise, little attention is paid to the hydrogeological aspects and their relationship with water quality, focusing mostly on quantity. Fairness and sustainability concerns, in this particular case, are central to the robustness of this long-standing case of successful water management.

Water Footprinting Frame

The focal problem addressed by the water footprinting frame is to identify where productive uses of water are wasteful. It seeks to reveal where water consumption in production is not accounted for or visible to the consumer, which leads to overconsumption.

The solution promoted within this frame is to calculate and display the amount of consumptive use of water embedded in a product or service. Further developments include water accounting and vulnerability evaluation (WAVE), which incorporates basin-level evaporation recycling, and corporate supply chain accounting (value chain capture of water use).

Key Aspects

1. Intellectual history
   - Emerged out of the virtual water concept by Tony Allan and is related to the ecological footprint (Hoekstra 2003).
2. Basic assumptions
   - More consumption is negative (no sustainability thresholds).

- More efficient production will automatically lead to systemic water savings.
3. Values emphasized in the frame
   - Sustainability: Emphasizes efficiency in the consumption of water.
   - Justice: Not developed.
   - Diversity: Not developed.
4. Additional values promoted through implementation
   - Economic efficiency: Optimizes economic output per drop of water.
5. Representational accuracy
   - Strengths: Contributes information currently hidden, as prices seldom reflect water use in production.
   - Weaknesses: A drop of water is not comparable across sites. More information may not lead to behavioral change if consumers respond only to prices. Efficiency gains may not lead to water savings because production expands.
6. Political effectiveness
   - Attractive to the corporate sector due to its simple, biophysical quantification approach.
   - Susceptible to "greenwashing."

Example: Mitigating the Water Footprint of Export Cut Flowers from the Lake Naivasha Basin, Kenya

The water footprint of a crop (m^3/ton) is calculated as the ratio of the volume of water (m^3/ha) consumed or polluted during the entire period of crop growth to the corresponding crop yield (ton/ha). Water consumption has green and blue components: green refers to the volume of rainwater consumed whereas blue refers to the volume of fresh water from rivers, lakes, or aquifers needed to produce crops. A third component, gray water, refers to the fresh water required to assimilate the load of pollutants, based on existing water quality standards.

Mekonnen et al. (2012) illustrate how the water footprint could be used to inform water policy decisions through an elaborate calculation of the water footprint of export cut flowers in Kenya. Economically, the export of Kenyan cut flowers is a success: from 1996–2005, it contributed to an annual average of $141 million foreign exchange (7% of Kenyan export value), with about $352 million in 2005 alone. The industry also provides employment, income, and infrastructure (e.g., schools, hospitals) for a large population around Lake Naivasha. Estimates of the net benefits of the cut flower industry, however, do not include the value (or "price") of water in their calculation and may thus be overly optimistic. Decreasing water levels in the lake, as well as complaints about pollution and a reduction in the lake's biodiversity, have caused concern—expressed among others by environmental NGOs—that the economic profits of this industry are realized at the expense of the longer-term health of the lake, and by implication at the expense of all those (plants, animals, and human beings) who depend on a healthy lake for their survival or well-being. To quantify the amount of virtual water being exported, Mekonnen et al. (2012) present a meticulous and precise calculation of the water footprint of the cut-flower industry around Lake Naivasha. The water footprint of one rose flower is estimated to be seven to thirteen liters; this is the amount of water needed to produce one rose. To put this differently: for every rose exported, seven to thirteen liters of water are also virtually exported from Kenya to retailers and consumers overseas. From 1996–2005, the total virtual water export related to the export of cut flowers from the Lake Naivasha Basin was 16 Mm^3/yr, with further division into green (22%), blue (45%), and gray water (33%). The calculations also show that six big farms account for more than half (56%) of this water footprint.

The calculation of the water footprint at this scale is data intensive. It makes use of a combination of available statistical and remote-sensing data about areas cropped or irrigated and quantities exported, combined with (among others) soil moisture, precipitation, and evapotranspiration data (making use of the standard FAO CROPWAT approach). The results give an interesting aggregate indication and quantification of how much water is needed to produce something and what fraction of that might be ending up in exports.

To understand whether, and to what extent, this footprint is environmentally problematic requires, however, further investigation of the hydro-ecology of Lake Naivasha, the amount of blue water that may be considered "available," the gray water assimilative capacity, and ultimately "how much a drop in lake level is socially and politically acceptable" (Mekonnen and Hoekstra 2010). This cannot be easily or unambiguously estimated. The authors assumed that the current use was above the acceptable threshold, further acknowledged that pricing of irrigation water supplied to farmers would not work, and focused on identifying consumer-end solutions; namely, a labeling and certification scheme that would enable customers to pay a "sustainable water premium" for sustainably grown flowers. They speculated that the extra money earned would be used by producers for investment in more sustainable ways—consuming less water or polluting less—of producing flowers.

The water footprint is an attractive and effective tool to improve water consciousness, as it creates awareness about the water costs associated with producing goods and services. An important hope of the water footprint is that better information on how much water it costs to produce something can be used to inform consumers who would then put pressure (through consumption choices) on producers to change their water practices. This hope is founded on several assumptions which may not always hold: (a) that consumers actually care about how the flowers are produced; (b) that transaction costs in labeling and certification are low; and (c) that the premium generated will translate seamlessly into changes in production practices by the farmers. Further insight into the nature and processes which shape water behaviors of different actors along the water-value chain are thus needed to translate improved water consciousness effectively into more water-wise practices. Also, a better understanding of the site-specific nature of actual environmental implications of water consumption—including a detailed identification of how costs and benefits are distributed (or the equity question)—is needed to identify realistic solution pathways. More fundamentally, perhaps, the water footprint framing and the solutions proposed sidestep some of the fundamental questions about the drivers of water overconsumption. By suggesting that water costs can and should be incorporated into the price of water, the idea that water can be protected through market mechanisms—and that consumption levels can continue increasing—go unchallenged.

Hydrosocial Cycle Frame

The hydrosocial cycle frame identifies unfairness and inequality in the distribution of water among humans as a core water problem. Biophysical causes identified by this frame include the ways by which water flows are mediated, interrupted, and diverted by technology and labor. Social causes are that powerful people appropriate water using this technology and labor at the cost of nature and less powerful people. Further, the frame highlights that water problems also occur because the dominant understandings of water are themselves reflections of the needs of the powerful, leaving out other concerns or problem framings.

Coevolution of biophysical and social dynamics is revealed through the hydrosocial cycle frame: changes in water flow alter society and vice versa. This frame, however, is not prescriptive and does not provide clear solutions. It is often associated with and informed by social movements that aim to alter the status quo and challenge existing hierarchical power relations.

Key Aspects

1. Intellectual history
   - Grew out of political ecology, eco-Marxism, and feminist scholarship.
   - Emphasizes reflexivity in scholarship.
2. Basic assumptions
   - There is no such thing as "natural" water; nature and society coevolve.
   - Power differentials always exist and are always used exploitatively.
3. Values emphasized in the frame
   - Sustainability: Not emphasized because when water is distributed equitably, the frame assumes that sustainability will follow.
   - Justice: Constitutes the primary concern.

- Diversity: This frame includes (bio)diversity in the sense that mis-allocation includes taking water away from nature. It also engages with social and cultural diversity by demanding attention to life styles beyond the mainstream to understand water.
4. Additional values promoted through implementation
   - Critique and resistance.
5. Representational accuracy
   - Strengths: By treating water as a flowing resource and highlighting power, it explains unequal appropriations well.
   - Weaknesses: Does not explain resource overuse per se. Accuracy of "equity implies sustainability" assumption is debatable. Causality is hard to trace, and reasoning can become circular.
   - Biophysical properties of water not fully considered. Its social constructivist stance precludes the possibility of a common scientific understanding of hydrology.
6. Political effectiveness
   - Does not speak to water experts and is not accessible (because of dense language) to practitioners or other disciplines. It is restricted to a small academic community.
   - Assumes that marginalized or "oppressed" peoples oppose inequality, ignoring the possibility of dependency relations which may be highly unequal yet lasting.

Example: The Case of Mollepata, Peru

The framing of the hydrosocial cycle (Swyngedouw 2009) aims to reveal how water governance, deeply permeated by power relations that are often hierarchical, produces highly uneven "waterscapes." The case of Mollepata in the Peruvian Andes has been analyzed using this framework to highlight the interactions between water, power, and cultural politics (Boelens 2014). By tracing how contemporary water arrangements in the Andes have evolved through a historical series of contestations over water between indigenous communities and state or private actors, Boelens (2014) demonstrates how water and nature are sociopolitical constructs. The problem of water scarcity is seen as rooted in existing power asymmetries, whereby powerful elites appropriate water, and this has a negative impact on ecosystems and less powerful people (Boelens 2014:234):

> Since ancient times, elites have striven to reinforce subjugation over Andean peoples by creating "convenient histories" and "socionatural order" [...] that support water hierarchies and legitimize particular distribution, extraction and control practices, as if these were entirely natural.

His analysis explicitly includes an examination of the politics of the very process of knowledge production to show how struggles over water partly play out through contestations about what is the best or most accurate way to represent water. Here, the label of science, through claims of efficiency or modernity, works to legitimize expropriations, while also performatively categorizing farmers into those (the potentially efficient ones) that deserve recognition and support, and those (the backward and inefficient ones) that should instead be left to their own devices.

Human Right to Water Frame

The core problem addressed by the human right to water frame is that some people do not have sufficient, safe, acceptable, and physically accessible and affordable water for personal and domestic use. This frame identifies the lack of legal and institutional protections as a major cause of the problem, and is also frequently used as a reaction against the trend toward water privatization and the emphasis on efficiency as a core value in water management (Murthy 2013). This frame does not take biophysical limitations into much consideration, given the relatively small amount of daily per capita water allocations typically involved. The immediate solutions suggested by this frame call for the enshrinement of a human right to water through international agreements (e.g., a dedicated UN resolution) and national legal frameworks (e.g., Bolivian constitution). In some cases, this has led to a quantitative operationalization of the minimum individual water allotment to survive (estimates range between 7–50 liters per day). Its larger purpose is to facilitate institutional reforms that would improve basic water access among underserved human populations. In 2011, the United Nations Human Rights Council Resolution A/HRC/RES/18/1 and World Health Organization Resolution 64/2 called for the development of strategies and solutions to realize the human right to water (e.g., financing for water and sanitation infrastructure).

Key Aspects

1. Intellectual history
   - In 1977, the UN Water Conference Action Plan identified the human right to water.
   - In 2010, the UN approved Resolution 64/292 to secure the human right to water. The resolution was introduced by Bolivia, where concern emerged from a local, indigenous framing of "water is life" (the Andean *vivir bien* worldview) and the Cochabamba Water War against water privatization.
2. Basic assumptions
   - Every human being has a right to sufficient, safe, physically accessible and affordable water for personal and domestic uses.
   - Exclusive focus on human well-being.

3. Values emphasized in the frame
   - Sustainability: Not emphasized as goal.
   - Justice: Core focus is on distributional and, to a lesser extent, procedural justice. More recent approaches argue for interactional justice (recognition).
   - Diversity: Includes the capacity to accommodate cultural, linguistic, and institutional diversity. Could have a deleterious effect on existing diversity by putting all under state or international framework. Does not consider biological diversity.
   - The moral or value orientation is that clean water is essential to accomplish all other human rights and dignity.
4. Additional values promoted through implementation
   - Promotes community management, commons, indigenous values.
   - Sometimes aligns with anti-capitalist views.
5. Representational accuracy
   - Strengths: Exposes the human health costs of inadequate water management and reveals structural inequities in water access within and across societies.
   - Weaknesses: Very narrow focus. Addresses only human water needs and is arguably inaccurate in representing these needs, as it rarely reflects cross-cultural variation in biological (e.g., cultural adaptations to water availability or scarcity) and symbolic needs for water (e.g., for ritual ablution).
6. Political effectiveness
   - Very effective in the sense that it emerged from successful social movements and has become enshrined in national and local law.
   - Its success in providing the basis for the reform of water management institutions and practices has yet to be determined.
   - Often used by communities to challenge the uses and allocation of water (e.g., rural vs. urban drinking water).

Example: The Case of Cochabamba, Bolivia

After its Water War in 2000, Cochabamba, Bolivia was celebrated internationally as a site of anti-privatization resistance. Many of the protesters who opposed the privatization of the water system lived in squatter settlements on the outskirts of the city of Cochabamba. Ironically, after the Water War was won, and control of the water system reverted to the municipal authority, these squatters were denied access to municipal water, just as they had been before, and remained dependent on informal water vendors for the bulk of the households' water supply. The research we discuss here (Wutich et al. 2016) asks: Do water vendors have a role to play in achieving the human right to water in Cochabamba?

Using a human right to water frame, this research examines three dimensions of justice (distributive, procedural, and interactional) in informal water vending from the perspective of the water vendors and of their clients. This frame explicitly emphasizes justice as a value and does not include sustainability or biodiversity as values. Wutich et al. (2016) find that informal water vendors adopt (institutional) rules and norms that are designed to improve distributional justice, but that their clients are much more concerned with procedural and interactional injustices in water delivery. They also find that unionized vendors (a small minority of vendors) are much more effective in designing and enforcing rules to protect distributive, procedural, and interactional justice than nonunionized vendors (the vast majority of vendors). These conclusions are very accurate in representing human views of justice, but they only address environmental sustainability (e.g., water quality) and economic sustainability (e.g., pricing) briefly, in terms of their relevance to distributive justice. They do not consider diversity at all. In its recommendations, the research suggests vendor unionization and community consultation with vendors as possible pathways to achieving the human right to water in communities that are dependent on informal water vendors.

Ecosystem Services Frame

The core problem identified by the ecosystem services frame is the degradation of ecosystems and decline of benefits they provided to society. Increasing degradation or the nonsustainable use of ecosystems causes significant harm to human well-being and represents a loss of natural assets or wealth of a country. According to the ecosystem services frame, the primary cause is that the contributions of ecosystems to humans (especially through indirect, regulatory, and/ or cultural services) are not well captured by current markets, nor are they well understood or recognized by society (especially policy makers), which leads to their neglect and deterioration. This frame points to the physical, quantitative, or qualitative assessment of ecosystem functions and services as a solution, often accompanied by valuation exercises that inform decision makers, who then take policy actions that respond to or capture the value of ecosystem services (thus, trying to internalize current market externalities). The frame assumes that nonrecognition can be addressed by simply assessing ecosystem services (to make them visible), estimating their benefits and values (monetary or otherwise), or setting up market-based instruments and other governance structures to get beneficiaries to transfer economic value (e.g., payments for ecosystem services).

Key Aspects

1. Intellectual history
   * Documented as early as Plato.
   * Term and current analytical tools broadly established by the

Millennium Ecosystem Assessment (2005) and TEEB (2008), based on previous and parallel discourses in landscape planning, agriculture, forestry, and ecological economics (e.g., Daily 1997).

2. Basic assumptions
    *   Natural (biotic) ecosystems and humanly transformed ecosystems (e.g., cultural landscapes) contribute substantially to human well-being.
    *   Time horizon of policy makers is long enough, it is only a matter of not "seeing" certain service flows.

3. Values emphasized in the frame
    *   Sustainability: Includes social, economic, and ecological dimensions.
    *   Justice: Recognition of all values is core to the frame.
    *   Diversity: Biological diversity (fostering or enabling many other services) and some aspects of cultural diversity (when different cultural groups hold different values for functioning ecosystems).
    *   Anthropocentric concept featuring a fairly broad understanding of human well-being. In practice, focus is on economic values and respective valuation methods; largely perceived to be a concept to "put a price on nature," yet, some initiatives (e.g., IPBES) are trying to go beyond this. Conceptually highly integrative in terms of including stakeholders at all levels.

4. Additional values promoted through implementation
    *   Some implementations promote multi-stakeholder participation and partly address distributional justice. Economic valuation or market-based instruments promoted in many cases.

5. Representational accuracy
    *   Strengths: Improves accuracy of understanding the relationship between ecosystem and human well-being by emphasizing (previously ignored) services and scale effects. Recognizes the relationship between ecosystem structures, functions, and service flows. Captures how management decisions create trade-offs between elements of human well-being to alter the benefit distribution among different groups of people.
    *   Weaknesses: Ignores disservices of ecosystems. Assumes that accounting for or capturing of ecosystem services will benefit nature broadly, although a causal linkage between biodiversity and ecosystem service provision is only established for a small number of services.
    *   Original formulation ignored role of human labor, technology, and capital in transforming "natural ecosystems" into benefits (MEA 2005). Recent methods (e.g., ecological production functions) recognize these elements more accurately.
    *   Emphasizes the supply side, but not the demand side, of ecosystems.

- • Assumption that lack of recognition is cause of degradation is often incorrect.
6. Political effectiveness
    - • Economic valuation is attractive to policy makers. Many national ecosystem assessments are currently ongoing or in planning.
    - • Some mainstreaming in policies at national and other levels, including impact evaluation requirements of multilateral lenders.
    - • Many commitments are being made (e.g., Natural Capital Protocol, WAVES); action to alter policies in response still not widespread.
    - • Strongly applied in European context for monitoring.
    - • IPBES to strengthen the science–policy interface for biodiversity and ecosystem services.

Example: The Case of the EU OpenNESS Project on the Lower Danube River Wetlands System in Romania

The research presented here was carried out within the EU research project OpenNESS, which aims to translate the concepts of natural capital and ecosystem services into operational frameworks that provide tested, practical, and tailored solutions for integrating ecosystem services into land, water, and urban management decision making. OpenNESS examines how the concepts link to, and support, wider EU economic, social, and environmental policy initiatives and scrutinizes the potential and limitations of their integration at national, regional, and local scales. It is a transdisciplinary project that works in close cooperation with decision makers and other stakeholders, as natural capital and ecosystem services concepts (or elements thereof) are applied to concrete management and decision-making situations, such as integrated river-basin management, in a set of real-world case studies (Furman et al. 2018).

One case study used the ecosystem services frame to facilitate the design and implementation of an adaptive management plan for the Lower Danube River Wetlands System in Romania, which is characterized by a complex network of wet meadows, alluvial forests, agricultural polders, and fish ponds. In particular, a long-standing intensive and monofunctional agricultural production system reduced many of the wetlands' major functions and services, including fish catches, nutrient retention, water quality regulation, and river pulse regulation as well as recreational services and biodiversity. This study used the ecosystem services concept to (a) assess the relationships between biophysical structure and functions, and the supply of ecosystem services in the region (mapping); (b) identify and assess the trade-offs between sectoral policy objectives (e.g., inland navigation, hydropower production, food production, water quality, flood protection, biodiversity conservation) and policy instruments (e.g., NATURA 2000, Water Framework Directive, Common Agricultural Policy) at multiple scales (local to EU) for effective conflict management; and (c) enhance the operational capacity of stakeholders involved in the development of the adaptive management plan for assessment and valuation of ecosystem services (including biophysical methods, mapping tools, monetary and nonmonetary valuation tools).

The underlying research questions were framed by the ecosystem services concept, informed by the case study characteristics and needs. Stakeholders were involved in the research process via a Case Study Advisory Board. This board decided on the respective methods and tools to use as well as on the specific objects/areas of application (particular ecosystems, ecosystem services, policies, or subregions), and did not have a political mandate. Various aspects of sustainability were considered and a wide range of stakeholder organizations, institutions, and perceptions were acknowledged. Only key stakeholders, however, were represented on the Board.

Preliminary results indicate that stakeholders involved in the planning process thought that the ecosystem services frame helped improve the effectiveness of the management and that it fostered an integrative understanding of linkages between ecosystem functions provided and the trade-offs between relevant sectoral policy objectives (Grizzetti et al. 2016). The Board embraced the inclusive approach of this frame, in principle, yet its practice focused on a subset of identification and mapping tools (e.g., QuickScan) and multi-criteria decision analysis methods provided by the researchers. Substantial efforts are, however, needed to generate and compile required data and information. Some justice issues (representational equity) were addressed through the inclusion of a broad range of stakeholders, both as Board members and as active participants, in the regional project workshops. Further, one major long-term goal was to enhance the operational capacity of all stakeholders involved in the development and implementation of the river-basin management plans (Dick et al. 2018).

Assessing and Comparing the Frames

The seven frames described above provide a reasonable cross-section of the different ways in which water problems tend to be framed in research and/or action. These frames vary widely, both in terms of their theoretical and normative inclusiveness as well as in terms of their representational accuracy; namely, how well, or in what way, they represent particular water realities.

At the outset, it is important to note that particular frames are often used in, and developed for, particular contexts. This means that the values they emphasize and the representations they produce may be adequate in these contexts. For instance, the uncertainty introduced by climate change may be the biggest source of stress on the water system in a Central or Northern European context, where basic water needs have been amply met across households and sectors, and thus adaptability or resilience gain more importance. In contrast, in the context of a low-income country, the challenge of meeting basic water needs for different users and uses may take priority over goals of adaptability and resilience. Similarly, where the allocation of rights between upstream and downstream users is well-defined and accepted, a focus on collective action *within* a user group becomes relevant. Related to this, we note that various actors may frame problems differently, depending on the context, their own viewpoint, or their association with others (Gyawali and Dixit 1999; Moench et al. 1999).

In the subset of frames we detailed more thoroughly, the most common normative goal in different framings of water seems to be sustainability, which four of the frames (IWRM, AWM, common-pool resource, and ecosystem services) emphasize. This is somewhat surprising given the point made by Lele et al. (2017) that since water is a flow resource, fairness and equity should be the key concerns rather than intertemporal sustainability. Closer examination, however, indicates that "sustainability" means different things in each frame. In the case of AWM and IWRM, sustainability means adaptability or resilience of water service delivery in the face of uncertainty or external shocks. In the case of common-pool resources, sustainability means avoiding a decline in groundwater levels or in the surface water delivery infrastructure. In the ecosystem services frame, sustainability means maintaining natural capital intact on the assumption that it will generate necessary water flows.

Few of the frames explored here emphasize justice, and fewer still diversity. Even where explicitly mentioned, such as in the IWRM frame, its conceptualization is merely process based, expressing justice in terms of whether or not all stakeholders are equally represented. Here, one sees the interplay between theory and values: theory explicitly recognizes power differentials, as in the frame of the hydrosocial cycle frame, and justice is also emphasized more comprehensively.

Our conclusion is that no frame explicitly front-pages all three dimensions: sustainability, justice, and diversity. Most of the frames can become more inclusive in their values, but within limits: the principle of water, for instance, as an economic good (implicit in the IWRM framing) is inherently biased in favor of those who can pay for water.

It is when we examine the theory behind each frame that we see the bigger barriers to inclusiveness. The assumption of power differentials as fundamental and ubiquitous in one case (hydrosocial cycle frame) is difficult, if not impossible, to reconcile with the methodological individualism of water-as-a-common-pool resource, or the process and planning emphasis of IWRM. The deep cleavages between political economy, coevolutionary thinking, and rational choice models become apparent here, as do the differences between framings aimed at descriptive understandings and those aimed at guiding planning or interventions. Similarly, if water is considered part of a cycle and its uses are contextual (hydrosocial cycle frame, IWRM frame), then the assumptions of linearity and universality of water footprints, or even water as a human right, become problematic.

Finally, there are some clear correlations between theoretical positions and normative ones. Ecologists, for example, tend to be more receptive to the ecosystem services frame than to, say, the frame of hydrosocial cycles, because the former places a greater normative focus on biodiversity, whereas sociologists are more receptive to the hydrosocial cycle and water as a human right frame than say economists or hydrologists.

The Use of Inclusive Frames in Academic Contexts

While most of the frames we discussed have clear limitations in terms of our definition of inclusiveness, it is possible to expand these frames pragmatically to make them more inclusive. Here, we explore cases in which two frames, ecosystem services and IWRM, were expanded to include a broader range of normative and theoretical positions. In doing so, we seek to address the following question: How can scholars make framings more inclusive while ensuring they remain representationally accurate in research on complex water problems?

Expanding an Ecosystem Services Frame to Embrace a Wider Range of Values

In Northern Kenya, wildlife, pastoralists, and private ranches exist in one large landscape. Much of the landscape is not fenced, and thus decisions made about grazing-land management have implications for pastoralist well-being, private revenue generation, and wildlife populations. A group of transdisciplinary researchers (disease ecologists, agronomists, ecologists, economists, social scientists) analyzed how different forms of cattle and grazing-land management affect wildlife and human well-being in this landscape (Allan et al. 2017). Initially they used an ecosystem services frame to develop a conceptual hypothesis of the connections acting across social, economic, and ecological elements of the landscape. Thereafter they conducted field experiments, observational surveys, integrated model development, and scenario analyses to explore this hypothesis.

The ecosystem services frame captured some connections in the system between elements such as forage and livestock production, nutrition, income, tourism, wildlife populations, and disease risk. Each of these connections exists through some environmental change. However, the frame did not capture several other important connections that link rangeland management to human well-being directly, rather than through an environmental change. For example, one mode of cattle grazing, called "bunched herding," requires keeping cattle close together and moving them around the landscape differently: a practice that necessitates more herders per head of cattle. Implementing this grazing mode directly creates jobs because of the higher labor requirement.

As the environmental system does not need to change for the creation of these jobs to be viable, this is not an ecosystem service, but it is still an important linkage between landscape management and human well-being. The research team also faced challenges in integrating ecological, agronomic, economic, disease ecology, and social models, as few dominant modeling frames in these disciplines incorporate elements of key interactions in this system. In addition, the ecosystem service frame emphasizes the flow of nature's benefits to people, often identifying those people as "beneficiaries." However, the frame is not very explicit about distributional effects and vague about how beneficiary groups should be specified. Ecosystem services analyses often use land-use classifications as the basis for modeling, but in this system, there can be multiple beneficiary groups (e.g., ranch owners, resident ranch staff, day workers) receiving different ecosystem services on one property (land-use type). Thus the researchers needed to expand the frame to capture this complexity and ensure more direct treatment of the distributional effects of landscape management decisions.

Expanding an Integrated Water Resources Management Frame to Address Multiple Concerns and Stressors

In analyzing water management in two regions of Southern India, and the potential impacts of climate change on it, Lele et al. (2018) devised an approach using the basic descriptive aspects of IWRM. This included: (a) clarifying linkages between upstream and downstream (using basin as the scale), (b) clarifying linkages between groundwater and surface water, (c) clarifying linkages across sectors and stakeholders, and (d) the core normative idea of representing all stakeholders. Over time, the frame was expanded to add the following elements:

1. Multiple concerns: adequacy, quality, fairness across sectors and equity within sectors, sustainability and democratic governance
2. Multiple stressors: climate change is not the only source of stress on the system, many other stressors already exist, including land-use change, cropping-pattern change, population growth, industrialization, so their relative impact has to be assessed
3. Clearer role of infrastructure and institutions: water is distributed through built infrastructure and rules associated with it (both in supply and in effluent disposal)
4. Participatory research: scientific monitoring was done in tandem with participatory monitoring and water-literacy programs at the grassroots level and continuous dialogue with water agencies to build a somewhat common understanding of the "system"

The implementation of this framework (Figure 12.1) in research has confronted multiple challenges. First, the absence of biophysical knowledge

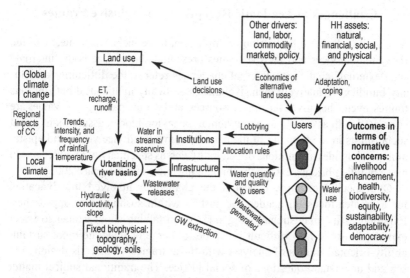

Figure 12.1 Framework for analyzing water issues in urbanizing basins in developing countries that delineates multiple concerns and stressors at multiple scales (Lele et al. 2017). CC: climate change; ET: evapotranspiration; GW: groundwater; HH: household.

and methods to understand the behavior of hard rock aquifers at 1000 feet of pumping, and the links between groundwater and surface water meant large amounts of primary research had to be conducted. Second, the limitations and contradictions in social science knowledge became quickly apparent. For instance, farmers' behavior is complex and coevolves with penetration of markets, but models of farmer decision making are usually unidirectional and incorporate limited parameters. Understanding or modeling the decision making of water agencies, (e.g., whether they will simply import more water, making the basin-level model meaningless) or the lobbying by farmers for electricity subsidies in pumping, or the level of corruption in pollution regulation and how it might change, was found to be very difficult. Third, involving farmers as partners in the research was challenging as they were at least partly complicit in groundwater depletion and feared that monitoring for research might result in them having to pay for water extraction. Overcoming these challenges was possible due to a combination of factors: a team consisting of both disciplinary and interdisciplinary scholars, a funding agency actively promoting inter- and transdisciplinarity, and sustained efforts at multilevel outreach.

As these two cases show, it is possible to build, from a base of somewhat less inclusive frames, more inclusive approaches to research on complex water problems. Implementing research based on such frames will, typically, require large interdisciplinary teams, substantial resources, and contextualized adaptation of existing theories.

Challenges in Academic Receptivity to Inclusive Frames

Academics who wish to develop or implement more inclusive frames, like the ecosystem services and IWRM frames presented above, face both "internal" and "external" challenges. Internal challenges refer to the difficulties of actually building normatively inclusive and theoretically multicausal but rigorous frames, even when given the time, resources, and willing participants. Values are correlated with assumptions about human behavior, hidden assumptions about social limits in the thinking of the natural scientist and vice versa, and epistemological divides and notions of generality versus specificity of knowledge can make building bridges extremely challenging (Lele and Norgaard 2005).

One specific example concerns the challenges involved in advocating for a greater focus on gender and justice within sustainability approaches. Experiences in "mainstreaming" gender, or in talking about gender to water experts, illustrate how difficult it can be to merge or integrate more sustainability-oriented frames of analysis with those frames of analysis designed to see and understand questions of social justice. The traditional subject matter of water analysts is "nonsocial": the physical, biological, and chemical characteristics of water. Although efforts are increasingly made to also include social questions in the analysis of water problems, preferred or dominant scientific languages and methods continue to be derived from the physical sciences.

These, however, are not well suited for understanding the behavior of human beings and their interactions. Gender is a deeply contextual phenomenon: What gender is, and what it means to act or identify as a specific gender, is dependent on time and place. It is also variable, depending on class, caste, religion, or ethnicity. This realization makes it difficult to make general statements about gender in relation to water and to reconcile with a desire for generic truths and universally applicable solutions. Analyzing gender and analyzing water not only seem to require different ways of ordering and making abstractions about reality, the levels and units of analysis may also be difficult to reconcile. Manifestations of gendered inequities and injustices in water occur, or are most clearly visible, at the level of the end users. If the unit of analysis is a river basin or a large surface irrigation system, the group of end users is so large that it becomes conceptually difficult to do justice to all diversities and differences, including those based on gender, between stakeholders and actors. This is even more so because water interests and needs are usually not clearly gendered; although women may have specific water interests, they are usually not a homogeneous group in terms of water.

The difficulty in recognizing gender issues that affect the framing of water, and probably more broadly in combining justice frameworks with sustainability frameworks, is linked to the irreconcilability of epistemic traditions in knowing water and in knowing or thinking about (gendered) injustices. Importantly, to understand (and act on) problems of justice in water management requires active efforts to change normal ways of knowledge production related to water.

This importantly hinges on the forging of new alliances between critical stakeholders (e.g., feminists, political ecologists) and water scholars. These alliances need to go beyond the latter studying and criticizing the former, and should instead concentrate on the active co-creation of different water knowledges.

Problems in overcoming theoretical understandings of water and society are compounded by the institutional challenges faced in academia. The full deployment of multiple value dimensions and theoretical perspectives in a research frame involves significant additional time and resources as compared to deploying the narrower, preexisting disciplinary frames. However, academia tends to value productivity over breadth. Over time, the institutionalization of interdisciplinary scholarship is also stymied, because young scholars see that engaging with such frameworks poses a handicap as they build a career in an academic world that still puts a premium on disciplinary scholarship (Bruce et al. 2004).

These difficulties increase when the frame to be developed and articulated in a specific case is not just interdisciplinary but also transdisciplinary; namely, integrating stakeholders in all steps of research (Jahn et al. 2012). Transdisciplinarity increases the appropriateness of research and its potential to contribute to resolving actually occurring water-related societal problems. However, it often does not receive full academic credit, since, by definition, it is not meant to follow disciplinary guidelines or to contribute to disciplinary development (Defila and Di Giulio 2015). As the primary instrument to guarantee scientific excellence, peer review poses a unique challenge for transdisciplinary publications, due to the lack of specific standards and journals.

Exploring How Inclusive Frameworks Work (or Not) in Practice

Using three illustrative examples, we explore how inclusive frames facilitate (or not) societal relevance, impactful research, and concerted action. We reflect on researchers' struggles to use different inclusive frames in applied contexts by describing examples that demonstrate how inclusive frameworks work in practice. While these examples also show that the inclusion of values other than those put forward by the research framing can be made through the transdisciplinary engagement of researchers, a more detailed analysis on this point would go beyond the scope of our chapter.

Success and Failure of Adopting a More Inclusive Frame: Revisiting the Tisza River Case of Adaptive Water Management

The frame adopted by the flood management research project in Hungary embraced a quite inclusive approach. Justice may not have been spelled out, but the fate of marginalized groups was explicitly addressed. In the spirit of participatory action research and transdisciplinarity, the project engaged with an ongoing process in the region with the goal of both analyzing and supporting it. Given the apparent success of innovative approaches, the initial research design did not explicitly address power structures and points of view of opposing groups. The choice of interview partners and people included in the action research processes had a clear bias toward representatives of the shadow network (i.e., the network of informal actors).

However, the promising initial development toward integrated flood management practices stated in new national flood policy experienced a backlash caused by a weakening influence of the shadow network and the increasing dominance of supporters of a technocratic approach and traditional flood management paradigm in the formal policy process. The influence of the shadow network was never formalized but was triggered by the presence of powerful and charismatic individuals. A subsequent analysis of the process revealed this weakness more clearly (Sendzimir et al. 2010; Pahl-Wostl et al. 2013).

This example illustrates that certain frames resonate and are supported by different groups, and paradigm shifts cause resistance. As we have long known (and still often ignore), introducing AWM does not just imply a change in some procedural aspects of water management: it requires a real transformation (Allan and Curtis 2005; Pahl-Wostl 2015).

Making an Inclusive Frame Even More Inclusive: Revisiting the Ecosystem Services Frame in the Kenya Case Example

The Northern Kenya analyses were developed in collaboration with a local NGO, and results will be shared with that NGO and local and regional decision makers. The local NGO, the Northern Rangelands Trust, has an interconnected set of goals that they state as "resilient community conservancies, transforming lives, securing peace, and conserving natural resources." Given the broad set of interests held by these groups, the inclusive framing by these transdisciplinary researchers has been relatively well received, with some challenges. This is a case where there is already discussion and recognition of the importance of multiple values on the landscape, held by a diverse set of groups. These groups have asked the transdisciplinary researchers to be even more inclusive in the analyses, to represent and consider elements of governance (particularly grazing-related cultural norms) and other interests, such as security (especially related to cattle raiding and grazing incursions) and social cohesion (including elements of trust and engagement). In that sense, even the expanded ecosystem service frame of the transdisciplinary researchers has not been inclusive enough to reflect the myriad interests from the social groups that interact within this landscape.

At the same time, the researchers have needed to develop specific metrics for each of the stakeholder groups so that these groups can readily translate research findings in terms they find useful to their thinking and decision making. Without this tailoring, a broad frame is not accessible and would likely have produced results that were not very relevant to stakeholders. For example, employment in cattle management is reflected in the broad frame. However, each group is interested in employment for different reasons, and thus they are interested in different measures. Managers at the Northern Rangelands Trust are interested in the cost of employing staff, donors are interested in distribution of employment by county and by gender, local and national government leaders are interested in the total number of people employed, and community and local leaders are interested in distribution of employment among tribal groups. Discussing these metrics revealed that there are further distributional justice interests held by different groups that need to be included in the frame for it to be accepted.

Adopting a Less Inclusive Frame to Understand a Water Problem in a More Inclusive Way: The Case of the Adaptive Management Frame in Arizona Water Decision Making

Here, we introduce a final case to illustrate an interesting phenomenon: sometimes the adoption of a *less* inclusive frame as a starting point can facilitate the development of a more inclusive frame later. This case involves the problem of water scarcity in Phoenix, Arizona. Phoenix is a large city located in a desert; future projections indicate that the climate will become warmer and drier over time, even as the human population continues to grow. Researchers at Arizona State University's Decision Center for a Desert City developed WaterSim, an interactive model to address water scarcity that is designed to be a "boundary object" spanning science and policy-making processes (Gober and Wheater 2014; Larson et al. 2015). Initially researchers adopted an adaptive management frame, which placed a strong emphasis on sustainability (including environmental, economic, and societal dimensions) as a core value. In emphasizing sustainability, this frame was politically appropriate and a good fit with local values. Discussions of justice and diversity, however, can be politically sensitive and, critically, impede discussion on genuine points of mutual interest and shared values in the Arizona water decision-making context (Wutich et al. 2010). As with all boundary objects, the WaterSim model was revised as the result of many conversations, critiques, and contributions from both the scientists and policy makers (Wutich et al. 2010).

Over time, this process allowed scientists and policy makers to build trust in each other and, by engaging in discussions around WaterSim, to explore issues that most academics would identify as relevant to justice (e.g., fairness in the distribution of water across sectors) and diversity (e.g., using artificially constructed wetlands to provide tertiary treatment of wastewater, and also enhance biodiversity). This case clearly highlights the importance of resisting the academic impulse to build a highly inclusive frame that emphasizes all normative values and goals at all times. Rather, it is important to consider carefully cases in which it may be better to embrace a frame that is less inclusive or emphasizes only those normative values that are most shared. Such a frame, while less overtly inclusive, can play an important role in accomplishing other normative goals by allowing these goals to be pursued in the background or for less-shared values to be built into the frame over time as a consensus slowly emerges. This case also illustrates the value of having boundary concepts and border zones where people who do not necessarily share normative concerns can meet in a way that is not politically polarizing (cf. Schleyer et al. 2017).

Some Final Reflections

If environmental problems are inherently problems of sustainability, justice, and diversity, then it may be argued that the analysis of environmental problems (including water problems) should routinely emphasize all three groups of concerns, in addition to some form of life or livelihood enhancement. It could further be argued that theoretical explanations should speak to all these concerns. This is easier said than done.

At the theoretical level, integration is often difficult because different experts use different conceptual metaphors, have different methodological preferences, and come from different epistemic traditions. Theoretical integration can perhaps best happen when scientists and researchers engage in joint research projects that run long enough to arrive at shared problem analyses and to co-develop frames of understanding and problem solving. This type of integration will perhaps necessarily happen at lower levels of abstraction than what normal disciplinary scientific rigor prescribes, but may eventually result in the types of framings that allow seeing the connections between social, environmental, and political questions.

Perhaps integration can also happen at the more practical level, by bringing together experts and practitioners (and their respective frames) to help jointly solve an environmental or justice problem. This kind of process-based integration brings its own set of challenges: Who is invited? Is "consensus" the best solution? How will issues of power and tradition affect those who participate? Overcoming such problems could entail compromises and lead to a watering down of the original objectives that may be unacceptable to some participants.

One important role for researchers involves the more instrumental and practical collection and provision of information to support decision making as well as the development and implementation of plans. Researchers can also provide critical reflection: the detailed documentation and critical assessment (against objectives of sustainability, justice, and diversity) of the messy bargains, dilemmas, choices, and compromises that any water intervention project or policy development inevitably entails. Such critical reflections could provide an interesting starting point for rethinking environmentalism.

Many changes will only happen when actively demanded and struggled for by social movements or activists, who may be much more effective when forcefully emphasizing one value or concern instead of integrating many. In particular, concerns of social justice are often articulated most forcefully by social movements. For example, the "water warriors" network of activists (hosted by the Blue Planet Project of the Council of Canadians) has been successful in a cross-national reframing of water as a commons; this frame has been used to challenge power effectively *across scale*. It is worth noting that the difficulty of inserting justice concerns into water frameworks is itself a reflection of dominant power–knowledge networks and epistemic traditions. Accepting this may allow researchers to occupy yet another valuable role, one that comes with different requirements: researchers can support single-issue or value movements by providing them with information and analyses in support of their cause. For example, instead of attempting to arrive at integrative frameworks, the role of researchers concerned with justice can also be to uncover how the seemingly technical solutions proposed in the name of diversity or sustainability imply deeply political decisions in that they redistribute water responsibilities, rights, benefits, and risks.

Researchers who choose to adopt a less inclusive frame, with the goal of facilitating action, must have a strong identification with the movements or activists they are supporting. These researchers may also document and analyze the strategies of social movements or activists in an effort to help identify which strategies are more effective and under which conditions. In cases where the goal is not to produce impactful research or concerted action, some researchers may find that more inclusive frames are more appropriate for achieving representational accuracy in the understanding of complex water problems.

Concluding Thoughts

Water problems are clearly framed in different ways: some normatively narrow or analytically simplistic, some normatively broad or analytically deliberately fuzzy, some speaking to managers, and others highly academic. Assessing these frames on a common set of criteria, as we have done here, can help us understand points of tension, overlap or disconnect between them, as well as the contexts in which some work better than others. In the current academic culture, however, we sense that there is some interest in developing a frame that is even more inclusive than those currently used in interdisciplinary or even transdisciplinary scholarship.

When developing more inclusive frames, there is naturally a tendency for disciplinary experts to desire representation of their own familiar theoretical and normative positions. This leads to frames that "front-page" or clearly highlight multiple values. For example, an environmental justice expert is more likely to be supportive of a frame where justice is clearly recognized as a normative interest, rather than one where justice is implicit or could be built into the frame. However, such overtly inclusive frames may not always be acceptable to all actors because they appear to elevate values not shared by all, or preference some values that seem outside the scope or interest of the context. While it may be tempting to conclude that frames which do not explicitly emphasize all values are not inclusive, we believe that *the more important consideration in determining their inclusivity is how open a frame is to accommodating other values* (even if these values are not highlighted).

While frames that are flexible in allowing for multiple values can become more inclusive, they will not necessarily do so. It is important to acknowledge that values that are not emphasized in a frame can easily be overlooked or treated shallowly. This leads us to consider another reason why we should perhaps avoid the impulse to develop highly inclusive frames for understanding water problems. Ontological and methodological diversity is important for advancing scholarship as well as for the public debates that can result from scholarly analysis. Water systems or realities are always complex. We suggest that adopting any single language or logic in attempts to know water may be too constraining or limited. There are different versions of "water": its meanings, its uses, its management, and so on. Too much emphasis on integration, equivalence or commensuration will inevitably hide gaps, slippages, and frictions between these differences. Acknowledging these frictions instead of, or in addition to, trying to solve them or gloss over them through more inclusivity can be valuable when it helps us see the fault lines of potential conflicts that stem from human struggles over water.

These tensions—between explicit and implicit inclusivity, and between inclusiveness as viewed by researchers, practitioners, and community members—are very significant ones that need to be addressed if more inclusive frames are to be used to advance rigorous science that is socially relevant and impactful.

References

Allan, B. F., H. Tallis, R. Chaplin-Kramer, et al. 2017. Can Integrating Wildlife and Livestock Enhance Delivery of Ecosystem Services in Central Kenya? *Front. Ecol. Environ.* **15**:328–325.

Allan, C., and A. Curtis. 2005. Nipped in the Bud: Why Regional Scale Adaptive Management Is Not Blooming. *Environ. Manage.* **36**:414–425.

Biswas, A. K. 2008. Integrated Water Resources Management: Is It Working? *Int. J. Water Res. Develop.* **24**:5–22.

Boelens, R. 2014. Cultural Politics and the Hydrosocial Cycle: Water, Power and Identity in the Andean Highlands. *Geoforum* **57**:234–247.

Bruce, A., C. Lyall, J. Tait, and R. Williams. 2004. Interdisciplinary Integration in Europe: The Case of the Fifth Framework Programme. *Futures* **36**:457–470.

Butterworth, J., J. Warner, P. Moriarty, S. Smits, and C. Batchelor. 2010. Finding Practical Approaches to Integrated Water Resources Management. *Water Altern.* **3**:68–81.

Daily, G. C. 1997. Nature's Services. Washington, D.C.: Island Press.

Defila, R., and A. Di Giulio. 2015. Integrating Knowledge: Challenges Raised by the "Inventory of Synthesis." *Futures* **65**:123–135.

Dick, J., F. Turkelboom, W. Verheyden, et al. 2018. Users' Perspectives of Ecosystem Service Concept and Tools in Openness Case Studies. *Ecosyst. Serv.*, in press.

Furman, E., K. Jax, H. Saarikoski, et al. 2018. The Openness Approach: From Real World Problems to Concepts and Back to Real World Solutions. *Ecosyst. Serv.*, in press.

Gober, P., and H. S. Wheater. 2014. Socio-Hydrology and the Science–Policy Interface: A Case Study of the Saskatchewan River Basin. *Hydrol. Earth Syst. Sci.* **18**:1413–1422.

Grizzetti, B., C. Liquete, P. Antunes, et al. 2016. Ecosystem Services for Water Policy: Insights across Europe. *Environ. Sci. Pol.* **66**:179–190.

Gyawali, D., and A. Dixit. 1999. Fractured Institutions and Physical Interdependence: Challenges to Local Water Management in the Tinau River Basin, Nepal. In: Rethinking the Mosaic: Investigations into Local Water Management, ed. M. Moench et al., pp. 57–122. Kathmandu: Nepal Water Conservation Foundation and Institute for Social and Environmental Transition.

Hoekstra, A. Y., ed. 2003. Virtual Water Trade. In: Value of Water Research Report Series. Proc. Intl. Expert Meeting on Virtual Water Trade, Dec. 12–13, 2002, vol. 12. Delft: IHE.

Jahn, T., M. Bergmann, and F. Keil. 2012. Transdisciplinarity: Between Mainstreaming and Marginalization. *Ecol. Econ.* **79**:1–10.

Larson, K. L., D. D. White, P. Gober, and A. Wutich. 2015. Decision-Making under Uncertainty for Water Sustainability and Urban Climate Change Adaptation. *Sustainability* **7**:14761–14784.

Lele, S., and R. Norgaard. 2005. Practicing Interdisciplinarity. *Bioscience* **55**:967–975.

Lele, S., V. Srinivasan, B. K. Thomas, and P. Jamwal. 2018. Adapting to Climate Change in Rapidly Urbanizing River Basins: A Multiple-Concerns, Multiple-Stressors and Multi-Level Approach. *Water Int.*, in press.

Lenton, R., and M. Muller, eds. 2009. Integrated Water Resources Management in Practice: Better Water Management for Development. London: Earthscan.

Medema, W., B. S. McIntosh, and P. J. Jeffrey. 2008. From Premise to Practice: A Critical Assessment of Integrated Water Resources Management and Adaptive Management Approaches in the Water Sector. *Ecol. Soc.* **13**:29.

Mekonnen, M. M., and A. Y. Hoekstra. 2010. Mitigating the Water Footprint of Export Cut Flowers from the Lake Naivasha Basin, Kenya. Value of Water Research Report Series No. 45. Delft: UNESCO-IHE.

Mekonnen, M. M., A. Y. Hoekstra, and R. Becht. 2012. Mitigating the Water Footprint of Export Cut Flowers from the Lake Naivasha Basin, Kenya. *Water Resour. Manage.* **26**:3725–3742.

Millennium Ecosystem Assessment. 2005. Ecosystems and Human Well-Being: Synthesis. Washington, D.C.: Island Press.

Moench, M., E. Caspari, and A. Dixit, eds. 1999. Rethinking the Mosaic: Investigations into Local Water Management. Kathmandu: Nepal Water Conservation Foundation, & Institute for Social and Environmental Transition.

Molden, D., R. Sakthivadivel, and Z. Habib. 2001. Basin-Level Use and Productivity of Water: Examples from South Asia. Colombo: Intl. Water Management Institute.

286 A. Wutich et al.

Murthy, S. 2013. The Human Right(s) to Water and Sanitation: History, Meaning, and the Controversy over Privatization. *Berkeley J. Int. Law* **31**:89–147.

Ortega-Reig, M., G. Palau-Salvador, M. J. Cascant i Sempere, J. Benitez-Buelga, and P. Trawick. 2014. The Integrated Use of Surface, Ground and Recycled Waste Water in Adapting to Drought in the Traditional Irrigation System of Valencia. *Agric. Water Manage.* **133**:55–64.

Ostrom, E. 1990. Governing the Commons: The Evolution of Institutions for Collective Action. New York: Cambridge Univ. Press.

Pahl-Wostl, C. 2015. Water Governance in the Face of Global Change: From Understanding to Transformation. Cham: Springer.

Pahl-Wostl, C., G. Becker, J. Sendzimir, and C. Knieper. 2013. How Multilevel Societal Learning Processes Facilitate Transformative Change: A Comparative Case Study Analysis on Flood Management. *Ecol. Soc.* **18**:58.

Petit, O., and C. Baron. 2009. Integrated Water Resources Management: From General Principles to Its Implementation by the State: The Case of Burkina Faso. *Natural Res. Forum* **33**:49–59.

Schleyer, C., A. Lux, M. Mehring, and C. Görg. 2017. Ecosystem Services as a Boundary Concept: Arguments from Social Ecology. *Sustainability* **9**:1107.

Sendzimir, J., Z. Flachner, C. Pahl-Wostl, and C. Knieper. 2010. Stalled Regime Transition in the Upper Tisza River Basin: The Dynamics of Linked Action Situations. *Environ. Sci. Pol.* **13**:604–619.

Sendzimir, J., P. Magnuszewski, Z. Flachner, et al. 2007. Assessing the Resilience of a River Management Regime: Informal Learning in a Shadow Network in the Tisza River Basin. *Ecol. Soc.* **13**:11.

Suhardiman, D., F. Clement, and L. Bharati. 2015. Integrated Water Resources Management in Nepal: Key Stakeholders Perceptions and Lessons Learned. *Int. J. Water Res. Develop.* **31**:284–300.

Swyngedouw, E. 2009. The Political Economy and Political Ecology of the Hydro-Social Cycle. *J. Contemp. Water Res. Educ.* **142**:56–60.

TEEB. 2008. An Interim Report. European Communities. Cambridge: Banson Production.

Werners, S. E., P. Matczak, and Z. Flachner. 2009. The Introduction of Floodplain Rehabilitation and Rural Development into the Water Policy for the Tisza River in Hungary. In: Water Policy Entrepreneurs. A Research Companion to Water Transitions around the Globe, ed. D. Huitema and S. Meijerink, pp. 250–271. Cheltenham: Edward Elgar.

Wutich, A., M. Beresford, and C. Carvajal. 2016. Can Informal Water Vendors Deliver on the Promise of a Human Right to Water? Results from Cochabamba, Bolivia. *World Dev.* **79**:14–24.

Wutich, A., T. Lant, D. D. White, K. L. Larson, and M. Gartin. 2010. Comparing Focus Group and Individual Responses on Sensitive Topics: A Study of Water Decision Makers in a Desert City. *Field Methods* **22**:88–110.

Further Titles in the Strüngmann Forum Report Series[1]

Better Than Conscious? Decision Making, the Human Mind, and Implications For Institutions
edited by Christoph Engel and Wolf Singer, ISBN 978-0-262-19580-5

Clouds in the Perturbed Climate System: Their Relationship to Energy Balance, Atmospheric Dynamics, and Precipitation
edited by Jost Heintzenberg and Robert J. Charlson, ISBN 978-0-262-01287-4

Biological Foundations and Origin of Syntax
edited by Derek Bickerton and Eörs Szathmáry, ISBN 978-0-262-01356-7

Linkages of Sustainability
edited by Thomas E. Graedel and Ester van der Voet, ISBN 978-0-262-01358-1

Dynamic Coordination in the Brain: From Neurons to Mind
edited by Christoph von der Malsburg, William A. Phillips and Wolf Singer, ISBN 978-0-262-01471-7

Disease Eradication in the 21st Century: Implications for Global Health
edited by Stephen L. Cochi and Walter R. Dowdle, ISBN 978-0-262-01673-5

Animal Thinking: Contemporary Issues in Comparative Cognition
edited by Randolf Menzel and Julia Fischer, ISBN 978-0-262-01663-6

Cognitive Search: Evolution, Algorithms, and the Brain
edited by Peter M. Todd, Thomas T. Hills and Trevor W. Robbins, ISBN 978-0-262-01809-8

Evolution and the Mechanisms of Decision Making
edited by Peter Hammerstein and Jeffrey R. Stevens, ISBN 978-0-262-01808-1

Language, Music, and the Brain: A Mysterious Relationship
edited by Michael A. Arbib, ISBN 978-0-262-01962-0

Cultural Evolution: Society, Technology, Language, and Religion
edited by Peter J. Richerson and Morten H. Christiansen, ISBN 978-0-262-01975-0

Schizophrenia: Evolution and Synthesis
edited by Steven M. Silverstein, Bita Moghaddam and Til Wykes, ISBN 978-0-262-01962-0

Rethinking Global Land Use in an Urban Era
edited by Karen C. Seto and Anette Reenberg, ISBN 978-0-262-02690-1

Trace Metals and Infectious Diseases
edited by Jerome O. Nriagu and Eric P. Skaar, ISBN 978-0-262-02919-3

Translational Neuroscience: Toward New Therapies
edited by Karoly Nikolich and Steven E. Hyman, ISBN: 9780262029865

---

[1] available at https://mitpress.mit.edu/books/series/str%C3%BCngmann-forum-reports-0

The Pragmatic Turn: Toward Action-Oriented Views in Cognitive Science
edited by Andreas K. Engel, Karl J. Friston and Danica Kragic
ISBN: 978-0-262-03432-6

Complexity and Evolution: Toward a New Synthesis for Economics
edited by David S. Wilson and Alan Kirman, ISBN: 9780262035385

Computational Psychiatry: New Perspectives on Mental Illness
edited by A. David Redish and Joshua A. Gordon, ISBN: 9780262035422

Investors and Exploiters in Ecology and Economics: Principles and Applications
edited by Luc-Alain Giraldeau, Philipp Heeb and Michael Kosfeld
Hardcover: ISBN: 9780262036122, ebook: ISBN: 9780262339797

The Cultural Nature of Attachment: Contextualizing Relationships and Development
Edited by Heidi Keller and Kim A. Bard
Hardcover: ISBN: 9780262036900, ebook: ISBN: 9780262342865

Printed in the United States
By Bookmasters